Springer Series in Information Sciences 9

Editor: T. S. Huang

Springer Series in Information Sciences

Editors: King-sun Fu Thomas S. Huang Manfred R. Schroeder

L. P. Yaroslavsky

Digital Picture Processing

An Introduction

With 87 Figures

Springer-Verlag
Berlin Heidelberg New York Tokyo

Dr. Leonid P. Yaroslavsky

Institute for Information Transmission Problems,
Yermolovoy Str. 19, SU-101447 Moscow, USSR

Series Editors:

Professor King-sun Fu

Professor Thomas S. Huang

Department of Electrical Engineering and Coordinated Science Laboratory,
University of Illinois, Urbana, IL 61801, USA

Professor Dr. Manfred R. Schroeder

Drittes Physikalisches Institut, Universität Göttingen, Bürgerstraße 42–44,
D-3400 Göttingen, Fed. Rep. of Germany

Title of the original Russian edition:
Vvedyenie v Tsifrovuyu Obrabotku Izobrazheny

ISBN 3-540-11934-5 Springer-Verlag Berlin Heidelberg New York Tokyo
ISBN 0-387-11934-5 Springer-Verlag New York Heidelberg Berlin Tokyo

Library of Congress Cataloging in Publication Data. I͡Aroslavskiĭ, L.P. (Leonid Pinkhusovich) Digital picture processing. (Springer series in information sciences ; v. 9) Translation of: Vvedenie v t͡sifrovui͡u obrabotku izobrazheniĭ. Includes bibliographical references and index. 1. Image processing–Digital techniques. I. Title. II. Series.
TA1632.I1713 1985 621.36'7 85-14808

Offset printing: Beltz Offsetdruck, 6944 Hemsbach/Bergstr.
Bookbinding: J. Schäffer OHG, 6718 Grünstadt
2153/3130-543210

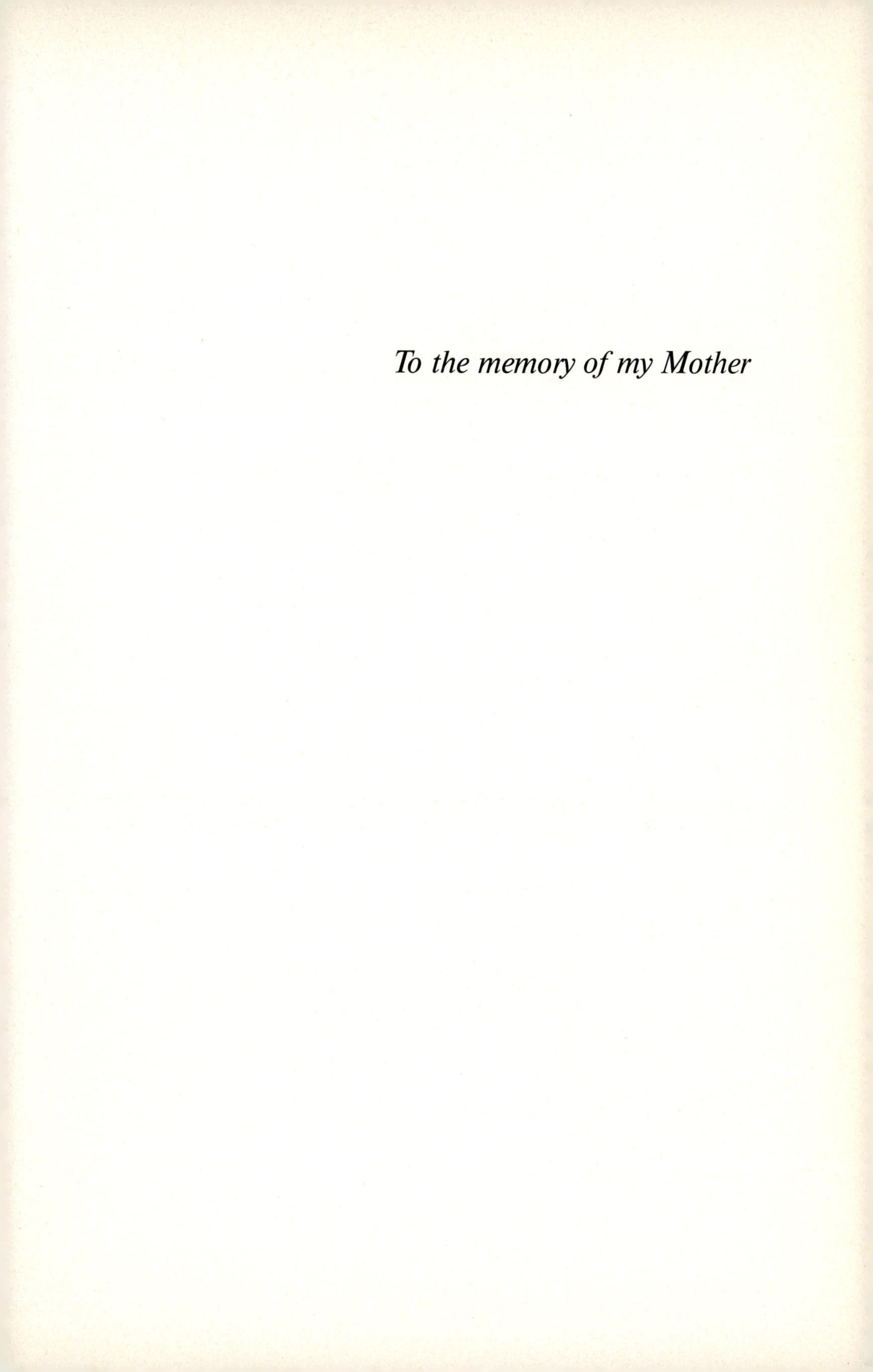

To the memory of my Mother

Preface

The text has been prepared for researchers involved in picture processing. It is designed to help them in mastering the methods at the professional level. From the viewpoint of both signal theory and information theory, the treatment covers the basic principles of the digital methods for the processing of continuous signals such as picture signals. In addition, it reviews schemes for correcting signal distortion in imaging systems, for the enhancement of picture contrast, and for the automatic measurement of picture details.

The text contains new results on digital filtering and transformation, and a new approach to picture processing. The main applications, as documented by numerous examples, are in space research, remote sensing, medical diagnostics, nondestructive testing. The material has been tested extensively in class-room use with students of both computer science and electrical engineering at the senior undergraduate and the first-year graduate level.

The present edition is not a translation of the original Russian book, but it has been extended substantially as well as updated.The author is grateful to Dr. H. Lotsch of Springer-Verlag for his proposal to prepare this text and for many helpful suggestions. He likes to thank Dr. P. Hawkes for a careful copy-editing of the manuscript, and acknowledges numerous critical comments by Professors S.L. Gorelik, T.S. Huang, A.W. Lohmann,and A.M. Trakhtman.

Moscow, September 1984 *L.P. Yaroslavsky*

Contents

Part 2 Picture Processing

1 Introduction

Digital Picture Processing considers a picture above all as a two-dimensional signal. The characteristics of signals as carriers of information are defined both by the receiver of the information and the nature of the information transferred. Among the broad class of two-dimensional signals, those intended for visual perception are unique and perhaps the most difficult to treat. These signals are the pictures under investigation.

Sight is one of man's most important senses. Through it he receives a large proportion of his information about the external world, and it is for this reason that pictures are so important in science and technology, and in our daily life.

Pictures are formed by imaging systems. The purpose of these systems is to make available information that cannot be directly perceived by sight. Imaging systems of whatever type, whether they function by means of optics, photography, television, X-rays, etc., consist of three types of devices: a video signal sensor, a device for transforming and transmitting the video signal, and a picture synthesizer (Fig.1.1)[1].

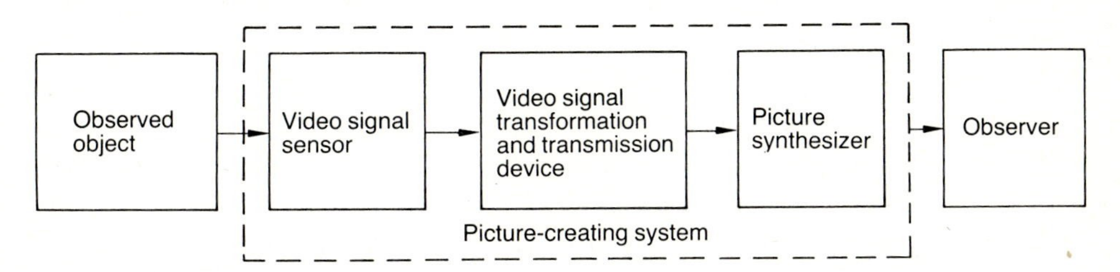

Fig. 1.1 Observation of an object by means of an imaging system

[1] In television technology the term video signal is understood to refer to the electrical signal which is formed by scanning the screen of the pickup and subsequently generates the TV picture when the screen is scanned. In radar the term is sometimes used to denote signals appearing on a radar screen. In this book it will have the wider meaning of a signal of any physical type carrying visual information.

The sensor interacts directly with the observed object. The picture synthesizer forms the picture, making it accessible for direct visual perception. The video signal transforming and transmitting device matches and links the sensor and the picture synthesizer.

The signals are transformed in these devices in order to change their physical nature, as in the transformation of a light wave into an electrical signal, or of an electrical signal into electromagnetic radiation in a radio antenna, etc., and/or to change their structure, i.e., their mathematical nature. This division is to some extent arbitrary, but nonetheless convenient, because it allows a differentiation between transformations that are described mathematically and those that are described in physical terms (matter, energy, wavelength, etc.).

We shall understand picture processing to mean only the mathematical processing of the video signal.

The following categories of tasks can be distinguished in picture processing:

1) correction of the imaging system
2) picture preparation
3) information display
4) automatic picture measurements and picture understanding
5) picture coding
6) simulation of imaging systems

Correction of imaging systems consists of processing the video signal in a real picture generator in order to obtain pictures corresponding to an ideal picture generator. By ideal we mean a system in which the observer disregards any discrepancies between the observed object and the picture, i.e., a system which creates a picture that for the observer is equivalent to the object. In designing a picture generator the requirements for an ideal system are usually expressed in terms of certain technical properties, such as resolution, photometric accuracy, level of background noice, etc. Thus, correcting a picture generator means adjusting the characteristics of the system to meet specific demands. Examples are increasing the sharpness of an unfocussed picture, removing blur, and suppressing noise.

The ideal picture generator does not necessarily yield the picture that would best correspond to the specific task of visual analysis and extraction of information, since in meeting the demand for ideal characteristics a compromise must in practice be made among the demands of a wide class of tasks. Certain problems sometimes require supplementary transformation of the signal to facilitate visual analysis. This may involve emphasizing some features and details of the picture and suppressing others, changing spatial relationships, measuring and visualizing numerical characteristics, etc. Such transformations, which are an instrument of visual analysis, we shall call *picture enhancement and preparation*. Examples are the following methods used in artistic and scientific photography: solarization. The plotting of isodensities, contouring, and picture representation in pseudocolours.

By *information display* we mean the processing of signals into pictures for subsequent visual interpretation. Examples are the presentation of one value as a function of another in the form of a plot and the presentation of the functions of two variables in the form of a distribution of shadow on a photograph or the distribution of brightness on a display screen.

In principle, any signal can be presented to man in the form of a picture in one way or another. Nevertheless, many problems involved in the extraction of information from signals can and must be solved by automatic devices. The automatic processing of signals that are represented in the form of pictures and can be analyzed visually we call the *automatic measurement of pictures*. Examples are automatic measurement of the number and spatial location of objects, and their detection and classification.

Picture coding is the transformation of the video signal that is necessary in order for it to be stored in memory banks or transmitted through communication channels in digital form.

Finally, the *simulation of imaging systems* is the category of tasks in picture processing that is concerned with the development and investigation of new types of systems.

All of these problems are closely interrelated both in the methods by which they are solved, and in the technical means that are used.

This book is devoted to picture-processing methods using digital computer technology, universal digital computers and specialized digital processors. The use of digital technology is an alternative to the traditional analogue technology of picture processing by means of optics, photography, television, etc. It has great advantages over analogue technology when flexibility, programmability, interactive control, intervention in processing, and the accurate reproduction of results are required. Digital technology currently surpasses analogue technology both in reliability and unification of functions, i.e., in economic efficiency.

Digital picture processing was introduced in the late fifties and early sixties with the application of universal digital computers to simulate imaging systems and methods of picture coding [1.1-5], although the digital representation of pictures for transmission was first mentioned back in the twenties [1.6]. With the development of computer technology at the end of the sixties it became clear that digital computers could also be used to solve other problems arising in picture processing, in particular the correction of imaging systems, the picture preparation, the classification in space, in physics, in medical diagnostics and in industrial non-destructive testing. The display of information using digital computer technology has become a special branch of computer sciences, namely computer graphics [1.7-9]. Finally, by the mid-seventies, owing to the development of a microelectronic basis specialized digital instruments and picture-processing systems began to be developed using minicomputers and microprocessors. These systems found practical application mainly in the study of the earth's resources and in medicine [1.10-19].

Nowadays digital picture processing has become a profession for many people. The present book is an attempt to bring together the information necessary to master this profession. For this reason it is divided into two parts: the first (Chaps.2-6) treats fundamentals of the theory, and the second (Chaps.7-9) presents concrete examples of problem-solving in digital picture processing.

Chapter 2 is devoted to the elements of signal theory used in the subsequent treatment. Chapters 3 - 5 are the heart of Part 1, and deal with the discretization and quantization of pictures, the discrete representation of signal transformations, and algorithmic methods of digital transformation with minimum expenditure of computing resources. These chapters are thus devoted to the central problem of the digital processing of signals, namely how to construct the digital representation of signals and their transformation in digital processing.

Despite the importance of this problem, many of its aspects, particularly those concerned with the processing of two-dimensional signals (such as pictures), have been inadequately discussed in the literature. For this reason, Chaps.3-5 present, together with known results, useful new results from both the methodological and practical standpoints. They include an analysis of the features of picture sampling (Sect.3.4); the characteristic of digital filters (Sect.4.4); shifted discrete Fourier transforms (Sect.4.7); a representation of orthogonal transforms in the form of layered Kronecker matrices that allows a unified approach to these transforms and their corresponding fast algorithms with the use of matrix algebra (Sects.5.1-5); a new algorithm of the combined discrete Fourier transform for even sequences (Sect.5.6); and the two-dimensional recursion algorithm for calculating local discrete Fourier spectra (Sect.5.7).

As the coding of pictures is not discussed in detail in the book, a short survey of picture-coding methods currently being used or studied is presented in Sect.3.11 on the basis of the concepts presented in Chap.3.

Chapter 6 is devoted to digital statistical methods of measuring the statistical characteristics and random variations of pictures, and to the generation of pseudorandom sequences and pictures. A knowledge of the statistical characteristics of pictures is usually required when coding pictures and correcting imaging systems. Pseudorandom sequences are also used to simulate imaging systems.

As already mentioned, Part 2 contains descriptions of specific examples of picture processing. Chapter 7 is devoted to the problem of correcting imaging systems. It presents methods of correcting linear and nonlinear picture distortion and combatting interference. Especial attention is given to the adaptive approach to the synthesis of correction algorithms, in which distortion is evaluated directly on the basis of the distorted signal and the algorithm parameters are adjusted from the observed picture. Methods that have proved their effectiveness on real pictures are described in detail. Chapter 7 draws primarily on results obtained in processing pictures from the space missions "Mars-4" and "Mars-5".

Chapter 8 looks at some methods of picture preparation, focusing on nonlinear and linear algorithms with automatic adjustment of parameters. The geometrical transformation of pictures is considered briefly, as is its relation to the problem of interpolation for the transfer from discrete to continuous picture representation.

Chapter 9 is devoted to a problem of measurement in pictures, namely the determination of the coordinates of particular objects. As this important practical question has been inadequately discussed in the literature, Sects. 9.2-4 take an analytical look at it. The conclusions arrived at are supported by the results of simulation on a digital computer. Section 9.5 describes one practical application of these conclusions.

Part 2 is, with the exception of Chap.9, of a more descriptive character than Part 1. It can be considered independently of Part 1 but for a thorough understanding of the processing methods it describes and, more importantly, for an appreciation of the unity of the methods used in various problems and the reasons for their selection, a reading of Part 1 is indispensable.

Part 1

Fundamentals of the Theory of Digital Signal Processing

2 Elements of Signal Theory

In picture processing, pictures are regarded as two-dimensional signals. This chapter introduces and explains the basic concepts of signal theory. The treatment of signals as mathematical functions is discussed in Sect.2.1, and signal space and related notions, including linear representation of signals, are introduced in Sect.2.2. Section 2.3 reviews the most common systems of basis functions used for the linear representation of signals, while Sect.2.4 deals with the integral representation of signals, with special emphasis on Fourier transform. Section 2.5 is devoted to ways of describing signal transformations as well as to concepts of linear and nonlinear transforms. The discrete representation of linear transformation corresponding to linear representation of a signal with respect to a discrete basis is described in Sect.2.6, while Sect.2.7 treats the representation of linear transforms corresponding to integral representation of signals. Finally, Sect.2.8 reviews examples of the most popular linear operators.

2.1 Signals as Mathematical Functions

Mathematical models are used to describe signals and transformations of them analytically. Signals are considered above all to be functions defined in physical coordinates. It is in this sense that signals are referred to as one-dimensional (e.g., functions of time), two-dimensional (defined on a plane; e.g. a picture), or three-dimensional (characterizing spatial objects, for example). Normally, scalar functions are used as mathematical models of signals. It is sometimes necessary, however, to employ more complicated models such as complex functions or vector functions. For example, to describe electromagnetic fields it is convenient to use complex functions. In order to describe colour pictures, three-component vector functions are used; and four- to six-component vector functions are applied to describe the data of multispectral surveys.

An important general characteristic of signals is the set of values that the signals themselves and their arguments can have. If a signal and its arguments can take any values on a finite or infinite interval, i.e., an infinite set of values, then the signal is termed *continuous* or *analogue* (being, as it were, an analogue representation of a natural object, which is usually continuous). A signal whose arguments can take only a finite set of values is termed *discrete*. A signal with a finite set of values is called *quantized*. Signals which are both, discrete and quantized are called *digital*.

The class of signals that is dual in the familiar sense to discrete signals is the class of *periodic* signals. They are described by periodic functions, defined in the one-dimensional case by

$$a(t + kT) = a(t) \quad , \tag{2.1}$$

where k is an integer.

When discussing two-dimensional and multidimensional signals, the arguments describing their functions can be considered to be vectors defined by a given system of coordinates. The choice of this coordinate system is usually determined by the essence of the task at hand, and on this the simplicity of the analytical description of signals largely depends. We shall illustrate this using an example of two-dimensional periodic signals.

The two-dimensional interval, the period of the periodic signal, is a plane figure and not a line segment. If a two-dimensional signal repeats itself on a plane with a period in the form of a rectangle, it is convenient to use the Cartesian coordinate system (t_1, t_2), in which the signal is described as

$$a(t_1, t_2) = a(t_1 + k_1T_1, t_2 + k_2T_2) \quad . \tag{2.2}$$

If the period is a parallelogram, a skewed coordinate system is preferable, in which the signal is also described as in (2.2). In the Cartesian coordinate system the signal looks more complicated

$$a(t_1, t_2) = a(t_1 + k_1T_{11} + k_2T_{21}, t_2 + k_1T_{12} + k_2T_{22}) \quad , \tag{2.3}$$

where $T_{11}, T_{21}, T_{12} - T_{22}$ are the projections of periods about the axes of the skewed coordinate system on the axes of the Cartesian coordinate system.

In the mathematical description of signals deterministic and probablistic descriptions are also considered. In deterministic descriptions signals are considered individually and mutually independently, and the value of a signal is considered to be defined at every point where it is determined. However, it is not always possible to treat physical objects individually and only a certain number of what can be called macroparameters characterizing the object on the average can be measured and calculated. In such cases probablistic description is used, i.e., signals are considered to be sample functions or the realization of an ensemble of signals, and a mathematical description is made not of each individual signal but of the ensemble as a whole.

Since this book is devoted to digital image processing, two-dimensional digital signals will be its main theme. However, in digital signal processing, digital signals are artificial objects, being the result of a transformation of continuous signals. In order to clarify the relationship between digital signals and the continuous signals from which they originate, we shall consider continuous signals. It should be noted that in order to simplify the formulae, the signals will be described whenever possible as functions of one variable, irrespective of whether these signals are one-dimensional or two-dimensional. For the latter case, this variable can be considered vectorial.

2.2 Signal Space

When describing signals mathematically, it is convenient to regard them as points or vectors in some function space. Signal transformation is considered to be a mapping in this space. In this context "space" is used to give the set of signals geometric meaning and thus render them easier to visualize.

The treatment of signals as elements of a linear normed metric space is simplest, and at the same time sufficiently meaningful in a physical sense.

A space is termed *metric* when a distance between its spatial elements (points) is defined, i.e., each pair of elements, say $\mathbf{a}_1$ and $\mathbf{a}_2$, can be placed in correspondence with a real non-negative number $d(\mathbf{a}_1, \mathbf{a}_2)$, satisfying the following rules

$$\begin{aligned} &d(\mathbf{a}_1, \mathbf{a}_2) = 0, \quad \text{if } \mathbf{a}_1 = \mathbf{a}_2 \quad ; \\ &d(\mathbf{a}_1, \mathbf{a}_2) = d(\mathbf{a}_2, \mathbf{a}_1) \quad ; \\ &d(\mathbf{a}_1, \mathbf{a}_3) \leqslant d(\mathbf{a}_1, \mathbf{a}_2) + d(\mathbf{a}_2, \mathbf{a}_3) \quad , \end{aligned} \tag{2.4}$$

where the real valued function $d(\cdot,\cdot)$ is called the *metric* or *distance function*.

The purpose of the first two conditions is obvious. The third condition, called the "triangle inequality", is the formal expression of the following natural requirement of a metric: if two points are close to a third point, they must be close to each other. Table 2.1 shows the metrics most commonly used in functional and signal analysis.

Table 2.1. Examples of metrics (*Note:* T denotes the range of the variable t; $\sup(x)$ denotes supremum of x, i.e., the least upper bound of x_i and $\oplus$ denotes addition modulo 2.)

Discrete signals	l_N	$d(a_1,a_2) = \sum_{n=0}^{N-1} \lvert a_{1n}-a_{2n}\rvert$
	l_N^2	$d(a_1,a_2) = \sqrt{\sum_{n=0}^{N-1} \lvert a_{1n}-a_{2n}\rvert^2}$
	m_N	$d(a_1,a_2) = \max_n \lvert a_{1n}-a_{2n}\rvert$
	h_n	$d(a_1,a_2) = \sum_{k=0}^{N-1} (a_{1k} \oplus a_{2k})$
Continuous signals	L_T	$d(a_1,a_2) = \int_T \lvert a_1(t)-a_2(t)\rvert dt$
	L_T^2	$d(a_1,a_2) = \sqrt{\int_T \lvert a_1(t)-a_2(t)\rvert^2 dt}$
	M_T	$d(a_1,a_2) = \sup_T \lvert a_1(t)-a_2(t)\rvert$

The metric ℓ_N^2 and its continuous analogue metric L_T^2, as well as its generalization when $N\to\infty$ and $T\to\infty$ are termed Euclidean, since ℓ_3^2 corresponds to the Euclidean metric of real physical space.

The concept "distance" in signal theory is used to measure the difference between one signal and another, or the error in representing one signal by another. Therefore in order to characterize signal space, the metric must be chosen which best describes this difference with one number, i.e., the distance.

At least two examples can be given in which the metrics are chosen unambiguously. The first is when the difference between one signal and another is the result of additive non-correlated Gaussian noise. For simplicity let us take the case in which two discrete signals are distinguished:

$$a_{2_k} = a_{1_k} + n_k \quad , \quad k = 0 \quad , \quad 1, \ldots , \quad N - 1 \quad , \tag{2.5}$$

where n_k are samples of independent random numbers with normal probability density function

$$p(n) = (1/\sqrt{2\pi}\,\sigma) \exp(-n^2/2\sigma^2) \tag{2.6}$$

and variance σ^2. Clearly the entire difference between a_1 and a_2 is contained in the signal $\mathbf{n} = \{n_k\}$, and it can be fully described statistically by its multidimensional probability density function

$$p(\mathbf{n}) = p(n_0)\, p(n_1) \,\dots\, p(n_{N-1}) = (2\pi)^{-N/2}\sigma^{-N}$$
$$\times \exp\left[-\frac{1}{2}\,\sigma^2 \sum_{k=0}^{N-1} (a_{2k} - a_{1k})^2\right] \quad , \tag{2.7}$$

which in turn is wholly defined by the value

$$d(\mathbf{a}_2, \mathbf{a}_1) = \sqrt{\sum_{k=0}^{N-1} (a_{2k} - a_{1k})^2} \quad , \tag{2.8}$$

i.e., the Euclidean distance between a_2 and a_1. The Euclidean metric is generated in this way.

The Euclidean metric is popular in signals theory for two reasons. First, it is very convenient for calculations and was a definite physical meaning (the metric is proportional to the difference between the energy of the two signals). Second, it is sufficiently accurate for tasks in which the difference between signals is generated by the combined effect of a large number of errors.

Often the Euclidean metric is also called a mean-square metric, because it gives the squared difference between signals, averaged over the region in which they are defined. In this sense its generalization is a weighted mean-square metric defined for the discrete case by

$$d(\mathbf{a}_1, \mathbf{a}_2) = \sqrt{\sum_{k=0}^{N-1} w_k (a_{2k} - a_{1k})^2} \quad , \tag{2.9}$$

where $\{w_k\}$ is a set of weight coefficients. Such metrics would be used, for example, if in (2.5) we assumed that $\{n_k\}$ has various values σ_k^2 for the variance.

Let us now look at another example. This time we consider the space of binary digital signals whose samples take only two values and which change from one to another as a result of the inversion of the value of randomly chosen samples. Let two such signals be a_1 and a_2. The total difference between them can then be described by the binary signal n and the bitwise addition, of samples $\{a_{1k}\}$ and $\{n_k\}$ modulo 2 :

$$a_{2k} = a_{1k} \oplus n_k \quad . \tag{2.10}$$

By definition signal n is random, and its samples can take only the values 1 or 0, subject to some probability distribution. If these samples are independent, in a statistical sense the signal n is completely described by the number of samples that are 1; in other words, the number of non-identical samples of $\mathbf{a}_1$ and $\mathbf{a}_2$ equals

$$d(\mathbf{a}_1, \mathbf{a}_2) = \sum_{k=0}^{N-1} (a_{1k} \oplus a_{2k}) \quad . \tag{2.11}$$

This is how Hamming metric is derived.

A signal space is called a *linear* or *vector* space if it satisfies the following conditions:

1) Any two of its elements $\mathbf{a}_1$ and $\mathbf{a}_2$ uniquely determine a third element $\mathbf{a}_3$, which is called their sum and denoted by $\mathbf{a}_1 + \mathbf{a}_2$, provided the summation obeys the commutative law

$$\mathbf{a}_1 + \mathbf{a}_2 = \mathbf{a}_2 + \mathbf{a}_1 \tag{2.12}$$

and the associative law

$$\mathbf{a}_1 + (\mathbf{a}_2 + \mathbf{a}_3) = (\mathbf{a}_1 + \mathbf{a}_2) + \mathbf{a}_3 \quad . \tag{2.13}$$

2) There is an element ϕ, such that

$$\mathbf{a} + \phi = \mathbf{a} \tag{2.14}$$

for all elements of the space.

3) Each element $\mathbf{a}$ of the space can be placed in correspondence with its negative element $-\mathbf{a}$ such that

$$\mathbf{a} + (-\mathbf{a}) = \phi \quad . \tag{2.15}$$

4) For any number α and any element $\mathbf{a}$ of the space an element $\alpha\mathbf{a}$ is defined such that

$$\begin{aligned} &\alpha_1(\alpha_2\mathbf{a}) = (\alpha_1\alpha_2)\mathbf{a} \quad ; \quad 1 \cdot \mathbf{a} = \mathbf{a} \quad ; \\ &(\alpha_1 + \alpha_2)\mathbf{a} = \alpha_1\mathbf{a} + \alpha_2\mathbf{a} \quad ; \\ &\alpha(\mathbf{a}_1 + \mathbf{a}_2) = \alpha\mathbf{a}_1 + \alpha\mathbf{a}_2 \quad . \end{aligned} \tag{2.16}$$

The elements of a linear or vector space are called *vectors*. A vector formed summing several vectors multiplied by scalar coefficients is called a *linear combination vector*:

$$\mathbf{a} = \sum_{k=0}^{N-1} \alpha_k \mathbf{a}_k \quad . \tag{2.17}$$

The set of vectors $\{\mathbf{a}(k)\}$ is said to be linearly independent if there exists no set of non-zero constants α_k such that

$$\sum_{k=0}^{N-1} \alpha_k \mathbf{a}_k \neq \phi \quad . \tag{2.18}$$

Consequently, a linearly independent set is one in which no vector can be expressed as a linear combination of the remainder.

A space $\mathbf{A}_N$, which is fully described by linear combinations of N linearly independent vectors $\{\varphi_k\}$, k=0, 1, ..., N-1, is called an *N-dimensional vector space*. The set of linearly independent vectors $\{\boldsymbol{\varphi}_k\}$ is known as a *discrete basis* of this space, and $\mathbf{A}_N$ is thought of as generated by this basis. Any set of N linearly independent vectors in $\mathbf{A}_N$ can serve as its basis.

Any vector in the space $\mathbf{A}_N$ corresponds to a unique linear combination of the vectors $\{\boldsymbol{\varphi}_k\}$ and hence to a unique set of scalar coefficients $\{\alpha_k\}$. The set of (ordered) scalar coefficients $\{\alpha_k\}$ is the representation of the vector for the given basis.

The physical principle of superposition is mathematically described by the concept of a linear space. Therefore the property of linearity is ascribed to the signal space whenever the principle of superposition is applicable for physical signals.

The term *normalized linear space* is used to describe a linear space in which a vector norm that satisfies the following conditions is defined:

$$\begin{aligned} &\text{a)} \quad \|\mathbf{a}\| \geqslant 0 \quad ; \quad \|\mathbf{a}\| = 0 \quad , \quad \text{if and only if} \quad \mathbf{a} = \phi \\ &\text{b)} \quad \|\mathbf{a}_1 + \mathbf{a}_2\| \leqslant \|\mathbf{a}_1\| + \|\mathbf{a}_2\| \quad ; \\ &\text{c)} \quad \|\alpha\mathbf{a}\| = |\alpha| \; \|\mathbf{a}\| \quad . \end{aligned} \tag{2.19}$$

The geometrical analogue of the vector norm is the length. A norm satisfying condition (2.19) can be used as a metric:

$$d(\mathbf{a}_1, \mathbf{a}_2) = \|\mathbf{a}_1 - \mathbf{a}_2\| \quad , \tag{2.20}$$

in which case

$$\|\mathbf{a}\| = d(\mathbf{a}, \phi) \quad . \tag{2.21}$$

Thus if the norm of a signal space generates its metric in accordance with (2.20), the norm shows how much two vectors differ or how far a vector is from the zero vector.

In linear space the identity of norms and metrics is natural in view of the presence of the zero vector.

When using signal space to describe so-called linear processing systems, yet another geometric characteristic is used, namely the *inner product of two vectors*. This is usually a complex quantity having the following properties:

$$\begin{aligned} &\text{a)} \quad (\mathbf{a}_1, \mathbf{a}_2) = (\mathbf{a}_2, \mathbf{a}_1)^* \quad ; \\ &\text{b)} \quad (\alpha_1\mathbf{a}_1 + \alpha_2\mathbf{a}_2, \mathbf{a}_3) = \alpha_1(\mathbf{a}_1, \mathbf{a}_3) + \alpha_2(\mathbf{a}_2, \mathbf{a}_3) \quad ; \\ &\text{c)} \quad (\mathbf{a}, \mathbf{a}) \geqslant 0 \quad ; \quad (\mathbf{a}, \mathbf{a}) = 0 \quad , \quad \text{if and only if} \quad \mathbf{a} = \phi \end{aligned} \tag{2.22}$$

(the asterisk * indicates the complex conjugate).

The following definition of the inner product is the most commonly used:

$$(\mathbf{a}_1, \mathbf{a}_2) = \int_T a_1(t)\, a_2^*(t)\, dt \quad . \tag{2.23}$$

The concepts of inner product and vector norm can be linked and the norm can be defined as

$$\|\mathbf{a}\| = \sqrt{(\mathbf{a}, \mathbf{a})} \quad . \tag{2.24}$$

It follows from the properties of inner products (2.22) that this definition satisfies the requirements of the norm (2.19).

As stated above, the norm can in turn generate a metric. Thus, a space with an inner product can be converted into a normalized metric space.

A naturally important concept in a space with an inner product is the *orthogonality of vectors*. Two vectors $\mathbf{a}_1$ and $\mathbf{a}_2$ are orthogonal if

$$(\mathbf{a}_1, \mathbf{a}_2) = 0 \quad . \tag{2.25}$$

If vectors $\{\mathbf{a}_k\}$ are mutually orthogonal then they are linearly independent. Therefore orthogonal vectors can be used as the basis vectors of a linear space.

In a space in which an inner product is defined, a simple relationship can be established between the signal and its representation.

Let A_N be N-dimensional space spanned by the basis $\{\boldsymbol{\varphi}_k\}$, k=0, 1, ..., N-1, so that it consists of vectors having the form

$$\mathbf{a} = \sum_{k=0}^{N-1} \alpha_k \boldsymbol{\varphi}_k \quad . \tag{2.26}$$

Suppose that $\{\boldsymbol{\psi}_k\}$ are vectors that are mutually orthogonal to $\{\boldsymbol{\varphi}_k\}$ and normalized such that

$$(\boldsymbol{\varphi}_k, \boldsymbol{\psi}_l) = \delta(k, l) = \begin{cases} 1, & k = \ell \quad ; \\ 0, & k \neq \ell \quad . \end{cases} \tag{2.27}$$

The function $\delta(k,\ell)$ is known as the Kronecker delta function.

Then

$$(\mathbf{a}, \boldsymbol{\psi}_\ell) = \sum_{k=0}^{N-1} \alpha_k (\boldsymbol{\varphi}_k, \boldsymbol{\psi}_\ell) = \sum_{k=0}^{N-1} \alpha_k \delta(k, \ell) = \alpha_\ell \quad . \tag{2.28}$$

Equation (2.28) is the rule for calculating the coefficients α_k in (2.26). The basis $\{\boldsymbol{\psi}_k\}$ which satisfies (2.27), is said to be reciprocal to $\{\boldsymbol{\varphi}_k\}$. Clearly

$$\mathbf{a} = \sum_{k=0}^{N-1} (\mathbf{a}, \boldsymbol{\psi}_k)\boldsymbol{\varphi}_k = \sum_{\ell=0}^{N-1} (\mathbf{a}, \boldsymbol{\varphi}_\ell)\psi_\ell \tag{2.29}$$

for any pair of reciprocal basis in space A_N.

If a basis $\{\boldsymbol{\varphi}_k\}$ consists of normalized and mutually orthogonal vectors, i.e., is reciprocal to itself:

$$\{\boldsymbol{\varphi}_k, \boldsymbol{\varphi}_\ell\} = \delta(k, \ell) \quad , \qquad \text{for all } k, \ell \tag{2.30}$$

it is termed orthonormal. In this basis

$$\mathbf{a} = \sum_{k=0}^{N-1} (\mathbf{a}, \boldsymbol{\varphi}_k)\boldsymbol{\varphi}_k \quad . \tag{2.31}$$

Given the projections of vectors on the orthonormal basis, it is easy to calculate their norms

$$\|\mathbf{a}\| = (\mathbf{a}, \mathbf{a}) = \sum_{k=0}^{N-1} |\alpha_k|^2 \tag{2.32}$$

and scalar products

$$(\mathbf{a}, \mathbf{b}) = \left(\sum_{k=0}^{N-1} \alpha_k \boldsymbol{\varphi}_k, \sum_{k=0}^{N-1} \beta_k\boldsymbol{\varphi}_k\right) = \sum_{k=0}^{N-1} \sum_{\ell=0}^{N-1} \alpha_k\beta_\ell^*(\boldsymbol{\varphi}_k, \boldsymbol{\varphi}_\ell) = \sum_{k=0}^{N-1} \alpha_k\beta_k^* \quad . \tag{2.33}$$

The representations $\{\alpha_{1,k}\}$, $\{\beta_{1,k}\}$, $\{\alpha_{2,k}\}$, $\{\beta_{2,k}\}$ of any pair of signals **a**, **b** using orthonormal bases $\{\boldsymbol{\varphi}_{1,k}\}$ and $\{\boldsymbol{\varphi}_{2,k}\}$ are linked by the equation

$$\sum_{k=0}^{N-1} \alpha_{\ell,k}\beta_{\ell,k}^* = \sum_{k=0}^{N-1} \alpha_{2,\ell}\beta_{2,\ell}^* \quad , \tag{2.34}$$

which is called *Parseval's theorem* [2.1].

The linear representation of signals as elements of a linear metric space spanned by a finite-dimensional basis is convenient because it makes it possible to describe any signal by a set of a few standard basis functions and a set of numbers. The choice of the basis is determined by the convenience of finding signal representations and, of course, by the nature of the task at hand.

2.3 The Most Common Systems of Basis Functions

2.3.1 Impulse Basis Functions

In the one-dimensional case these functions take the form

$$\varphi_k(t_1) = \text{rect}\,[(t_1 - kT_1)/T_1] \quad , \tag{2.35}$$

where

$$\text{rect}(x) = \begin{cases} 1, & 0 \leqslant x < 1 \quad ; \\ 0, & x < 0,\ x \geqslant 1 \end{cases} \quad .$$

The functions $\varphi_k(t_1)$ are orthogonal on the entire axis t_1 [1]. The space spanned by this basis is composed of step functions. The reciprocal basis for this function system consists of the functions

$$\psi_k(t_1) = \frac{1}{T_1} \operatorname{rect}\left(\frac{t_1 - kT_1}{T_1}\right) \quad . \tag{2.36}$$

Two-dimensional impulse functions in Cartesian coordinates can be determined analogously

$$\varphi_{k,\ell}(t_1, t_2) = \operatorname{rect}\left(\frac{t_1 - kT_1}{T_1}\right) \operatorname{rect}\left(\frac{t_2 - \ell T_2}{T_2}\right) \quad . \tag{2.37a}$$

The reciprocal function system is

$$\psi_{k,\ell}(t_1, t_2) = \frac{1}{T_1T_2} \operatorname{rect}\left(\frac{t_1 - kT_1}{T_1}\right) \operatorname{rect}\left(\frac{t_2 - \ell T_2}{T_2}\right) \quad . \tag{2.37b}$$

The signals are represented on these bases by their average values over the corresponding intervals:

$$\alpha_{k,\ell} = \frac{1}{T_1T_2} \int_{kT_1}^{(k+1)T_1} dt_1 \int_{\ell T_2}^{(\ell+1)T_2} a(t_1, t_2)dt_1dt_2 \quad . \tag{2.38}$$

2.3.2 Harmonic Functions

In the one-dimensional case these functions, which are orthogonal on the interval $(0, T_1)$, take one of three forms:

$$\varphi^s k(t_1) = \sin(\pi k t_1/T_1) \quad ; \tag{2.39a}$$

$$\varphi^c k(t_1) = \cos(\pi k t_1/T_1) \quad ; \tag{2.39b}$$

$$\varphi^e k(t_1) = \exp(i2\pi k t_1/T_1) \quad . \tag{2.39c}$$

The most commonly used are the complex exponential functions (2.39c). The space spanned by the basis (2.39c) is composed of signals defined on an interval of length T_1 and of periodic signals with a period T_1. The expansion of signals in terms of this basis is called expansion in Fourier series.

The reciprocal basis is formed by the functions

$$\psi_k(t_1) = (1/T_1) \exp(i2\pi k t_1/T_1) \quad . \tag{2.40}$$

---

[1] Here and elsewhere it is assumed that the functions are orthogonal in the metrics L^2_T, see Table 2.1 and (2.23).

The coefficients of the Fourier series are calculated from

$$\alpha_k = (\mathbf{a}, \boldsymbol{\psi}_k) = \frac{1}{T_1} \int_0^{T_1} a(t_1) \exp\left(-i2\pi \frac{kt_1}{T_1}\right) dt_1 \quad . \tag{2.41}$$

Complex exponential functions form a so-called *multiplicative* function system: the product of two functions again gives a member of the system. The value of k for this resulting function is equal to the sum of the values for the original functions, and the product of two functions with the same value of k but with different arguments gives the function with the argument equal to the sum of the individual arguments.

The two-dimensional basis, which is composed of complex exponential functions, is usually defined in Cartesian coordinates as the product of one-dimensional functions:

$$\varphi_{k,\ell}(t_1, t_2) = \exp\left[i2\pi\left(\frac{kt_1}{T_1} + \frac{\ell t_2}{T_2}\right)\right] \quad . \tag{2.42}$$

With this definition, two-dimensional integration in calculating the Fourier coefficients (2.23) reduces to two one-dimensional ones.

The space spanned by this basis is composed of functions defined in the rectangle (T_1, T_2) and periodic two-dimensional signals with periods in the form of that rectangle.

2.3.3 Walsh Functions

Walsh functions are interesting in that they take only two values. Let us first consider the one-dimensional Walsh functions. They are derived from the Rademacher functions

$$\mathrm{rad}_k(t_1) = \mathrm{sign}[\sin(2^k \pi t_1 / T_1)] \quad . \tag{2.43}$$

Plots of the first four Rademacher functions are shown in Fig.2.1. Any two Rademacher functions are mutually orthogonal but the set of functions $\{\mathrm{rad}_k(t_1)\}$ is not complete: on the segment $(0, T_1)$ there are also other functions orthogonal to the Rademacher functions (2.43). Examples are the functions $\{\mathrm{sign}[\cos(2^k \pi t_1 / T_1)]\}$.

The Walsh functions are an extension of the Rademacher function system into a complete system. They are defined by [2.2]:

$$\mathrm{wal}_k(t_1) = \frac{1}{\sqrt{T_1}} \prod_{m=0}^{\infty} [\mathrm{rad}_{m+1}(t_1)]^{k_m^G} \quad , \tag{2.44}$$

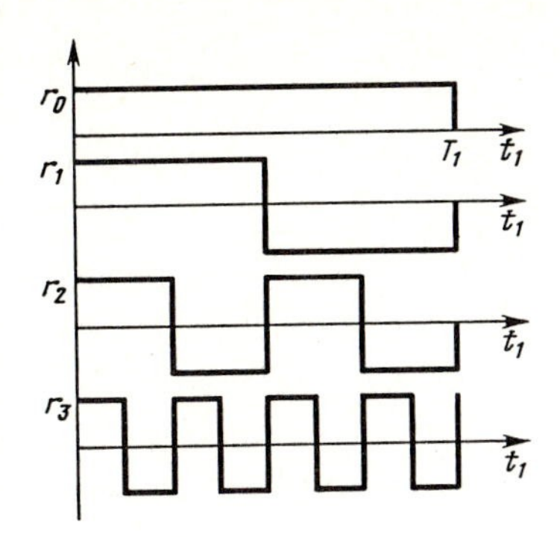

Fig. 2.1 The first four Rademacher functions

where k_m^G is the m digit of the Gray code of the number k. The Gray code is formed from the direct binary code of k by applying the following rule:

$$k_m^G = k_m \oplus k_{m+1} \quad . \tag{2.45}$$

where m is the position of the binary digit (numbered from right to left); k_m is the binary digit in the binary representation of number k;

$$k = \sum_{m=0}^{\infty} k_m 2^m \quad ,$$

and the sign $\oplus$ denot s addition modulo 2.

Plots of the first eight Walsh functions are shown in Fig.2.2. Equation (2.44) helps us to understand Walsh functions. When calculating them by computer, it is more convenient to present the Walsh functions in another form, namely through the values of the digits $\{\xi_\ell\}$ of the binary code of the normalized value of the argument $\xi = t_1/T_1$:

$$\mathrm{wal}_k(\xi) = \frac{1}{\sqrt{T_1}} \prod_{m=0}^{\infty} (-1)^{\xi_{m+1} k_m^G} = \frac{1}{\sqrt{T_1}} (-1)^{\sum_{m=0}^{\infty} k_m^G \xi_{m+1}} \quad , \text{ where} \tag{2.46}$$

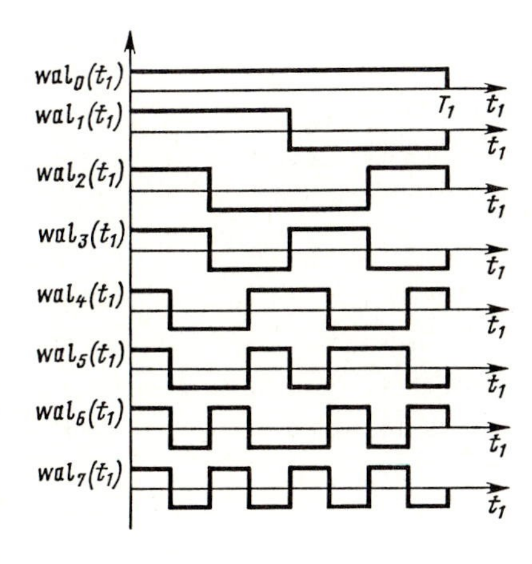

Fig. 2.2 The first eight Walsh functions

$$\xi = \sum_{m=0}^{\infty} \xi_{m+1} \, 2^{-(m+1)} \quad . \tag{2.47}$$

The Walsh functions are orthonormal on the interval $(0, T_1)$. Like complex exponential functions, they form a multiplicative function system. The product of two Walsh functions is also a Walsh function:

$$\begin{aligned} \mathrm{wal}_k(t_1)\,\mathrm{wal}_\ell(t_1) &= \mathrm{wal}_{k\oplus\ell}(t_1) \quad ; \\ \mathrm{wal}_k(T_1\xi)\,\mathrm{wal}_k(T_1\zeta) &= \mathrm{wal}_k[T_1(\xi \oplus \zeta)] \quad , \end{aligned} \tag{2.48}$$

where $\oplus$ denotes logical addition modulo 2. Multiplying two Walsh functions leads to a shift (no carry) of the index or argument, known as a dyadic shift to distinguish it from an arithmetical shift [2.2 - 4].

Two-dimensional Walsh functions are usually defined as the products of one-dimensional ones, i.e.,

$$\mathrm{wal}_{k\ell}(t_1, t_2) = \mathrm{wal}_k(t_1)\,\mathrm{wal}_\ell(t_2) \quad , \tag{2.49}$$

where t_1 and t_2 are considered to be Cartesian coordinates.

The first 16 two-dimensional Walsh functions are shown in Fig.2.3.

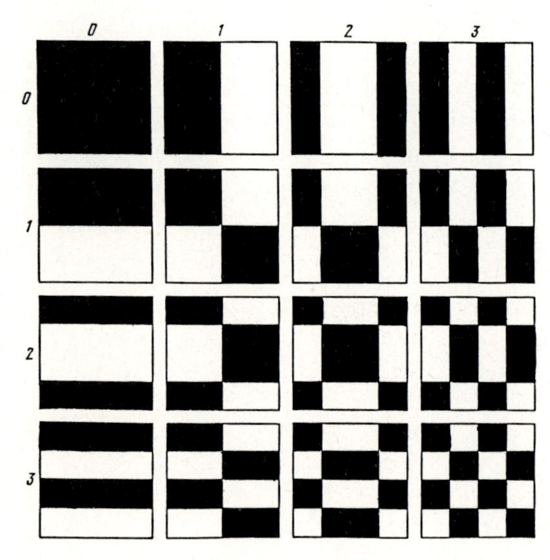

Fig. 2.3 The first 16 two-dimensional Walsh functions

2.3.4 Haar Functions

The combination of impulse functions rect(t) and Rademacher functions generates another system of orthogonal functions, interesting for picture processing, namely the Haar functions. One-dimensional Haar functions are defined on the interval $(0, T_1)$ in the following way:

$$\mathrm{har}_k(t_1) = \frac{2^{m_s/2}}{\sqrt{T_1}} \mathrm{rad}_{m_s+1}(t_1)\, \mathrm{rect}\left[\frac{t_1}{T_1} 2^{m_s} - (k)\mathrm{mod}\, 2^{m_s}\right] \quad , \tag{2.50}$$

where $\mathrm{rad}_{m_s+1}(t_1)$ is the Rademacher function (2.43); rect(t) is defined by (2.35); m_s is the position of the most significant non-zero digit in the binary representation of k; and $(k)\mathrm{mod}\, 2^{m_s}$ is the value of k modulo 2^{m_s}.

Haar functions are orthonormal on the interval $(0, T_1)$. The numerical value of Haar functions at every point can be found and expressed, as with Walsh functions, in terms of the argument written in binary code:

$$\begin{aligned} \mathrm{har}_k(t_1) &= \frac{2^{m_s/2}}{\sqrt{T_1}} (-1)^{\xi_{m_s+1}} \prod_{m=0}^{m-1} \delta(\xi_{m_s-1}, k_m) \\ &= \frac{2^{m_s/2}}{\sqrt{T_1}} (-1)^{\xi_{m_s+1}} \delta([\xi]_{m_s}, (k)\mathrm{mod}\, 2^{m_s}) \quad , \end{aligned} \tag{2.51}$$

where $\delta(\cdot)$ is the Kronecker delta, and $[\xi]_{m_s}$ is the binary number obtained by retaining the m_s highest binary digits of the number ξ. Figure 2.4 shows the first eight Haar functions.

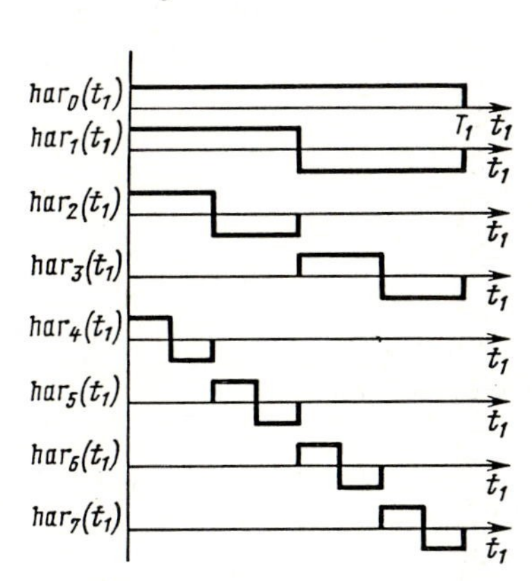

Fig. 2.4 The first eight Haar functions

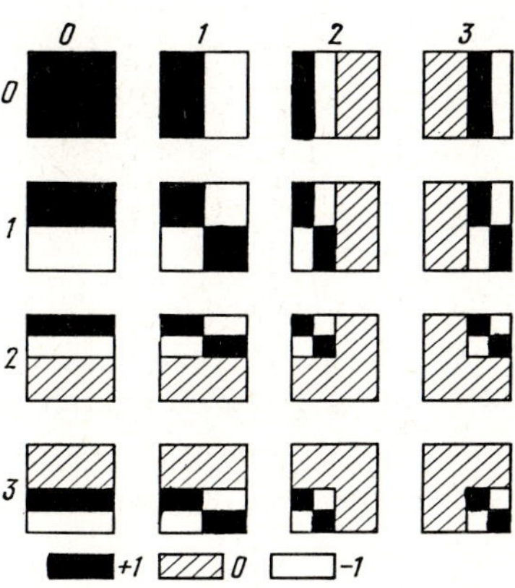

Fig. 2.5 The first 16 two-dimensional Haar functions

The two-dimensional (and multidimensional) Haar functions are usually defined as the product of one-dimensional ones:

$$\mathrm{har}_{k\ell}(t_1, t_2) = \mathrm{har}_k(t_1)\, \mathrm{har}_\ell(t_2) \quad . \tag{2.52}$$

Figure 2.5 depicts the first 16 two-dimensional Haar functions.

2.3.5 Sampling Functions

Sampling functions are defined in the one-dimensional case by

$$\varphi_k(t_1) = \mathrm{sinc}[2\pi F_1(t_1 - k/2F_1)] = \frac{\sin[2\pi F_1(t_1 - k/2F_1)]}{2\pi F_1(t_1 - k/2F_1)} \quad . \tag{2.53}$$

These functions are orthogonal on the entire axis $(-\infty, \infty)$:

$$\begin{aligned} &\int_{-\infty}^{\infty} \mathrm{sinc}\left[2\pi F_1\left(t_1 - \frac{k}{2F_1}\right)\right] \mathrm{sinc}\left[2\pi F_1\left(t_1 - \frac{\ell}{2F_1}\right)\right] dt_1 \\ &= \frac{1}{2F_1} \mathrm{sinc}\left[2\pi(k - \ell)\right] = \begin{cases} 1/2F_1, & k = \ell \quad ; \\ 0, & k \neq \ell \quad . \end{cases} \end{aligned} \tag{2.54}$$

From this it is clear that the reciprocal basis consists of functions of the form $2F_1 \mathrm{sinc}[2\pi F_1(t_1 - k/2F_1)]$.

Sampling functions are usually used for the discrete representation of signals as indicated by the sampling theorem (Sect.3.3). They were given this name because for signals with a limited Fourier spectrum (Sects.2.4 and 3.3) the coefficients α_k corresponding to the basis functions are simply the values (or samples) of signals at $t = k/2F_1$:

$$\alpha_k = 2F_1 \int_{-\infty}^{\infty} a(t_1) \mathrm{sinc}[2\pi F_1(t_1 - k/2F_1)] dt_1 = a(k/2F_1) \quad . \tag{2.55}$$

Two-dimensional sampling functions are usually defined as products of one-dimensional functions.

2.4 Continuous Representations of Signals

So far we described the representation of signals as elements of a finite-dimensional Euclidean or countable-dimensional Hilbert space. This representation, which for a given basis places the signal in relation to a set of numbers, can be called *discrete*. It is the basic digital description of continuous signals. To understand how such a description is obtained it is convenient to regard the discrete representation of signals

$$a(t) = \sum_{k=0}^{N-1} \alpha_k \varphi_k(t) \tag{2.56}$$

as the extreme case of a continuous representation, derived by replacing the number labelling the basis function k by a continuous variable $f \in F$, where F is a finite or infinite interval [2.1]. Then the analogue of (2.56) is

$$a(t) = \int_F \alpha(f) \varphi(t, f) df, \quad t \in T \quad . \tag{2.57}$$

It is also natural to apply such an approach to methods of determining $\alpha(f)$ from $a(t)$, by introducing reciprocal functions $\psi(f, t)$ or the conjugate-basis kernel

$$\alpha(f) = \int_T a(t)\, \psi^*(f, t)dt \quad . \tag{2.58}$$

The function $\alpha(f)$ is called the *integral transformation* of signal $a(t)$ or its *spectrum relative to the continuous basis* $\varphi(t, f)$.

The condition under which functions $\varphi(t, f)$ and $\psi(f, t)$ are reciprocal can be obtained by substituting (2.58) into (2.57):

$$\begin{aligned} a(t) &= \int_F \left[\int_T a(\tau)\, \psi^*(f, \tau)\, \varphi(t, f)d\tau \right] df \\ &= \int_T a(\tau) \left[\int_F \varphi(t, f)\, \psi^*(f, \tau)df \right] d\tau = \int_T a(\tau)\, \delta(t, \tau)d\tau \quad , \end{aligned} \tag{2.59}$$

where

$$\delta(t, \tau) = \int_F \varphi(t, f)\, \psi^*(f, \tau)df \quad . \tag{2.60}$$

Thus functions $\varphi(t, f)$ and $\psi(f, t)$ are reciprocal if the integral (2.60) of their product satisfies the condition

$$a(t) = \int_T a(\tau)\, \delta(t, \tau)d\tau \quad . \tag{2.61}$$

The function $\delta(t, \tau)$, defined by (2.61), is called the δ function. Equation (2.60) can be considered a generalization of condition (2.27) for the reciprocity of basis functions of a finite-dimensional or countable-dimensional signal space, as well as a version of the continuous representation of signals (2.57) when the δ function is the basis.

By substituting (2.57) into (2.58), condition (2.61) can also be made applicable to the second argument f of the reciprocal basis functions:

$$\delta(f, p) = \int_T \varphi(t, f)\, \psi^*(p, t)dt \quad .$$

It is often convenient to consider the δ function as a function of one argument, by assuming that $\delta(t, \tau) = \delta(t - \tau)$. Then (2.61) takes the form

$$a(t) = \int_T a(\tau)\, \delta(t - \tau)d\tau \quad . \tag{2.62}$$

Using these concepts of continuous representation, the discrete representation of signals (2.56) can be written as

$$a(t) = \sum_k \alpha_k \varphi(t, f_k) \quad , \tag{2.63}$$

where $\{f_k\}$ are discrete values of the continuous arguments of the basis function $\varphi(t, f)$, which correspond to the basis functions $\varphi_k(t)$ in (2.56).

Let us now link the discrete signal representation $\{a_k\}$ with its continuous representation, the spectrum $\alpha(f)$, by substituting (2.63) into (2.58):

$$\alpha(f) = \int_T a(t)\, \psi^*(f, t)dt = \sum \alpha_k \int \varphi(t, f_k)\, \psi^*(f, t)dt = \sum \alpha_k \delta(f - f_k) \quad . \tag{2.64}$$

This equation can be regarded as a method of expressing discrete spectra in terms of continuous functions, considering that the discrete representation $\{\alpha_k\}$ of a signal (where such a representation is possible) corresponds to the spectrum $\alpha(f)$, a function which according to (2.64) is an element of space spanned by basis functions $\{\delta(f - f_k)\}$. Clearly, (2.64) can also be regarded as a method of expressing a discrete signal in the form of a continuous one.

Using (2.60) as an analogue of (2.27), the concept of an orthonormal basis (2.30) can also be generalized:

$$\int_F \varphi(t, f)\, \varphi^*(\tau, f)df = \delta(t - \tau) \quad , \tag{2.65a}$$

$$\int_T \varphi(t, f)\, \varphi^*(p, t)dt = \delta(f - p) \quad . \tag{2.65b}$$

The basis, or kernel, of a transformation $\varphi(t, f)$ that satisfies (2.65a,b) is termed selfconjugate. The following equation is true for a self-conjugate basis:

$$\int_T a_1(t)\, a^*_2(t)dt = \int \alpha_1(f)\, \alpha^*_2(f)df \quad . \tag{2.66}$$

This is the continuous analogue of Parseval's theorem (2.34). It is not difficult to check using (2.65) and the spectrum definition (2.58). In the particular case when $a_1(t) = a_2(t)$, expression (2.66) takes the following form:

$$\int_T |a(t)|^2 dt = \int_F |\alpha(f)|^2 df \quad . \tag{2.67}$$

All the basis functions described in the preceding section can also be used as continuous bases, generating the corresponding integral transformations. The most important of these for signal theory is the *Fourier transform* [2.1,5,7], defined in the one-dimensional case by

$$\alpha(f_1) = \int_{-\infty}^{\infty} a(t_1)\, \exp(i2\pi f_1 t_1)dt_1 \quad . \tag{2.68}$$

According to the Fourier integral theorem [2.6], $a(t_1)$ can be recovered from $\alpha(f_1)$ by the inverse Fourier transform:

$$a(t_1) = \lim_{F_1 \to \infty} \int_{-F}^{F_1} \alpha(f_1) \exp(-i2\pi f_1 t_1) df_1 \quad . \tag{2.69}$$

Hence it follows that the kernel of $\exp(i2\pi f_1 t_1)$ is self-conjugate, and the δ function corresponding to this kernel may be written as

$$\begin{aligned} \delta(t - \tau) &= \lim_{F_1 \to \infty} \int_{-F_1}^{F_1} \exp(i2\pi f_1 t_1) \exp(-i2\pi f_1 \tau_1) df_1 \\ &= \lim_{F_1 \to \infty} 2F_1 \frac{\sin[2\pi F_1(t_1 - \tau_1)]}{2\pi F_1(t_1 - \tau_1)} = \lim 2F_1 \operatorname{sinc}[2\pi F_1(t_1 - \tau_1)] \quad . \end{aligned} \tag{2.70}$$

Bearing this in mind, we can define the inverse Fourier transform

$$a(t_1) = \int_{-\infty}^{\infty} \alpha(f_1) \exp(-i2\pi f_1 t_1) df_1 \quad . \tag{2.71}$$

The Fourier transform of two-dimensional signals is usually defined in the Cartesian system of coordinates:

$$\alpha(f_1, f_2) = \int_{-\infty}^{\infty} \int_{-\infty}^{\infty} a(t_1, t_2) \exp[i2\pi(f_1 t_1 + f_2 t_2)] dt_1\, dt_2 \quad . \tag{2.72a}$$

The use of Cartesian coordinates results in the decomposition of the two-dimensional Fourier transform into two one-dimensional ones:

$$\alpha(f_1, f_2) = \int_{-\infty}^{\infty} \exp(i2\pi f_2 t_2) dt_2 \int_{-\infty}^{\infty} a(t_1, t_2) \exp(i2\pi f_1 t_1) dt_1 \quad . \tag{2.72b}$$

Tables 2.2,3 contain the formulae most frequently in describing the properties of one-dimensional and two-dimensional Fourier transforms.

2.5 Description of Signal Transformations

It is convenient to treat the transformations of signals mathematically as a mapping. In general, to describe such a representation it is necessary to specify all possible pairs of input and output signals, i.e., to list all input-output pairs. But this is impractical. Since the basic operations of existing processors are operations on separate numbers, it is impossible to construct such a description for them. We are therefore limited to "hierarchic" descriptions, in which the desired transformation is expressed as a reasonably simple combination of "elementary" transformations, each of which can be described using a subset, not very large, of all possible input and output pairs.

The most important of these "elementary" transformations are the so-called linear transformations and element-by-element nonlinear transformations.

Table 2.2. Properties of the one-dimensional Fourier transform

Function	Fourier transform
1) $a(t_1) = \int_{-\infty}^{\infty} \alpha(f_1)\exp(-i2\pi f_1 t_1)\,df_1$	$\alpha(f_1) = \int_{-\infty}^{\infty} a(t_1)\exp(i2\pi f_1 t_1)\,dt_1$
2) $a^*(t_1)$	$\alpha^*(-f_1)$
3) $a(-t_1)$	$\alpha(-f_1)$
4) $a(t_1) = a^*(t_1)$	$\alpha(f_1) = \alpha^*(-f_1)$
5) $a(t_1) = -a^*(t_1)$	$\alpha(f_1) = -\alpha^*(-f_1)$
6) $a(t_1) = a(-t_1)$	$\alpha(f_1) = \alpha(-f_1)$
7) $a(t_1) = -a(-t_1)$	$\alpha(f_1) = -\alpha(-f_1)$
8) $[a(t_1)+a(-t_1)]/2$; $a(t_1) = a^*(t_1)$	$\mathrm{Re}\{\alpha(f_1)\} = \dfrac{\alpha(f_1)+\alpha^*(f_1)}{2}$
9) $[a(t_1)-a(-t_1)]/2$; $a(t_1) = a^*(t_1)$	$\mathrm{Im}\{\alpha(f_1)\} = \dfrac{\alpha(f_1)-\alpha^*(f_1)}{2i}$
10) $a(t_1-\tau_1)$	$\alpha(f_1)\exp i2\pi f_1 \tau_1$
11) $a(t_1)\exp(-i2\pi p_1 t_1)$	$\alpha(f_1-p_1)$
12) $\delta(t_1)$	1
13) 1	$\delta(f_1)$
14) $\exp(-i2\pi p_1 t_1)$	$\delta(f_1-p_1)$
15) $\cos 2\pi p_1 t_1$	$[\delta(f_1-p_1)+\delta(f_1+p_1)]/2$
16) $\int_{-\infty}^{\infty} a(\tau_1)b(t_1-\tau_1)\,d\tau_1$	$\alpha(f_1)\beta(f_1)$
17) $\int_{-\infty}^{\infty} a(t_1)b^*(t_1)\,dt_1 = \int_{-\infty}^{\infty} \alpha(f_1)\beta^*(f_1)\,df_1$	$\int_{-\infty}^{\infty} \alpha(f_1)\beta^*(f_1)\,df_1 = \int_{-\infty}^{\infty} a(t_1)b^*(t_1)\,dt_1$

Table 2.2 (continued)

Function	Fourier transform
18) $a(t_1)\cos 2\pi p_1 t_1$	$[\alpha(f_1+p_1)+\alpha(f_1-p_1)]/2$
19) $\sum_{k=-\infty}^{\infty} a(t_1-kT_1)$; $a(t_1)=a(t_1)\,\mathrm{rect}\,\frac{t_1+T_1/2}{T_1}$	$\frac{1}{T_1}\sum_{l=-\infty}^{\infty}\alpha\left(\frac{l}{T_1}\right)\delta\left(f_1-\frac{l}{T_1}\right)$
20) $\sum_{k=-\infty}^{\infty}\delta(t_1-kT_1)$	$\frac{1}{T_1}\sum_{l=-\infty}^{\infty}\delta\left(f_1-\frac{l}{T_1}\right)$ $=\sum_{k=-\infty}^{\infty}\exp(\mathrm{i}2\pi kf_1T_1)$
21) $b(t_1)\sum_{k=-\infty}^{\infty}a(t_1-kT_1)$	$\frac{1}{T_1}\sum_{l=-\infty}^{\infty}\alpha\left(\frac{l}{T_1}\right)\beta\left(f_1-\frac{l}{T_1}\right)$
22) $\sum_{k=-\infty}^{\infty}a(kT_1)\delta(t_1-kT_1)$	$\frac{1}{T_1}\sum_{l=-\infty}^{\infty}\alpha\left(f_1-\frac{l}{T_1}\right)$ $=\sum_{k=-\infty}^{\infty}a(kT_1)\exp(\mathrm{i}2\pi kf_1T_1)$
23) $\mathrm{rect}\,\frac{t+T/2}{T}$	$T\,\mathrm{sinc}\,\pi Tf_1$
24) $a(t_1/\varkappa)$	$\varkappa\alpha(\varkappa f_1)$
25) $m_n=\int_{-\infty}^{\infty}t_1^n a(t_1)\,dt_1$	$\alpha(f_1)=\sum_{n=0}^{\infty}\frac{(2\pi\mathrm{i})^n}{n!}m_n f_1^n$
26) $\frac{d^n a(t_1)}{dt_1^n}$	$(-2\pi\mathrm{i}f_1)^n\alpha(f_1)$
27) $\int_{-\infty}^{\infty}a(t_1)\,dt_1$	$\alpha(0)$
28) $\int_{-\infty}^{t}a(\tau)\,d\tau$	$-\frac{\alpha(f_1)}{\mathrm{i}2\pi f_1}+\int_{-\infty}^{\infty}a(t_1)\,dt_1$ $\times\lim_{T\to\infty}2T\,\mathrm{sinc}\,2\pi f_1T$ $=-\frac{\alpha(f_1)}{\mathrm{i}2\pi f_1}+\alpha(0)\,\delta(f_1)$

Table 2.3. Properties of the two-dimensional Fourier transform

Function	Fourier transform
1) $a(t_1, t_2) = \int_{-\infty}^{\infty}\int_{-\infty}^{\infty} \alpha(f_1, f_2)$ $\times \exp[-i2\pi(f_1 t_1 + f_2 t_2)]\, df_1 df_2$	$\alpha(f_1, f_2) = \int_{-\infty}^{\infty}\int_{-\infty}^{\infty} a(t_1, t_2)$ $\times \exp[i2\pi(f_1 t_1 + f_2 t_2)]\, dt_1 dt_2$
2) $a(-t_1, -t_2)$	$\alpha(-f_1, -f_2)$
3) $a(t_1, t_2) = a^*(t_1, t_2)$	$\alpha(f_1, f_2) = \alpha^*(-f_1, -f_2)$
4) $a(t_1, t_2) = a(-t_1, -t_2)$	$\alpha(-f_1, -f_2) = \alpha(f_1, f_2)$
5) $[a(t_1, t_2) + a(-t_1, -t_2)]/2;$ $a(t_1, t_2) = a^*(t_1, t_2)$	$\mathrm{Re}\{\alpha(f_1, f_2)\}$
6) $a(t_1, t_2) = a^*(t_1, t_2)$ $= a(-t_1, t_2)$	$\alpha(f_1, f_2) = \alpha^*(-f_1, -f_2)$ $= \alpha(-f_1, f_2)$
7) $a(t_1 - \tau_1, t_2 - \tau_2)$	$\alpha(f_1, f_2) \exp[i2\pi(f_1\tau_1 + f_2\tau_2)]$
8) $a(t_1, t_2) \exp[i2\pi(t_1 p_1 + t_2 p_2)]$	$\alpha(f_1 + p_1, f_2 + p_2)$
9) 1	$\delta(f_1, f_2) =. \delta(f_1)\delta(f_2)$
10) $\delta(t_1 - \tau_1)$	$\exp(i2\pi f_1 \tau_1)\delta(f_2)$
11) $\cos 2\pi(p_1 t_1 + p_2 t_2)$	$[\delta(f_1 - p_1)\delta(f_2 - p_2)$ $+ \delta(f_1 + p_1)\delta(f_2 + p_2)]/2$
12) $\int_{-\infty}^{\infty}\int_{-\infty}^{\infty} a(\tau_1, \tau_2) b(t_1 - \tau_1,$ $t_2 - \tau_2)\, d\tau_1 d\tau_2$	$\alpha(f_1, f_2)\beta(f_1, f_2)$
13) $\sum_{k=-\infty}^{\infty}\sum_{k=-\infty}^{\infty} a(t_1 + k_1 T_1, t_2 + k_2 T_2);$ $a(t_1, t_2) = a(t_1, t_2)\,\mathrm{rect}\,\frac{t_1 + T_1/2}{T_1}$ $\times\,\mathrm{rect}\,\frac{t_2 + T_2/2}{T_2}$	$\frac{1}{T_1 T_2}\sum_{l_1=-\infty}^{\infty}\sum_{l_2=-\infty}^{\infty} \alpha\left(\frac{l_1}{T_1}, \frac{l_1}{T_2}\right)$ $\times\,\delta\left(f_1 - \frac{l_1}{T_1}\right)\delta\left(f_2 - \frac{l_2}{T_2}\right)$

Table 2.3 (continued)

Function	Fourier transform
14) $\sum_{k_1=-\infty}^{\infty}\sum_{k_2=-\infty}^{\infty}\delta(t_1+k_1T_1) \times\delta(t_2+k_2T_2)$	$\frac{1}{T_1T_2}\sum_{l_1=-\infty}^{\infty}\sum_{l_2=-\infty}^{\infty}\delta\left(f_1-\frac{k_1}{T_1}\right) \times\delta\left(f_2-\frac{k_2}{T_2}\right) = \sum_{k_1=-\infty}^{\infty}\sum_{k_2=-\infty}^{\infty}\exp(\mathrm{i}2\pi) \times(k_1f_1T_1+k_2f_2T_2)$
15) $b(t_1,t_2)\sum_{k_1=-\infty}^{\infty}\sum_{k_2=-\infty}^{\infty}a(t_1+k_1T_1, t_2+k_2T_2)$	$\frac{1}{T_1T_2}\sum_{l_1=-\infty}^{\infty}\sum_{l_2=-\infty}^{\infty}\alpha\left(\frac{l_1}{T_1},\frac{l_2}{T_2}\right) \beta\left(f_1-\frac{l_1}{T_1},f_2-\frac{l_2}{T_2}\right)$
16) $\sum_{k_1=-\infty}^{\infty}a(t_1+k_1T_1,t_2);\quad a(t_1\notin T_1,t_2)=0$	$\frac{1}{T_1}\sum_{l_1=-\infty}^{\infty}\alpha\left(\frac{l_1}{T_1},f_2\right)\delta\left(f_1-\frac{l_1}{T_1}\right)$
17) $\sum_{k_1=-\infty}^{\infty}\delta(t_1+k_1T_1)$	$\frac{1}{T_1}\sum_{l_1=-\infty}^{\infty}\delta\left(f_1-\frac{l_1}{T_1}\right)\delta(f_2)$
18) $\sum_{k_1=-\infty}^{\infty}\sum_{k_2=-\infty}^{\infty}a(t_1+k_1T_{11}+k_2T_{21}; t_2+k_1T_{12}+k_2T_{22})$	$\frac{1}{T_1T_2}\sum_{l_1=-\infty}^{\infty}\sum_{l_2=-\infty}^{\infty}\alpha\left(\frac{l_1}{T_1},\frac{l_2}{T_2}\right) \times\delta\left(p_1-\frac{l_1}{T_1}\right)\delta\left(p_2-\frac{l_2}{T_2}\right);$ $T_1=\sqrt{T_{11}^2+T_{12}^2};$ $T_2=\sqrt{T_{22}^2+T_{21}^2};$ $p_1=\frac{T_{22}f_1+T_{21}f_2}{T_{11}T_{22}-T_{12}T_{21}}T_{11};$ $p_2=\frac{T_{11}f_2+T_{12}f_1}{T_{11}T_{22}-T_{12}T_{21}}T_{22}$
19) $a(t_1/\varkappa_1, t_2/\varkappa_2)$	$\varkappa_1\varkappa_2\alpha(\varkappa_1f_1,\varkappa_2f_2)$
20) $\frac{\partial^2a(t_1,t_2)}{\partial t_1\partial t_2}$	$-4\pi^2f_1f_2\alpha(f_1,f_2)$

Table 2.3 (continued)

Function	Fourier transform
21) $\dfrac{\partial a(t_1, t_2)}{\partial t_1}$	$-i2\pi f_1 \alpha(f_1, f_2)$
22) $\int_{-\infty}^{\infty} \int_{-\infty}^{\infty} a(t_1, t_2) dt_1 dt_2$	$\alpha(0, 0)$
23) $\int_{-\infty}^{\infty} a(t_1, t_2) dt_1$	$\alpha(0, f_2)$
24) $\int_{-\infty}^{t_1} \int_{-\infty}^{t_2} a(\tau_1, \tau_2) d\tau_1 d\tau_2$	$\alpha(0, 0)\delta(f_1)\delta(f_2) - (1/2\pi i f_1)$ $\times \alpha(f_1, 0)\delta(f_2) - (1/2\pi i f_2)$ $\times \alpha(0, f_2)\delta(f_1)$ $-(1/4\pi^2 f_1 f_2)\alpha(f_1, f_2)$
25) $\int_{-\infty}^{t_1} a(\tau_1, t_2) d\tau_1$	$\alpha(0, f_2)\delta(f_1) - (1/2\pi i f_1)\alpha(f_1, f_2)$

2.5.1 Linear Transformations

They are defined on a linear space and have the properties

$$L(\alpha \mathbf{a}_1 + \beta \mathbf{a}_2) = \alpha L\mathbf{a}_1 + \beta L\mathbf{a}_2 \tag{2.73}$$

for any vectors $\mathbf{a}_1$, $\mathbf{a}_2$ and scalars α, β, where $L\mathbf{a}$ denotes the transformation of $\mathbf{a}$, by the operator L. Clearly,

$$L\phi = \phi \quad ; \quad L(-\mathbf{a}) = -L\mathbf{a} \quad , \tag{2.74}$$

i.e., the set of linearly transformed vectors also forms a linear space. For linear operators it is convenient to introduce yet another operation, the product

$$L\mathbf{a} = L_1(L_2\mathbf{a}) \quad . \tag{2.75}$$

The physical equivalent of the product is the series connecting of blocks, representing the individual operators.

Owing to the linearity of the operators, the product is distributive with respect to addition

$$\begin{aligned} L_1(L_2 + L_3) &= L_1L_2 + L_1L_3 \quad ; \\ (L_1 + L_2)L_3 &= L_1L_3 + L_2L_3 \quad . \end{aligned} \tag{2.76}$$

If operator L generates a one-to-one mapping with the domain of application, then there exists an inverse operator L^{-1}, such that

$$LL^{-1}\mathbf{a} = L^{-1}L\mathbf{a} = \mathbf{a} \quad . \tag{2.77}$$

2.5.2 Nonlinear Element-by-Element Transforms

These are transforms of each individual coefficient of the signal representation on the given basis. The transformation law of every coefficient depends only on its label and not on the value of other coefficients.

The simplicity of these "elementary" transformations can be illustrated by considering the linear space of quantized signals. Suppose we have a linear N-dimensional signal space, spanned by a basis $\{\varphi_k\}$, $k = 0, 1, \ldots, N-1$, such that the coefficients of the signal representation on this basis are quantized on M levels, i.e., can take one of M discrete values. Clearly, the number of different signals in such a space is M^N. This is the number of rows that a table must have in the general case to describe the mapping of this space into itself.

Linear transformations are specific in that their effect can be described in terms of the results of only basis function transformations. For this N vectors must be assigned and, in the case where the vector bases are the so-called eigenvectors of the transformations, N numbers.

Nonlinear element-by-element transformations can be described by M numbers, i.e., the results of nonlinear transformation of signal representation coefficients on the given basis. Thus, to determine nonlinear element-by-element transformations it is sufficient to construct a look-up table of M numbers.

2.6 Representation of Linear Transformations with Respect to Discrete Bases

This and the following sections are devoted to an analysis of specific methods of describing linear transformations. Let us begin by discussing signal transformations in spaces spanned by discrete bases. These transformations have a direct bearing on digital signal processing, and by generalizing the results obtained for them, one can move straight on to continuous signal space.

Let A_N be a linear space with an inner product spanned by a linearly independent basis $\{\boldsymbol{\varphi}_k\}$, $k = 0, 1, \ldots, N-1$. There are several ways of representing the linear transformations of the elements of this space.

2.6.1 Representation Using Vector Responses

By virtue of the linearity of transformation L,

$$La = L \sum_{k=0}^{N-1} \alpha_k \boldsymbol{\varphi}_k = \sum_{k=0}^{N-1} \alpha_k L\boldsymbol{\varphi}_k = \sum_{k=0}^{N-1} \alpha_k \boldsymbol{\theta}_k \quad . \tag{2.78}$$

The set $\{\boldsymbol{\theta}_k\}$ contains all the responses of the transformation L to the basis functions $\{\boldsymbol{\varphi}_k\}$. Knowledge of this set is sufficient to find the response of the operator L to any signal from the representation of this signal $\{\alpha_k\}$. This set can therefore be considered to be a representation of L with respect to basis $\{\boldsymbol{\varphi}_k\}$. As distinct from the vector representation on this basis, which is a set of numbers, the representation of the linear transformation is a ranked set of vectors. These vectors are not necessarily linearly independent. If they are not linearly independent, this means that there exists a subspace (i.e., a set of signals in A_N) which is mapped by L to zero. This subspace is called the zero space of linear transformations. Transformations with a non-empty zero space are singular. For example, the transformation operator is singular in irradiation of a photodetector that is insensitive to radiation of certain wavelengths, or of a photodetector whose entrance is sealed with a grid, etc.

For singular transformations the dimensions of output signal space are smaller than the dimensions of input signal space, in the sense that the vectors $\{\boldsymbol{\theta}_k\}$ are not all linearly independent.

2.6.2 Matrix Representations

In operator representations using vector responses, the signal representation coefficients remain unchanged, and during transformations the spatial basis changes, as it were. However, it is usually more convenient to assign a basis to the space that contains the results of linear signal transformation. Let us find the operator representation in this case.

Let $\{\boldsymbol{\theta}_k\}$ be a set of linearly independent vectors, which span the space B_N of the results of the linear transformation of input space A_N. Any vector of B_N can be represented as

$$b = \sum_{k=0}^{N-1} \beta_k \boldsymbol{\theta}_k = \sum_{k=0}^{N-1} (b, \boldsymbol{\eta}_k) \boldsymbol{\theta}_k \quad , \tag{2.79}$$

where $\{\boldsymbol{\eta}_k\}$ is the basis that is reciprocal to $\{\boldsymbol{\theta}_k\}$.

Substituting $b = La = \sum_{\ell=0}^{N-1} \alpha_\ell L\boldsymbol{\varphi}_\ell$, we obtain

$$b = \sum_{k=0}^{N-1} \left(\sum_{\ell=0}^{N-1} \alpha_\ell L\boldsymbol{\varphi}_\ell, \boldsymbol{\eta}_k \right) \boldsymbol{\theta}_k = \sum_{k=0}^{N-1} \sum_{\ell=0}^{N-1} \alpha_\ell (L\boldsymbol{\varphi}_\ell, \boldsymbol{\eta}_k) \boldsymbol{\theta}_k \quad . \tag{2.80}$$

Hence the output signal representation coefficients of the operator L on the basis $\{\boldsymbol{\theta}_k\}$ are

$$\beta_k = \sum_{\ell=0}^{N-1} \alpha_\ell \lambda_{\ell k} \quad , \qquad \text{where} \tag{2.81}$$

$$\lambda_{\ell,k} = (L\boldsymbol{\varphi}_\ell, \boldsymbol{\eta}_k) \quad . \tag{2.82}$$

Equation (2.81) can be written in matrix form:

$$\mathbf{B} = \Lambda \mathbf{A} \quad , \tag{2.83}$$

where Λ is the matrix $\{\lambda_{\ell k}\}$ and $\mathbf{A}$, $\mathbf{B}$ are the column vectors representing the vectors $\mathbf{a}$ and $\mathbf{b} = L\mathbf{a}$ on the bases $\{\boldsymbol{\varphi}_k\}$ and $\{\boldsymbol{\theta}_k\}$, respectively.

2.6.3 Representation of Operators by Means of Their Eigenfunctions and Eigenvalues

Operators can be represented in yet another way by generalizing their representations using vector responses. This method is constructed on the basis of the eigenfunction and eigenvalue concepts of the operator.

The eigenfunctions (eigenvectors) $e_k(t)$ of the operator are functions that are transformed by the operator into themselves apart from a scalar coefficient E_k, which is known as the eigenvalue.

$$Le(t) = E_k e_k(t) \quad . \tag{2.84}$$

The set of vectors $\{e_k(t)\}$ and eigenvalues $\{E_k\}$ is fully defined by the operator and is a very convenient way of representing it. Indeed, if we select $\{e_k(t)\}$ as the basis, then the effect of operator L on signal a(t), which is represented in this basis as

$$a(t) = \sum_{k=0}^{N-1} \alpha^e{}_k e_k(t) \quad , \tag{2.85}$$

is clearly

$$b(t) = La(t) = \sum_{k=0}^{N-1} \alpha^e{}_k Le(t) = \sum_{k=0}^{N-1} \alpha^e{}_k E_k e_k(t) \quad . \tag{2.86}$$

In other words, the representation on the eigenbasis is equal to the product of representation coefficients $\{\alpha_k^e\}$ of signal a(t) times the corresponding eigenvalues $\{E_k\}$:

$$\beta_k = \alpha^e{}_k E_k \quad . \tag{2.87}$$

By substituting (2.84) into (2.82) it is easy to obtain the matrix representation of the operator on this basis:

$$\lambda^{e}_{k,\ell} = E_k \delta(k, \ell) \quad , \tag{2.88}$$

The matrix Λ^e, which represents the operator L relative to the basis consisting of its eigenfunctions, is a diagonal matrix with the eigenvalues $\{E_k\}$ on the diagonal.

2.7 Representing Operator with Respect to Continuous Bases

2.7.1 Operator Kernel

Signals $a(t)$ and the response $b(\tau)$ that linear operators have on them can be represented by their spectra $\alpha(f)$ and $\beta(f)$ with respect to a continuous basis kernel $\varphi(t, f)$:[2]

$$\begin{aligned} a(t) &= \int_F \alpha(f)\, \varphi(t, f)df \quad , \\ b(\tau) &= \int_F \beta(f)\, \varphi(\tau, f)df \quad . \end{aligned} \tag{2.89}$$

Because of the linearity of operator L we have the analogue of (2.78):

$$b(\tau) = La(t) = \int_F \alpha(f)\, L\varphi(t, f)df = \int_F \alpha(f)\, \theta(\tau, f)df \quad , \tag{2.90}$$

where $\theta(\tau, f) = L\varphi(t, f)$, and $\alpha(f)$ and $\beta(f)$ are linked by

$$\beta(f) = \int_F \alpha(p)\, H(f, p)dp \quad , \quad \text{where} \tag{2.91}$$

$$H(f,p) = \int_T \theta(\tau, p)\, \psi^*(f, \tau)d\tau \quad ; \tag{2.92}$$

$\psi(f, \tau)$ is the kernel that is reciprocal to $\varphi(\tau, f)$ in the sense of (2.60).

Equation (2.92) is the analogue of (2.81). It could be said that the linear operator with respect to the continuous basis is characterized by the kernel $H(f, p)$, which in turn is defined by the operator response to basis functions.

2.7.2 Description in Terms of Impulse Responses

A common method of describing a linear operator is to use impulse response. This corresponds to the basis of δ functions:

---

[2] The arguments of signals may not coincide with the result of the action of the operator on them.

$$\varphi(t, \tau) = \delta(t - \tau) \quad . \tag{2.93}$$

The *impulse response* is the operator response to the δ function:

$$h(t, \tau) = L\delta(t - \tau) \quad . \tag{2.94}$$

For such a basis the transformation kernel

$$H(f, p) = \int_{-\infty}^{\infty} \delta(p - t)\, h(f, t)dt = h(f, p) \tag{2.95}$$

coincide with the impulse response, and with the appropriate replacement of variables, (2.91) takes the form

$$\beta(t) = b(t) = \int_{-\infty}^{\infty} \alpha(\tau)\, h(t, \tau)d\tau \quad . \tag{2.96}$$

2.7.3 Description Using Frequency Transfer Functions

Another widely used basis is that defined by $\varphi(t, f) = \exp(-i2\pi ft)$, which leads to the frequency representation of signals and operators. In this case $\alpha(f)$ and $\beta(f)$ are the Fourier transforms of $a(t)$ and $b(t)$.

The response of the operator L to the exponential basis function is

$$\begin{aligned}\theta_{exp}(t, f) &= L \exp(-i2\pi ft) = L \int_{-\infty}^{\infty} \exp(-i2\pi f\tau)\, \delta(t - \tau)d\tau \\ &= \int_{-\infty}^{\infty} h(t, \tau) \exp(-i2\pi f\tau)d\tau\end{aligned} \tag{2.97}$$

The operator kernel with respect to this basis is therefore a two-dimensional Fourier transform of the impulse response

$$H(f, p) = \int_{-\infty}^{\infty} \theta(t, p)\, \psi^*(f, t)dt = \iint_{-\infty}^{\infty} h(t, \tau) \exp[i2\pi(ft - p\tau)]dtd\tau \tag{2.98}$$

– the so-called *operator transfer function*.

2.7.4 Description with Input and Output Signals Referred to Different Bases

It is sometimes convenient to refer the input and output signals of operators to different bases:

$$\begin{aligned}a(t) &= \int_F \alpha(f)\, \varphi_a(t, f)df \quad , \\ b(\tau) &= \int_F \beta(f)\, \varphi_b(\tau, f)df \quad .\end{aligned} \tag{2.99}$$

In this case

$$\beta(p) = \int_F \alpha(f)\, H(f, p)df \quad , \tag{2.100}$$

where

$$H(f, p) = \int_T \theta_a(t, p)\, \psi^*_b(f, t)dt \quad ; \tag{2.101}$$

$$\theta_a(t, p) = L\varphi_a(t, p) \quad . \tag{2.102}$$

Thus, in analyzing spatially non-homogeneous systems, for example, a frequency description of input signals, i.e., a description relative to the basis $\{\exp(i2\pi ft)\}$ is convenient, whereas it is preferable to describe the output signals relative to the basis $\delta(t-\tau)$.

In this case

$$H(f, p) = \int_F \theta_{exp}(t, p)\, \delta(t - f)dt = \theta_{exp}(f, p) \quad , \tag{2.103}$$

where $\theta_{exp}[(f, p)]$ is determined by (2.97).

2.7.5 Description Using Eigenfunctions

Operators are most conveniently described in terms of eigenfunctions. Let $e(t, f)$ be the family of eigenfunctions of operator L, and $E(f)$ be the family of its eigenvalues (operator spectrum):

$$Le(t, f) = E(f)\, e(t, f) \quad . \tag{2.104}$$

Then the kernel of the operator L with respect to $e(t, f)$ defined in accordance with (2.92) is

$$H_e(f, p) = \int_T E(p)\, e(t,p)\, e^*_{re}(f, t)dt = E(p)\, \delta(f - p) \quad , \tag{2.105}$$

where $e_{re}(f, t)$ are the functions reciprocal to $e(t, f)$ in the sense of (2.60). This clearly is the continuous analogue of (2.88).

It follows from (2.91) that the representation (spectrum) of the output signal relative to eigenbasis will be equal to the product of the spectra of the input signal and of the operator:

$$\beta(f) = \int_F \alpha(p)\, H_e(f, p)dp = \int \alpha(p)\, E(p)\, \delta(f - p)dp = \alpha(f)\, E(f) \quad . \tag{2.106}$$

2.8 Examples of Linear Operators

Linear operators are often known as linear filters in signal theory. Let us look at some of the more important classes of filters [2.1].

2.8.1 Shift-Invariant Filters

If the pulse response of a filter (the operator kernel in the δ function basis) is independent of the signal coordinates, that is, if

$$b(t + t_0) = \int_{-\infty}^{\infty} a(\tau + t_0)\, h(t, \tau) d\tau \tag{2.107}$$

for any t_0, then the filter is termed shift-invariant. It can easily be seen, for example, by changing the variables in (2.107), that the impulse response of a shift-invariant filter is a function of the difference between the arguments

$$h(t, \tau) = h(t - \tau) \quad , \tag{2.108}$$

so that

$$b(t) = \int_{-\infty}^{\infty} a(\tau)\, h(t - \tau) d\tau \quad . \tag{2.109}$$

This operation is known as a convolution and is denoted by $\circledast$; we write

$$b = a \circledast h \quad . \tag{2.110}$$

Linear systems with constant parameters (for signals as a function of time), and spatially homogeneous system that are invariant with respect to spatial shift have such an impulse response.

Let us establish the form of the kernel of such systems on an exponential basis. According to (2.98) (see also Table 2.2, line 16),

$$\begin{aligned} H(f, p) &= \int_{-\infty}^{\infty}\int_{-\infty}^{\infty} h(t, \tau) \exp[i2\pi(ft - p\tau)] d\tau\, dt \\ &= \int_{-\infty}^{\infty}\int_{-\infty}^{\infty} h(t - \tau) \exp[i2\pi(ft - p\tau)] d\tau\, dt \\ &= \int_{-\infty}^{\infty} h(t) \exp(i2\pi ft) dt \int_{-\infty}^{\infty} \exp[i2\pi(f - p)]\tau\, d\tau = H(f)\, \delta(f - p) \quad , \end{aligned} \tag{2.111}$$

where

$$H(f) = \int_{-\infty}^{\infty} h(t) \exp(i2\pi ft) dt \tag{2.112}$$

is the frequency transfer function of the shift-invariant filter.

Equation (2.111) shows that complex exponential functions are the eigenbasis of shift-invariant filters. Indeed, their filter response to complex exponential functions is

$$\begin{aligned} \int_{-\infty}^{\infty} h(t - \tau) \exp(-i2\pi f\tau) d\tau &= \exp(-i2\pi ft) \int_{-\infty}^{\infty} h(t) \exp(i2\pi ft) dt \\ &= H(f) \exp(-i2\pi ft) \quad . \end{aligned} \tag{2.113}$$

A remarkable feature of shift-invariant filters is that the successive effect on the signal of several such filters is equivalent to the effect of one filter whose frequency transfer function is equal to the product of the frequency transfer functions of these filters. This follows from (2.106), which links the signal spectrum on the filter eigenbasis with the filter frequency transfer function, as well as from the convolution theorem (Table 2.2, line 6):

$$\beta(f) = \alpha(f)\, H(f) \quad . \tag{2.114}$$

The result of successive signal transformations by several shift-invariant filters is not dependent on the positions of the filter in the sequence, and it is possible to combine these filters in arbitrary groups in any order.

2.8.2 The Identity Operator

This operator is described by

$$a(t) = \int_{\infty}^{\infty} a(\tau)\, h(t, \tau) d\tau \quad . \tag{2.115}$$

It is clear that this operator is shift invariant and its pulse response is a δ function.

2.8.3 Shift Operator

This operator is similar to the identity operator. It maps a signal $a(t)$ onto a shifted signal $a(t - t_0)$:

$$a(t - t_0) = \int_{-\infty}^{\infty} a(\tau)\, h(t, \tau) d\tau \quad . \tag{2.116}$$

For it

$$h(t, \tau) = \delta(t - t_0 - \tau) \quad , \tag{2.117}$$

and its frequency transfer function is

$$H(f) = \exp(-i2\pi f t_0) \quad . \tag{2.118}$$

2.8.4 Sampling Operator

This operator is related to the identity and shift operators and is described by the impulse response

$$h(t - \tau) = 2F \operatorname{sinc} 2\pi F(t - \tau) \tag{2.119}$$

and the frequency transfer function

$$H(f) = \operatorname{rect} \frac{f + F}{2F} \quad . \tag{2.120}$$

For signals whose Fourier spectra differ from zero only on the interval $(-F, F)$, i.e. signals with a limited bandwidth, this operator is clearly equivalent to the identity operator (and the shift operator):

$$a(t) = 2F \int_{\infty}^{\infty} a(\tau) \text{ sinc } 2\pi F(t - \tau)d\tau \quad . \tag{2.121}$$

As follows from (2.114,120), this operator converts all other signals into signals with limited bandwidths.

2.8.5 Gating Operator (Multiplier)

This operator effects transformations of the type

$$b(t) = k(t)\, a(t) \quad . \tag{2.122}$$

Its impulse response is

$$h(t, \tau) = k(t)\, \delta(t - \tau) \quad . \tag{2.123}$$

It is thus not a shift-invariant filter operator.

If the convolution theorem (Table 2.2, line 16) is applied to (2.122), it can be seen that the kernel of this operator with respect to the exponential basis is $K(f - p)$, where

$$K(f) = \int_{-\infty}^{\infty} k(t) \exp(i2\pi ft)dt \quad . \tag{2.124}$$

3 Discretization and Quantization of Signals

The first step in digital picture processing is the conversion of a continuous picture into a digital signal, i.e., a sequence of digits. This chapter is devoted to the problem of optimal conversion.

General ideas and theoretical limits are discussed in Sect.3.1, while Sect.3.2 presents the commonly used two-stage procedure of conversion, with discretization as the first stage and element-by-element quantization as the second. The sampling theorem as the basis of discretization is formulated and proved. Section 3.3, and Sect. 3.4 describe the peculiarities of the 2-D sampling theorem, together with some generalizations. Section 3.5 elaborates a method of estimating the inaccuracies connected with the sampling and restoration of continuous signals. In Sect.3.6, some practical considerations concerning picture discretization are discussed, and in Sect.3.7, general ideas relating to the dimensionality of signal space needed for discrete representation of signals are explained. Section 3.8 deals with optimum element-by-element quantization, examples of which are presented in Sect.3.9. Some useful practical aspects of picture quantization are discussed in Sect.3.10, and Sect.3.11 reviews methods of picture coding based on the ideas introduced.

3.1 Generalized Quantization

So far we have been concerned mainly with analogue signals. In digital processing these are usually the original signals, and the first stage of processing consists in converting them into digital form.

In a broad sense, a digital signal can be considered as a number. The conversion of an analogue signal into a digital signal consists in mapping analogue signal space into a finite set of signals. The signal space is divided into a finite set of subspaces, and in every subspace a representative signal is chosen such that all the other signals of the given subspace can, within a given degree of accuracy, be replaced by the representative signal. The subspace is indexed, and thus every analogue signal in the signal space, which is divided into cells or domains of equivalence, can be made to correspond to a number, i.e., the number of representative signal. Partitioning signal space in this way is called *generalized quantization* (as distinct from simple quantization, which will be discussed below) [3.1]. The subspaces, or cells, into which the signal space is divided, are called ε cells. In this way emphasis is given to the fact that for all signals belonging to a given subspace or cell there is a representative signal which does not differ from other signals in the cell by more than a small quantity ε. Both the metric, which is a measure of the difference between signals, as measured in terms of a certain metric, and the value ε itself are defined by the accuracy required in the digital representation; that is, the demands of the user and the characteristics of the digital processing procedure must be taken into account when they are assigned.

An important question in generalized quantization is the volume of the digital representation obtained. This volume can be measured in terms of the number of representative signals. On this the complexity (i.e., the feasibility and cost) of digital processing systems directly depends.

The most general method of evaluating the volume of discrete representations is provided by information theory and is based on the statistical description of signals and their transformations [3.2,3]. Depending on the particular statistical treatment, every element in the volume of signal space is preassigned a certain probability measure, i.e., the probability that in all the realizations (experiments, observations and similar actions) the signals belonging to it will be found.

The accuracy of signal reproduction is also treated statistically as the mean value of a certain distance between signals in the metric determined by the signal user and by the distinctive features of the processing being conducted.

Let $dP(\mathbf{a},\mathbf{b})$ be the joint probability that signals from a certain elementary volume of signal space $\{A\}$ surrounding a point $\mathbf{a}$ will be converted into signals in the elementary volume of the space of their transformations, $\{B\}$, surrounding point $\mathbf{b}$; $p(\mathbf{a})$, $p(\mathbf{b})$, $p(\mathbf{a},\mathbf{b})$ are the corresponding probability densities. Let us introduce the value

$$I(A,B) = \int_A \int_B \log \frac{p(\mathbf{a},\mathbf{b})}{p(\mathbf{a})\,p(\mathbf{b})}\, dP(\mathbf{a},\mathbf{b}) \quad , \tag{3.1}$$

which in information theory is called the mutual information in the signals from $\{A\}$ and $\{B\}$ [3.2].

Let $D(\mathbf{a},\mathbf{b})$ be a measure of the difference between signals $\mathbf{a}$ and $\mathbf{b}$ from the user's point of view. Then the value

$$\bar{D}(A,B) = \int_A \int_B D(\mathbf{a},\mathbf{b})\, dP(\mathbf{a},\mathbf{b}) \qquad (3.2)$$

is a measure of the difference between the signals and the results of their transformation averaged over the signal space. If for various transformations $\mathbf{a} \to \mathbf{b}$, the value of D remains below a certain tolerable limit ε, one can consider the signals in {B} to be on average indistinguishable from the signals in {A} as regards the prescribed criterion $D(\mathbf{a},\mathbf{b})$ and the given average value of the mapping error ε. Some of these transformations result in a minimal quantity of mutual information I(A,B). This minimal value I(A,B) is called the ε entropy of signal space [3.2]:

$$H_\varepsilon = \min I(A,B) \quad \text{for} \quad \bar{D}(A,B) \leqslant \varepsilon \quad . \qquad (3.3)$$

According to one of the basic results of information theory, the value H_ε determines the lower bound of the number of signals N which is necessary and sufficient for digital representation of signals from space {A} with the prescribed accuracy criterion $D(\mathbf{a},\mathbf{b})$ [3.2]:

$$N \geqslant 2^{H_\varepsilon} \quad . \qquad (3.4)$$

The numerical evaluation of H_ε for signals such as pictures is quite a complicated task, in that quantitative criteria for distinguishing one picture from another and adequately and exhaustively describing the viewer's perception of the picture are unknown. For a rough evaluation of the upper bound of H_ε for pictures (confirmed by results obtained in picture coding where a high reproductive quality was retained [3.4], the product of picture area $Q_{x,y}$ and the area of the corresponding spatial frequency spectrum, $Q_{fx,fy}$, can be used:

$$H_\varepsilon < Q_{x,y}\, Q_{fx,fy} \quad . \qquad (3.5)$$

3.2 Concepts of Discretization and Element-by-Element Quantization

Let us examine some concrete methods of digital signal description. It is usually difficult to implement general quantization in one go. In converting continuous signals into digital ones a two-step procedure is therefore adopted: discretization followed by element-by-element quantization.

3.2.1 Discretization

Discretization is the replacement of a continuous signal by a number sequence which is the representation of this signal on some discrete basis. In cases where the given signal is not an element of the linear space spanned by this basis, a signal is sought from that space which in a certain sense (specially defined for each task) is the best approximation of the given signal, and the representation of this approximate signal is then sought. The basis of the linear space can be chosen from consideration relating to convenience of implementation, accuracy of approximation, etc.

The best signal representation in a linear normalized space, with a metric generated by the norm and the norm generated by inner products (in Hilbert space), is the representation in which the norm of the difference between the signal and the vector representing it is least. It is known [3.5] that such a representation is obtained by projecting the signal on the N-dimensional space, spanned by the given basis; i.e., the coefficients of the representation are equal to the inner products of the signal and the corresponding basis functions[1]:

$$\alpha_k = (\mathbf{a}, \boldsymbol{\varphi}_k) \quad . \tag{3.6}$$

3.2.2 Element-by-Element Quantization

Element-by-element quantization is separate quantization of each of the numbers $\{\alpha_k\}$, which represent a given signal on the prescribed base. In other words, the continuous and, generally speaking, infinite scale of $\{\alpha_k\}$ is replaced by a discrete and finite scale of values. Discretization belongs to the class of linear signal transformations; whereas quantization is in the class of element-by-element nonlinear transformations.

Using a two-stage procedure in digital description means that in the signal space, cells of equivalence (quantized cells) are built which are, as it were, hypercubes with edges parallel to the base coordinate axes. The size of the edges is defined by the so-called quantizing step—the quantization interval at a given point on the number scale which corresponds to the given base functions. With this procedure the task of optimal quantization reduces

---

[1] This fact makes it easier to understand the concepts of norms, metrics, and inner products, as well as their relations and the methods of determining them. It shows that if the distinction between the signal and its linear representation $\sum_k \alpha_k \varphi_k(t)$ is characterized by their difference, then the metric of the signal space and inner products are determined by the norm, which is a means of measuring vector length (in this case the vector of the difference).

to making the optimal choice of basis function systems and achieving the optimal arrangement of quantization scales on the coordinate axes. This corresponds to choosing basis functions such that the resulting quantization hypercubes can be packed in the most favourable way with the minimum number of quantities in the ε domains corresponding to the given criteria for the accuracy of the digital representation.

3.3 The Sampling Theorem

The most common method of discretizing signals is based on the so-called sampling theorem: by means of interpolation, signals whose Fourier spectrum is zero beyond the interval $(-F, F)$ can be restored from samples taken at intervals $\Delta t = 1/2F$ apart:

$$a(t) = \sum_{k=-\infty}^{\infty} a(k\Delta t)\,\mathrm{sinc}\left[2\pi F\left(t - \frac{k}{2F}\right)\right] \cdot \tag{3.7}$$

Equation (3.7) represents the signal $a(t)$ as an expansion in the sampling functions $\mathrm{sinc}[2\pi F(t - k/2F)]$, and the signal representations relative to this basis are the samples of the signal itself:

$$a(k\Delta t) = 2F \int_{-\infty}^{\infty} a(t)\,\mathrm{sinc}\left[2\pi F\left(t - \frac{k}{2F}\right)\right]dt = \int_{-\infty}^{\infty} a(t)\,\delta\left(t - \frac{k}{2F}\right)dt \quad . \tag{3.8}$$

The validity of (3.7) for any signal with a limited bandwidth is derived as follows:

$$\begin{aligned}\sum_{k=-\infty}^{\infty} a(k\Delta t)\,\mathrm{sinc}\left[2\pi F\left(t - \frac{k}{2F}\right)\right] &= \sum_{k=-\infty}^{\infty} a(k\Delta t) \int_{-\infty}^{\infty} \mathrm{sinc}\left(2\pi F\tau\right)\delta\left(t - \tau - \frac{k}{2F}\right)d\tau \\ &= 2F\,\mathrm{sinc}(2\pi Ft) \circledast \frac{1}{2F}\sum_{k=-\infty}^{\infty} a(k\Delta t)\,\delta\left(t - \frac{k}{2F}\right) \quad ,\end{aligned} \tag{3.9}$$

where $\circledast$ is the convolution operation.

According to the convolution theorem (Table 2.2, line 16), the Fourier spectrum of a convolution (3.9) is equal to the product of the spectrum $\mathrm{rect}[(f+F)/2F]$ of the function $2F\,\mathrm{sinc}(2\pi Ft)$, see (2.120), and the spectrum

$$\sum_{\ell=-\infty}^{\infty} \alpha(f - 2F\ell) \quad \text{of the function} \quad \frac{1}{2F}\sum_{k=-\infty}^{\infty} a(k\Delta f)\,\delta\left(t - \frac{k}{2F}\right)$$

(Table 2.2, line 22), i.e. the spectrum $\alpha(f)$, repeated with period $2F$, of the signal $a(t)$. This product is obviously equal to $\alpha(f - 2F\ell)$ when $\ell = 0$, i.e., to the signal spectrum $\alpha(f)$, since $\mathrm{rect}[(f+F)/2F] = 0$ at all other ℓ. From this (3.7) follows.

If the signal spectrum $a(t)$ is limited to the interval (F_1, F_2), which is asymmetrical about the origin, then by a similar argument it is not difficult to show that the sampling theorem is valid in the following form

$$a(t) = \sum_{k=-\infty}^{\infty} a\left(\frac{k}{F_2 - F_1}\right) \times \operatorname{sinc}\left[\pi(F_2 - F_1)\left(t - \frac{k}{F_2 - F_1}\right)\right] \exp\left[i\pi(F_2 + F_1)t\right] \quad . \tag{3.10}$$

The method used to demonstrate the validity of (3.7) clearly illustrates what is meant by transforming the continuous signal into sample sequences and restoring a continuous signal from the samples by means of the sampling theorem. Corresponding to the discrete signal that has the form

$$\tilde{a}(t) = \frac{1}{2F} \sum_{k=-\infty}^{\infty} a\left(\frac{k}{2F}\right) \delta\left(t - \frac{k}{2F}\right) \tag{3.11}$$

and is composed of samples of a continuous signal, is the periodically multiplied spectrum

$$\tilde{\alpha}(f) = \sum_{t=-\infty}^{\infty} \alpha(f - 2F\ell) \quad , \tag{3.12}$$

of a continuous signal. The restoration of the continuous signal from its samples can be represented as the transformation of a discrete signal by the linear shift-invariant filter with rectangular frequency transfer function (2.120).

3.4 Sampling Theory for Two-Dimensional Signals

The discretization of two-dimensional signals and pictures is usually based on a generalization of the sampling theorem to the two-dimensional case, since for all practical purposes the picture spectrum can be considered non-zero only over a limited part of the frequency domain. However, two-dimensional sampling theory is much richer than its one-dimensional counterpart, since the two-dimensional interval is a substantially more complex mathematical object than the one-dimensional one [3.6].

Let the spectrum $\alpha(f_1, f_2)$ of a picture $a(t_1, t_2)$ differ from zero only within the limits of the figure **s**, which is enclosed by a line S in the spatial frequency plane with Cartesian coordinate system (f_1, f_2) (Fig.3.1a). With this knowledge the picture can be completely restored from the spectrum. Clearly, it can also be restored from the enlarged spectrum (Fig.3.1b) derived from the spectrum in Fig.3.1a by periodic extension along the coordinate axes f_1 and f_2, with steps F_1 and F_2 exceeding the longitudinal size (in direction f_1)

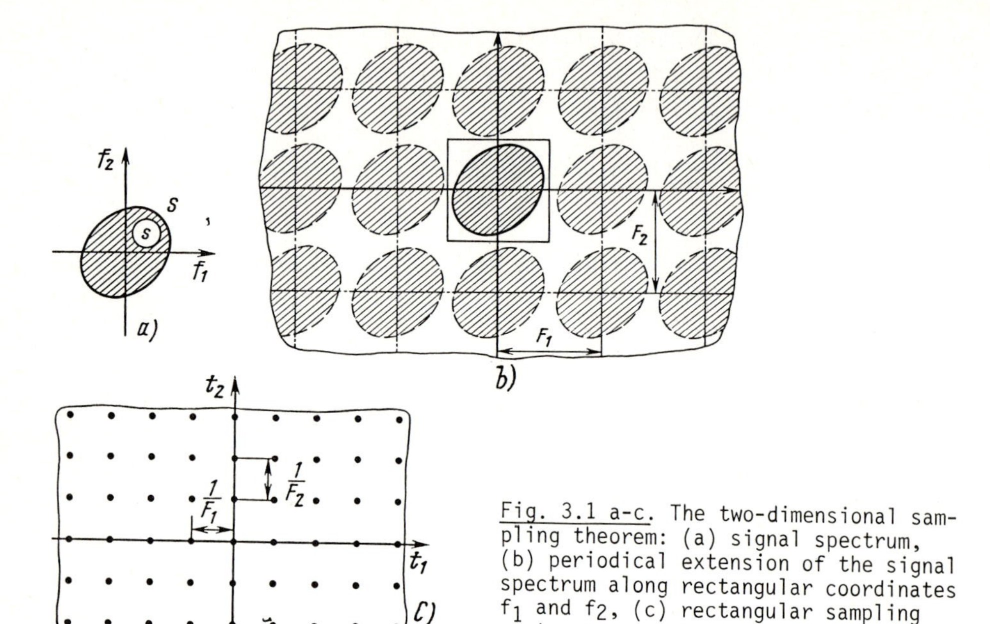

Fig. 3.1 a-c. The two-dimensional sampling theorem: (a) signal spectrum, (b) periodical extension of the signal spectrum along rectangular coordinates f_1 and f_2, (c) rectangular sampling raster

and the transverse size (in direction f_2) of the figure **s** (Fig.3.1a). To do this it is sufficient to pass a signal with a periodically multiplied spectrum (Fig.3.1b)

$$\tilde{\alpha}(f_1, f_2) = \sum_{\ell_1=-\infty}^{\infty} \sum_{\ell_2=-\infty}^{\infty} \alpha(f_1 + \ell_1 F_1, f_2 + \ell_2 F_2) \tag{3.13}$$

through a two-dimensional filter with a frequency transfer function

$$H(f_1, f_2) = \Pi_{F_1}(f_1)\, \Pi_{F_2}(f_2) \quad , \tag{3.14}$$

where

$$\Pi_F(f) = \operatorname{rect} \frac{f + F/2}{F} \quad . \tag{3.15}$$

In other words, the filter passes only those components of the spatial frequency spectrum that are within the rectangle shown in Fig.3.1b by a solid line.

As follows from the properties of the two-dimensional Fourier transform (Table 2.3, line 13), a discrete signal

$$\tilde{a}(t_1, t_2) = \frac{1}{F_1 F_2} \times \sum_{k_1=-\infty}^{\infty} \sum_{k_2=-\infty}^{\infty} a\left(\frac{k_1}{F_1}, \frac{k_2}{F_2}\right) \delta\left(t_1 - \frac{k_1}{F_1}\right) \delta\left(t_2 - \frac{k_2}{F_2}\right) \quad , \tag{3.16}$$

composed of the samples $a(k_1/F_1, k_2/F_2)$ of the initial picture, distributed on a rectangular raster with mesh-length $1/F_1$ and $1/F_2$ (Fig.3.1c), corresponds to spectrum (3.13). The initial picture, restored from the samples by filtering (3.14), can be written as

$$a(t_1, t_2) = \tilde{a}(t_1, t_2) \circledast h(t_1, t_2) = \sum_{k_1=-\infty}^{\infty} \sum_{k_2=-\infty}^{\infty} a\left(\frac{k_1}{F_1}, \frac{k_2}{F_2}\right) \times \operatorname{sinc}\left[2\pi F_1\left(t_1 - \frac{k_1}{F_1}\right)\right] \operatorname{sinc}\left[2\pi F_2\left(t_2 - \frac{k_2}{F_2}\right)\right] \quad , \tag{3.17}$$

where $h(t_1, t_2)$ is the impulse response of the restoring filter—the Fourier transform of the function $H(f_1, f_2)$ (3.14).

Equation (3.17) illustrates the two-dimensional sampling theorem for picture discretization on a rectangular raster.

The number of picture samples per unit area with such discretization is clearly F_1F_2, i.e., the area of the rectangle limiting the picture spectrum. In order to make this number as small as possible, the periodicity of the spectrum must be decreased. This can be done as long as the neighbouring periods do not overlap; otherwise it will be impossible to restore the picture without distortion by filtering which truncates these periods. Hence, decreasing the number of picture samples during discretization becomes a matter of packing the components of the spectrum densely during the periodic continuation in the spatial frequency plane.

A denser spectral packing than that of Fig.3.1b, and an accordingly less dense sample distribution during the discretization of pictures with the spectrum shown in Fig.3.1a can be achieved if periodic spectral continuation and the corresponding discretization are performed in a skewed coordinate system matched to the shape of a curve which restricts the picture spectrum (Fig.3.2a,b). If the line enclosing the spectrum is circular, the optimal

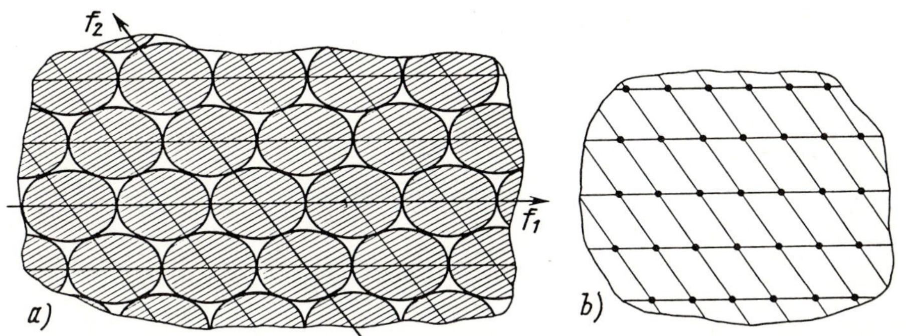

Fig. 3.2 a,b. The two-dimensional sampling theorem: (a) periodical extension of a signal spectrum in a skewed coordinate system, (b) skewed sampling raster

picture sample distribution will form the nodes of a regular hexagonal raster. This yields a saving of approximately 15 percent in the number of samples [3.1].

The relation between the spectrum continuation period in the skewed coordinate system and the discretization step is given in Table 2.3, line 18.

The choice of the optimal form for the raster is not the only method of dense spectral packing during discretization. Another possibilty, which again has no one-dimensional analogue, is rotation of the coordinate system. This results in a distinctive type of discretization that can be called "sampling with gaps" [3.7].

Let us consider the spectrum in Fig.3.3a, which differs from zero only in the shaded part. This part is inscribed within a square. Figure 3.3b shows the dense packing of a spectrum plane derived by superposing two initial spectra from Fig.3.3a that have been periodically continued in the Cartesian coordinate system and rotated by 90° relative to each other. Each of the periodically continued spectra corresponds to a discrete picture with the samples distributed at the nodes of a square raster, as shown in Fig.3.3c. Both of these discrete pictures, rotated as were their spectra by 90° relative to

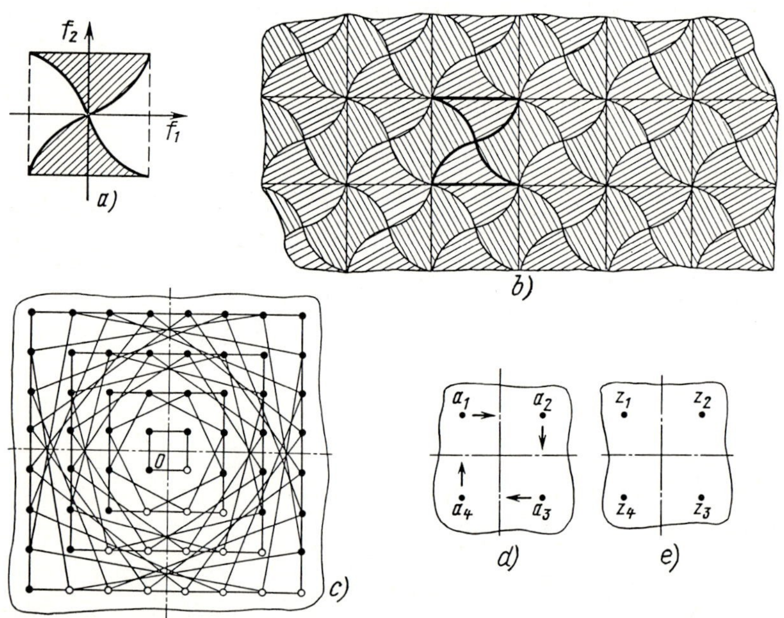

Fig. 3.3 a-e. Discretization with gaps: (a) signal spectrum, (b) pattern in frequency domain after superposing two initial spectra (a) rotated by 90° in relation to each other and periodically extending the result in a rectangular coordinate system, (c) sampling raster with open circles denoting redundant samples, (d) cyclical permutation of the raster nodes during rotation, (e) samples formed by the superposing of samples a_1, a_2, a_3, a_4 of the initial picture

each other (say around point 0, as in Fig.3.3c), are added during superposition. As a result a new discrete signal is obtained whose samples are clearly equal to the paired sums of the initial picture samples. The methods of forming these pairs during raster rotation around point 0 (Fig.3.3c) is presented in Figs.3.3c-e. Figure 3.3c shows the connecting lines of the raster nodes, which are superimposed during rotation by 90°. From the figure it is clear that the nodes form groups of four in which cyclical permutation of the nodes takes place during rotation, as shown by the arrows in Fig.3.3d.

Let z_1, z_2, z_3, z_4 (Fig.3.3e) be the samples formed in each group by the superposition of samples a_1, a_2, a_3, a_4 of the initial picture. From Fig.3.3d it can be seen that for clockwise rotation

$$z_1 = a_1 + a_4 \quad ; \quad z_2 = a_1 + a_2 \quad ;$$
$$z_3 = a_2 + a_3 \quad ; \quad z_4 = a_3 + a_4 \quad . \tag{3.18}$$

It is not difficult to see that of the four numbers z_1, z_2, z_3, z_4, only three are linearly independent, and the fourth can be written as a linear combination of the others. Thus, for example,

$$z_4 = z_1 - z_2 + z_3 \quad . \tag{3.19}$$

Hence, it follows that in each group of four samples (Fig.3.3c) one sample is redundant, since it can be calculated from (3.19) if the other three samples are known. Thus, owing to dense packing, rotation reduces the number of samples by 25 percent. These redundant samples are shown in Fig.3.3c as open circles. The figure clearly shows that they occupy the lower quarter of the plane, as if they had fallen out of the discretization of their own accord. The procedure for restoring continuous pictures after such "discretization with gaps" must take place in two stages: the restoration of the omitted samples by (3.19), and the restoration of the continuous signals by passing a discrete signal through the filter, which suppresses the frequency components beyond the field enclosed in Fig.3.3b by the thick line.

An interesting example of dense packing is a pattern by the Dutch artist M. Escher (Fig.3.4). It consists of three patterns constructed from the periodic continuation of lizard-like shapes in a hexagonal coordinate system. Two patterns are rotated 60°, clockwise and counterclockwise relative to the third pattern and shifted by half a period along one of the coordinate axes. This leads to dense packing of the plane. Each pattern, if it is considered the periodic continuation of the spectrum of a signal, has a corresponding discrete signal with the distribution of samples at the nodes of a hexagonal

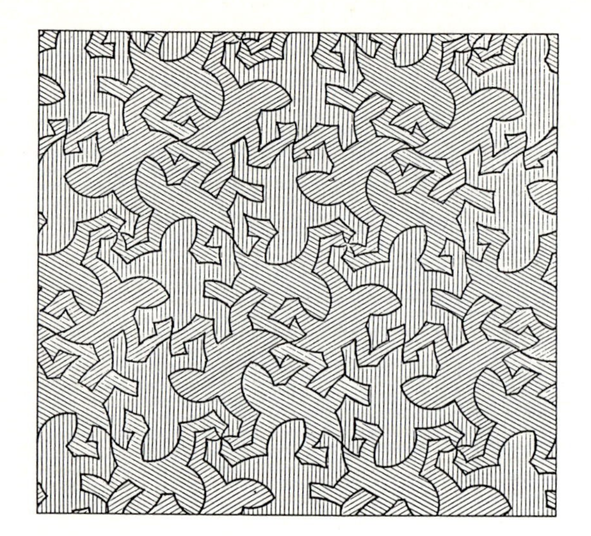

Fig. 3.4. Example of dense packing in a pattern by the artist M. Escher

lattice. Moreover, the patterns rotated by $\pm 60^{\circ}$ have corresponding signals whose samples are multiplied by a complex exponential function that depends on their coordinates (this is in accordance with the shift theorem; see Table 2.3, line 8). It can be shown that with the superposition of two rasters rotated by 60°, the samples separate into groups of six samples, taken in pairs. As a result, one out of every six is redundant; it can be calculated from the remaining five (as in the previous case of a square raster). Thus with the dense spectral packing shown in Fig.3.4, there is a saving of $1/6 + 1/6 = 1/3$ of the samples in the initial hexagonal rasters.

This discussion clearly shows that the saving in the number of samples is not as high with dense packing in the frequency domain after rotation as it is with the optimal choice of a skewed coordinate system. However, there is an even more general discretization procedure, which involves exploiting the topology of the spectral distribution, that in principle makes it possible to decrease the number of samples per unit of picture area to a minimum equal to the area of the spatial frequency spectrum of the picture. It consists in splitting the spatial frequency spectrum of the picture into zones, and reassembling them into a square of the very same area using rotations, shifts and mirror reflections. This corresponds to extracting separate components from the picture by means of filters, whose frequency transfer functions are constant within the area of each zone and zero beyond it; rotating them; multiplying them by a complex exponential function in conforming with the shift of the zone; and subsequently composing them into the auxiliary picture, which can now without loss be subjected to discretization on a square raster. The initial picture is restored in the usual way from samples, but in reverse order from the auxiliary picture.

Let us note in conclusion that in the digital processing of pictures and two-dimensional signals simple discretization along a rectangular raster is at present the most commonly used. It is the most universal and most suitable method, being well adapted to the one-dimensional structure of digital computers and processors. Hexagonal rasters are used in colour television and printing. Other forms of skewed rasters are sometimes used in special digital television systems.

3.5 Errors of Discretization and Restoration of Signals in Sampling Theory

Sampling theory is valid for signals whose spectra are exactly equal to zero beyond a certain frequency interval. Such idealized signals do not exist. However, to a known degree of accuracy the discretization method proposed by sampling theory can be applied to the discretization of signals whose spectrum falls more or less rapidly to a negligibly small level beyond a certain definite frequency interval.

Let us see how the signal distortion that occurs in such discretization can be estimated. Distortion can arise both during the transition from continuous to discrete representation and during the restoration of continuous signals from the discrete representation.

According to the sampling theorem the discrete representation of a signal with a limited spectrum given on the interval $(-F, F)$ is made up of the samples of that signal, with the step length $\Delta t = 1/2F$:

$$a(k\Delta t) = \int_{-\infty}^{\infty} a(t)\, \delta(t - k\Delta t)dt = 2F \int_{-\infty}^{\infty} a(t)\, \mathrm{sinc}[2\pi F(t - k\Delta t)]dt \quad . \tag{3.20}$$

According to (3.7), for signals whose spectrum is not strictly limited to the interval $(-F, F)$, only the first part of (3.20) is valid:

$$\alpha_k = 2F \int_{-\infty}^{\infty} a(t)\, \mathrm{sinc}[2\pi F(t - k\Delta t)]dt \neq a(k\Delta t) \quad . \tag{3.21}$$

With this the norm $||\varepsilon||^2$ of the difference between the restored

$$\tilde{a}(t) = \sum_{k=-\infty}^{\infty} \alpha_k\, \mathrm{sinc}[2\pi F(t - k\Delta t)] \tag{3.22}$$

and initial signal is minimal, and

$$||\varepsilon||^2 = \int_{-\infty}^{\infty} |a(t) - \tilde{a}(t)|^2 dt = \int_{\infty}^{F} |\alpha(f)|^2 df + \int_{F}^{\infty} |\alpha(f)|^2 df \quad . \tag{3.23}$$

However, values α_k will no longer be samples of the signal a(t), but rather samples of the signal

$$\tilde{a}(t) = 2F \int_{-\infty}^{\infty} a(\tau) \operatorname{sinc}[2\pi F(t - \tau)]d\tau \quad . \tag{3.24}$$

This is the result of the action on the initial signal of the linear operator filter, which passes only frequency components that fall within the interval (-F, F).

If the true samples are used in the discrete signal representation:

$$a(k\Delta t) = \int_{-\infty}^{\infty} a(t)\, \delta(t - k\Delta t)dt \quad , \tag{3.25}$$

the restoration error norm corresponding to

$$\tilde{a}_1(t) = \sum_{k=-\infty}^{\infty} a(k\Delta t) \operatorname{sinc}[2\pi F(t - k\Delta t)] \tag{3.26}$$

is greater than that given by (3.23). Figure 3.5 illustrates the reason for this increase in error. The solid line shows the one-dimensional signal spectrum (or the cross-section of a two-dimensional spectrum) periodically continued with step-length 2F, and the dashed line shows the shape of the frequency transfer function of the restoration filter. As the figure shows, the signal $\tilde{a}_1(t)$ (3.26) differs from the initial signal a(t) in that it does not contain spectral components of the signal a(t) beyond the interval (-F, F) (these determine the value of minimum error), but does contain "redundant" components owing to overlapping of the continued spectral components (shaded in the figure).

The inaccuracy in restoring the signal from samples is also called the strobe effect [3.9], since as a result of spectral overlap the components of the initial signal with frequency $F + \Delta$ appear in the restored signal with a

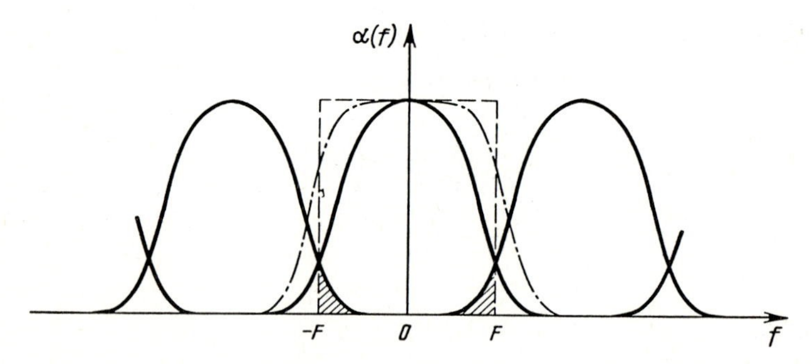

Fig. 3.5. Explanation of discretization and signal restoration errors. The solid line shows the signal spectrum; the dashed line the restoration filter frequency transfer function; and the dot-dashed line the spectrum of the weight function of the sensor

decreased frequency $F - \Delta$. In picture restoration these distortions are more visible in periodic structures of high spatial frequency, which as a result of beating on the discretization raster give low-frequency contributions [3.9,10].

To avoid such distortion it follows from (3.21) that the picture must be passed through a low pass filter before discretization so as to suppress the high-frequency components of the signal and to match the extent of the spatial spectrum to the discretization step adopted. In practice the strobing sensor of the discretization device usually plays the role of such a filter when measuring the signal samples. The result of such measurements can be described by

$$\tilde{\alpha}_k = \int_{-\infty}^{\infty} a(t)\, \varphi_d(t - k\Delta t)dt \quad , \tag{3.27}$$

where $\varphi_d(t)$ is the so-called aperture or weight function of the sensor.

This means that the signal spectrum is masked by the function (dot-dashed line in Fig.3.5), which is a Fourier transform of the function $\varphi_d(t)$. The mask weakens the highest spatial frequencies of the signal spectrum and hence attenuates the effect of overlap during the periodic continuation associated with discretization. To a certain extent it also distorts the relation between the spectral components of the signal within the "useful" interval [in Fig.3.5 the interval $(-F, F)$], since the weight function of the sensor is usually different from the functions $2F\ \mathrm{sinc}2\pi Ft$ required by sampling theory.

In modern picture discretization devices the sensor weight function is either approximately constant within a rectangular or square window, or varies as

$$\varphi_d(t_1, t_2) = h_0 \exp[-(t_1^2/\tau_1^2 + t_2^2/\tau_2^2)] \quad , \tag{3.28}$$

where τ_1, τ_2 are parameters defining the effective size of the aperture.

Like the discretization formula (3.21), equation (3.22) for the synthesis of the continuous signal from its discrete representation also cannot be precisely satisfied in practice, and restoration takes place by means of a linear synthesizer whose aperture differs from the required function $\mathrm{sinc}(2\pi Ft)$:

$$\tilde{a}_2(t) = \sum_k \tilde{\alpha}_k\, \varphi_s(t - k\Delta t) \quad . \tag{3.29}$$

There is another error associated with this. The frequency transfer function of the synthesizer $\Phi_s(f)$ [the Fourier transform of $\varphi_s(t)$] will, generally speaking, also fail to drop to zero beyond the interval $(-F, F)$; the components of the continued spectrum of the signal

$$\tilde{a}(t) = \sum_{k=-\infty}^{\infty} \tilde{\alpha}_k \, \mathrm{sinc}[2\pi F(t - k\Delta t)] \tag{3.30}$$

will thus end up in the spectrum of the restored signal.

If the signal $\tilde{a}(t)$ has a strong component with the frequency $F - \Delta$, an additional component with the frequency $F + \Delta$ will appear in the signal $\tilde{a}_2(t)$, albeit weakened to a certain degree by the filter $\varphi_s(t)$. Beating occurs between these two components and appears in pictures as a Moiré pattern. An example of such a pattern can be seen in the horizontal scan of a television test pattern [3.1].

These distortions can be reduced by choosing a synthesizer frequency transfer function $\Phi_s(f)$ for which the values $\Phi_s(f)$ beyond the useful interval $(-F, F)$ are minimal. However, this usually leads to changes in the values of $\Phi_s(f)$ within the useful interval and hence to distortion of the restored signal as compared with the original. This distortion, as well as the distortion in sample measurements, must be taken into account and corrected when the signals are digitally processed (Chap.7).

In modern devices for restoring pictures from a digital signal, and in discretization devices, rectangular (square) or Gaussian [as in (3.28)] weight functions $\varphi_s(t)$ are usually used.

The restoration of a continuous signal from sample sequences has yet another distinctive feature which is important for pictures. In real restoring devices it is impossible to maintain with absolute accuracy the distance Δt between separate picture samples during picture restoration. The fluctuation in this distance also leads to distortions in the restored picture. Let us consider these distortions in the case of devices with rectangular apertures equal in size to the discretization step in the corresponding directions. Let the picture samples have equal values. Then with unstable distances between samples, signal doubling takes place at certain points on the restored picture (owing to the overlapping of the apertures at neighbouring points on the raster), and at some of these points (where the distance between samples is larger than the aperture) the value of the restored signal will be zero.

Thus a small error in the distribution of the samples can result in substantial errors in reproducing the signal. However, in the restoration of the picture, it is not this which is decisive but rather the fact that raster instability is usually more or less regular. For example, errors in sample distribution are repeated from row to row. As a result false patterns appear in the picture which are noticeable even with small errors in the reproduction of signal values, owing to the high sensitivity of the eye to contours and lines.

The phenomena discussed in this section are very important when defining the requirements of picture restoration and discretization devices. A more detailed analysis of these requirements can be found in [3.8,10]..

3.6 Other Approaches to Discretization

The sampling theorem provides the rules for the discretization of signals with limited bandwidths. In order to use it for picture discretization, the form and size of the figure limiting the picture spectrum must be known. These data are sometimes directly linked with known constructional characteristics of the picture-creating system, such as the characteristics of the objective lens and of the film in photographic systems. But in many cases they are unknown or do not provide a useful description of the class of pictures whose discretization parameters have to be chosen. Moreover, picture distortions resulting from discretization cannot always be as conveniently described by distortion of their spectra as is the case when the sampling theorem is used.

In such cases the discretization parameters – the shape of the raster and the discretization step – are chosen on the basis of simpler criteria for evaluating discretization error and concrete methods for interpolating samples during the restoration of continuous pictures. One of these criteria for error evaluation is the maximum value by which the signal may deviate from the restoration based on the discrete representation. The real characteristics of the picture-reproducing device are taken into account by assuming for example that rectangular interpolation of the samples (with a square aperture) is employed during restoration. A square discretization raster is therefore chosen, and the distance between samples is selected so that the signal cannot change by more than the prescribed error magnitude in that distance. Clearly, this makes it necessary to set limits on the possible rate of change of the picture signal along the coordinates, i.e., on the maximum value of its derivatives. If this is unknown, the discretization device can be so designed that it automatically measures the difference between the current signal value and its value in the closest sample, and moves on to the next sample when this difference reaches the prescribed error threshold. This method is called adaptive discretization. In it, the distance between samples is clearly not constant, and therefore, although the method is sometimes used for economical picture coding for discrete communication channels, it is too inconvenient to be used for picture representation for digital processing. It can, however, be used for archival picture storage.

In certain cases it is impossible to find and formulate definite criteria for accuracy in picture discretization. The problem of choosing discretization parameters must then be solved on the basis of empirically obtained data concerning the required number of samples for an object of minimal area, or other similar indicators.

In the foregoing picture discretization methods, signal samples act as discrete representation that correspond to representation of the signal by sampling functions and δ functions, as already noted. Recently, methods of representing signals in terms of other basis functions have found application in converting radiation into a discrete electrical signal. In these methods the field of radiation to be measured is passed through interchangeable coding masks whose transmission functions correspond to each basis function. The radiation energy behind the mask, the coefficient of radiation field representation for the given basis functions, is measured. The most convenient masks are binary masks (transparent or opaque) which correspond to two-valued basis functions (e.g. Walsh functions). This discretization method is known as multiplex coding [3.11] or the coded aperture method [3.12]. It is used in measuring and discretization devices for weak radiation (radioactivity, X-rays, infrared radiation, etc.) to increase the sensitivity of receivers[2]. Short reviews of these methods can be found in [3.13,14].

3.7 Optimal Discrete Representation and Dimensionality of Signals

An important question in discretization theory has to do with the volume of the discrete signal description: the number N of basis functions used in the representation

$$a(t) = \sum_{k=0}^{N-1} \alpha_k \varphi_k(t) \quad , \tag{3.31}$$

or the volume of finite-dimensional space on which the signal is projected during discretization. For signals with a limited bandwidth, the description volume is the number of signal samples. As follows from the sampling theorem, the number of samples per unit area for two-dimensional signals is equal to the area of the rectangle limiting the signal spectrum in the frequency plane (in rectangular discretization). But measuring samples is not the only method

---

[2] Note that discretization by the measurement of samples can be considered a special case of the coded aperture method in which every mask is opaque to radiation apart from a small opening, the position of which changes from mask to mask depending on the raster chosen.

of discretization. It is logical to define the optimal method to be the one for which the basis volume is minimal for the prescribed accuracy of signal restoration. To find the optimal basis, it is necessary first to define the class of signals for which it is sought and to prescribe the restoration accuracy for that class. Two approaches are used in describing classes of signals: the deterministic method and statistical analysis.

In the deterministic approach, signals are considered to be the result of transforming arbitrary signals by a linear operator, so that the various classes are distinguished and described by the form of this operator alone. Signals with limited bandwidths form one of these classes. They have a corresponding linear filter or operator with a frequency transfer function of the form $\Pi_F(f)$ (3.15). With this approach, choosing the optimal basis becomes a matter of finding the links between the transfer functions of the linear operator defining the class, and the basis functions.

Let $\mathbf{b} = L\mathbf{a}$ be signals generated by the action of a certain linear operator L on arbitrary signals $\mathbf{a}$, and let

$$b_N = \sum_{k=0}^{N-1} \beta_k \varphi_k(t) \tag{3.32}$$

be their discrete representation in terms of the orthonormal basis $\{\varphi_k(t)\}$. If the accuracy of this discrete representation is measured by the norm of the difference

$$\|\varepsilon\| = \|\mathbf{b} - \mathbf{b}_N\| = \|\mathbf{b} - \sum_{k=0}^{N-1} \beta_k \varphi_k(t)\| \quad , \tag{3.33}$$

where

$$\boldsymbol{\beta}_k = (b, \boldsymbol{\varphi}_k) \quad , \tag{3.34}$$

then it follows [3.5] that the best function system, that is the system for which the maximum value of $\|\varepsilon\|$ (for various $\mathbf{a}$) is least, is defined by

$$\varphi_k(t) = \lambda_k^{-1/2} L\psi_k(t) \quad . \tag{3.35}$$

Here $\{\varphi_k(t)\}$ are the eigenfunctions of the composite operator L^*L, and L^* is the operator whose eigenfunctions are orthogonal to those of L, while its eigenvalues are the complex conjugates of those of L. The greatest relative error (3.32) is limited by the magnitude λ_N of the Nth eigenvalue of the operator $\mathbf{L^*L}$.

Thus, if the eigenvalues λ_k are arranged in decreasing order, the volume of signal spectra space generated by the operator $\mathbf{L}$ is determined by the index of the first eigenvalue that does not exceed the prescribed representation error. An example of such an approach is the evaluation of the volume of the space of signals of effectively finite length and bandwidths[3] [3.15].

Let a(t) be a signal satisfying the following conditions:

$$(P_t\mathbf{a}, \mathbf{a}) = 1 - \varepsilon_T^2 \quad ; (P_f\mathbf{a}, \mathbf{a}) = 1 - \eta_F^2 \quad ; (\mathbf{a}, \mathbf{a}) = \|\mathbf{a}\|^2 = 1 \quad , \tag{3.36}$$

where P_t is a gate operator, selecting a section of length T from the signal; P_f is an ideal band-pass filter passing only signal frequencies on the interval (-F, F); and ε_T^2 and η_F^2 are the errors of such truncation on the length and on the spectrum, respectively.

The best representation then becomes

$$a = \sum_{k=0}^{N-1} \alpha_k \varphi_k(t) \tag{3.37}$$

using the functions $\varphi_k(t)$, defined by

$$2F \int_{-T/2}^{T/2} \varphi_k(t) \operatorname{sinc}[2\pi F(t - \tau)]dt = \lambda_k \varphi_k(\tau) \quad . \tag{3.38}$$

These are known as prolate spheroidal wave functions, and

$$\| \mathbf{a} - \sum_{k=0}^{N-1} \alpha_k \varphi_k \|^2 \leqslant 12(\varepsilon_T + \eta_F)^2 + \eta_F^2 \tag{3.39}$$

if N is the smallest integer above 2TF [3.15].

When $T \to \infty$, the prolate spheroidal wave functions tend to the sample functions sinc $2\pi Ft$, and the expansion (3.37) reduces to the sampling theory expansion. For finite T, representation by (3.37) of signals satisfying (3.36) is better than expansion in the functions sinc $2\pi Ft$ with the same number of terms.

In the statistical approach, the optimum N-dimensional basis for the digital representation of individual signals is usually defined as the basis for which the error norm, averaged over the ensemble of realizations, is least. In that case a result analogous to (3.35) is obtained and is known as the Karhumen-Loève theorem [3.5,16]: the minimum value of the signal representation error norm on the interval length T is achieved by using as the basis the eigenfunctions of the operator whose kernel is the signal correlation function $R_a(t, \tau)$:

$$\int_{-T/2}^{T/2} R_a(t, \tau) \; \varphi_k(\tau)d\tau = \lambda_k \varphi_k(t) \quad . \tag{3.40}$$

[3] Note that a signal with a limited bandwidth cannot have a finite length, since the functions sinc $2\pi Ft$, used to represent such signals (2F is the spectrum width), continue to infinity.

The eigenfunctions corresponding to the N highest eigenvalues are retained.

In this case the error norm is

$$\| \varepsilon \|^2_{min} = \| a(t) - \sum_{k=0}^{N-1} \alpha_k \varphi_k(t) \|^2_{min} = \sum_{k=N}^{\infty} \lambda_k \quad . \tag{3.41}$$

This representation is called the *Karhunen-Loève expansion*. Its coefficients are uncorrelated [given the orthogonality of $\varphi_k(t)$] random values [3.5,16]. Note that for stationary processes, for which the correlation functions depend only on the difference between the arguments [$R_a(t,\tau) = R_a(t-\tau)$] with $T \to \infty$ [or when T becomes sufficiently large relative to the extension of $R(t-\tau)$], the eigenfunctions $\varphi_k(t)$ approach complex exponential functions with frequencies k/T (in the two-dimensional case k_1/T_2, k_2/T_2).

3.8 Element-by-Element Quantization

The stage following discretization in obtaining a digital signal from a continuous signal is the quantization of the signal expansion coefficients on the finite-dimensional orthogonal basis (element-by-element quantization). It consists in choosing a segment of finite length in the domain of values of the coefficients, which is then subdivided into a finite number of subintervals; the quantization steps and values falling in each interval are indexed with a single number (the interval number). During signal restoration it is replaced by a value that is representative of the given interval. This method of subdividing into intervals and the corresponding representative values are chosen so as to satisfy the requirements of accuracy in representing a continuous signal by a digital one. In view of the fact that the transformation from continuous digital signals takes place in two stages, i.e., discretization and element-by-element quantization, the accuracy requirements for digital representation must also be formulated separately for each stage.

Let α_k be the kth coefficient of the discrete representation of the signal and α_k^r the representative value of the rth quantization interval in the domain α_k. The quantization error can be characterized by the value

$$\varepsilon_k^r = \alpha_k - \hat{\alpha}_k^r \quad . \tag{3.42}$$

The requirements for accuracy of quantization are usually formulated by imposing restrictions on ε_k^r. The most general way of formulating these restrictions is to consider α_k and hence ε_k^r to be random, and to introduce a metric, i.e., a certain measure $D(\varepsilon_k^r)$ which distinguishes α_k from its quantized representation α_k^r. With this approach the accuracy of representation is characterized by the weighted mean of $D(\varepsilon_k^r)$, with probability distribution $p(\alpha_k)$:

$$Q = \sum_{r=0}^{M-1} \int_{\alpha_k^r}^{\alpha_k^{r+1}} p(\alpha_k)\, D(\varepsilon_k^r) d\alpha_k \quad , \tag{3.43}$$

where α_k^r, α_k^{r+1} are the edges of the r quantization interval, and M is the number of intervals, or levels of quantization.

The optimal choice of quantization intervals and their representatives is the one for which Q is smallest and does not exceed the prescribed limiting value. The problem of optimum element-by-element quantization is posed in this form in [3.17-18].

It is often expedient to distinguish between two types of quantization errors: limitation errors arising because of restrictions in the range of values of the quantized variables, and quantization errors within the chosen intervals. Indeed, if the distribution density $p(\alpha_k)$ has long "tails", restriction errors can be fairly significant in absolute values, while the quantization errors within the chosen interval are fairly limited. The distribution functions of these errors also differ greatly. It is therefore advisable to formulate the accuracy requirements for representing a continuous value by a quantized one separately for restriction errors

$$\begin{aligned} Q_{1,b} &= \int_{-\infty}^{\alpha_k^1} p(\alpha_k)\, D_{1,b}(\varepsilon_k^0) d\alpha_k \quad , \\ Q_{2,b} &= \int_{\alpha_k^{M-1}}^{\infty} p(\alpha_k)\, D_{2,b}(\varepsilon_k^{M-1}) d\alpha_k \end{aligned} \tag{3.44}$$

and for quantization errors within the restricted segment:

$$Q_0 = \sum_{r=1}^{M-2} \int_{\alpha_k^r}^{\alpha_k^{r+1}} p(\alpha_k)\, D_0(\varepsilon_k^r) d\alpha_k \quad , \tag{3.45}$$

where $D_{1b}(\varepsilon)$, $D_{2b}(\varepsilon)$, $D_0(\varepsilon)$ are measures of the difference for restriction errors and quantization errors within a limited segment.

With this formulation the optimal quantizations are those for which Q_{1b}, Q_{2b} and Q_0 are least, satisfying the conditions

$$Q_{1b,min} \leqslant \varepsilon_{1b} \quad ; Q_{2b,min} \leqslant \varepsilon_{2b} \quad ; Q_{0,min} \leqslant \varepsilon_0 \quad , \tag{3.46}$$

where ε_{1b}, ε_{2b}, ε_0 are the limiting values of the restriction and quantization errors.

The optimal values for the segment edges α_k^0, α_k^{M-1} (within which quantization takes place), for the edges of the quantization intervals $\alpha_k^r (r = 1, \ldots, M-2)$, and for the representative value α_k^r are defined by the following system of equations derived from (3.44,45) by differentiating over the desired values and setting the derivatives to zero:

$$\int_{-\infty}^{\alpha_k^1} p(\alpha_k)\, D_{1,b}(\alpha_k - \hat{\alpha}_k^0) d\alpha_k \leqslant \varepsilon_{1,b} \quad , \tag{3.47a}$$

$$\int_{-\infty}^{\alpha_k^1} p(\alpha_k)\, D'_{1,b}(\alpha_k - \hat{\alpha}_k^0) d\alpha_k \leqslant 0 \quad ; \tag{3.47b}$$

$$\int_{\alpha_k^{M-1}}^{\infty} p(\alpha_k)\, D_{2,b}(\alpha_k - \hat{\alpha}_k^{M-1}) d\alpha_k \leqslant \varepsilon_{2,b} \quad , \tag{3.48a}$$

$$\int_{\alpha_k^{M-1}}^{\infty} p(\alpha_k)\, D'_{2,b}(\alpha_k - \hat{\alpha}_k^{M-1}) d\alpha_k = 0 \quad ; \tag{3.48b}$$

$$D_0(\alpha_k^r - \hat{\alpha}_k^{r-1}) = D_0(\alpha_k^r - \hat{\alpha}_k^r) \quad , \tag{3.49a}$$

$$\int_{\alpha_k^r}^{\alpha_k^{r+1}} p(\alpha_k)\, D'_0(\alpha_k - \hat{\alpha}_k^r) d\alpha_k = 0 \quad , \tag{3.49b}$$

$$\sum_{r=1}^{M-2} \int_{\alpha_k^r}^{\alpha_k^{r+1}} p(\alpha_k)\, D_0(\alpha_k - \hat{\alpha}_k^r) d\alpha \leqslant \varepsilon_0 \quad . \tag{3.49c}$$

Equation (3.49a) implies that if $D_0(\varepsilon)$ is an even function, the borders of the quantization intervals must be chosen midway between the corresponding representative values:

$$\alpha_k^r = (\hat{\alpha}_k^{r-1} + \hat{\alpha}_k^r) / 2 \quad . \tag{3.50}$$

Equations (3.49a-c) can be solved to find optimum α_k^r and $\hat{\alpha}_k^r$ and minimum M by computer iteration. Some results of such calculations for quadratic loss functions $D_0(\varepsilon)$ and Gaussian distributions $p(\alpha_k)$ are presented in [3.18]. In [3.20] the results of calculations for the minimum criteria of fourth- and sixth-power quantization error are presented as well as Gauss, Rayleigh and Laplace distributions. If the quantized variables $\{\alpha_k\}$ are derived from digital computer calculations, this approach to the choice of optimal quantization is fully justified. In choosing optimum quantizations of a continuous signal for input into a digital processor, it is convenient to simplify the task somewhat so as to obtain an analytical solution, which can then be employed in devices converting the analogue signal into a digital one [3.10,21].

Let us consider the integral

$$Q_r = \int_{\alpha_k^r}^{\alpha_k^{r+1}} p(\alpha_k)\, D_0(\alpha_k - \hat{\alpha}_k^r) d\alpha_k \quad . \tag{3.51}$$

The distribution density $p(\alpha_k)$ and the metric $D_0(\varepsilon)$ can usually be considered smooth functions, and the number of quantization levels is sufficiently large for the interval width (step) of quantization $(\alpha_k^{r+1} - \alpha_k^r)$ to be small, and on the interval $(\alpha_k^r, \alpha_k^{r+1})$ the functions $p(\alpha_k)$ and $D_0(\alpha_k - \hat{\alpha}_k^r)$ can be represented in the form

$$p(\alpha_k) \approx p(\hat{\alpha}_{k_0}^r) + p'(\hat{\alpha}_{k_0}^r)\,(\alpha_k - \hat{\alpha}_{k_0}^r) \quad ; \tag{3.52}$$

$$D_0(\alpha_k - \hat{\alpha}_k^r) \approx D_0(\alpha_k - \hat{\alpha}_{k_0}^r) + D_0'(\alpha_k - \hat{\alpha}_k^r)\,(\hat{\alpha}_{k_0}^r - \hat{\alpha}_k^r) \quad , \tag{3.53}$$

where $\hat{\alpha}_{k_0}^r$ is the middle of the quantization interval;

$$\hat{\alpha}_{k_0}^r = (\alpha_k^r + \alpha_k^{r+1}) / 2 \quad . \tag{3.54}$$

By substituting (3.53) into (3.51) and taking (3.49b) into account, we obtain

$$Q_r \approx \int_{\alpha_k^r}^{\alpha_k^{r+1}} p(\alpha_k)\, D_0(\alpha_k - \hat{\alpha}_{k_0}^r) d\alpha_k \quad . \tag{3.55}$$

This result essentially means that the optimal representative value of the rth quantization interval $\hat{\alpha}_k^r$ differs little from the value $\hat{\alpha}_{k_0}^r$ in the middle of the interval.

By substituting (3.52) into (3.55) we obtain

$$Q_r \approx p(\alpha_{k_0}^r) \int_{\alpha_k^r}^{\alpha_k^{r+1}} D(\alpha_k - \hat{\alpha}_{k_0}^r) d\alpha_k + p'(\alpha_{k_0}^r) \int_{\alpha_k^r}^{\alpha_k^{r+1}} (\alpha_k - \alpha_{k_0}^r)\, D_0(\alpha_k - \alpha_{k_0}^r) d\alpha_k \quad . \tag{3.56}$$

In all practical problems $D_0(\varepsilon)$ can be considered an even function. The second integral in (3.56) is then zero, and

$$Q_r \approx p(\hat{\alpha}_{k_0}^r) \int_{\alpha_k^r}^{\alpha_k^{r+1}} D_0(\alpha_k - \hat{\alpha}_{k_0}^r) d\alpha_k = 2p(\hat{\alpha}_{k_0}^r) \int_0^{\Delta^r} D_0(\varepsilon) d\varepsilon \quad , \tag{3.57}$$

where Δ^r is the half-width of the rth interval;

$$\Delta^r = (\alpha_k^{r+1} - \alpha_k^r) / 2 \quad . \tag{3.58}$$

Let us write

$$\bar{D}_0(\Delta^r) = \left[\int_0^{\Delta^r} D_0(\varepsilon)d\varepsilon\right] / \Delta^r \quad . \tag{3.59}$$

By substituting (3.57,59) into (3.45) we obtain

$$Q_0 \approx \sum_{r=1}^{M-2} p(\tilde{\alpha}_{k_0}^r)\, \bar{D}_0(\Delta^r)\, (2\Delta^r) \quad . \tag{3.60}$$

In this way the problem of optimal quantization α_k in the restricted range $(\alpha_k^1, \alpha_k^{M-1})$ becomes a matter of minimalizing the sum (3.60) by dividing the segment $(\alpha_k^1, \alpha_k^{M-1})$ into intervals $2\Delta^r(r = 1, 2, \ldots, M - 2)$ of the optimal widths.

Existing quantization devices usually perform so-called uniform signal quantization, i.e., the quantization intervals are all equal. Optimum non-uniform quantization can be achieved with such devices if prior to uniform quantization the signal is subjected to non-linear transformation (predistortion), the form of which is chosen to satisfy the minimum condition (3.60).

Let $w(\alpha_k)$ be a function describing such a transformation. This is usually a smooth monotonic function, so that

$$\Delta_0 \approx w'(\tilde{\alpha}_{k_0}^r)\Delta^r \quad , \tag{3.61}$$

where Δ_0 is the half-width interval of uniform quantization;

$$\Delta_0 = \left[w(\alpha_k^{M-1}) - w(\alpha_k^1)\right] / 2(M - 2) \quad . \tag{3.62}$$

then (3.60) can be rewritten in the form

$$Q_0 = \int_{\alpha_k^1}^{\alpha_k^{M-1}} p(\alpha_k)\, \bar{D}_0\left(\frac{\Delta_0}{w'(\alpha_k)}\right)d\alpha_k \quad . \tag{3.63}$$

Assuming that the functions $p(\alpha_k)$ and $D_0(\varepsilon)$ are smooth, the minimum (3.63) can be found by seeking the functions $w(\alpha_k)$ that minimize the integral

$$Q_0 = \int_{\alpha_k^1}^{\alpha_k^{M-1}} p(\alpha_k)\, \bar{D}_0\left(\frac{\Delta_0}{w'(\alpha_k)}\right)d\alpha_k \quad . \tag{3.64}$$

In this way the problem of optimal quantization reduces to a standard variational task. The function $w(\alpha)$ that minimizes (3.64) is determined by the familiar Euler-Lagrange equation, which in the given case takes the form

$$\frac{\partial}{\partial w'}\left[p(\alpha_k)\ \bar{D}_0\left(\frac{\Delta_0}{w'(\alpha_k)}\right)\right] = \text{const} \quad . \tag{3.65}$$

Thus the form of optimum predistortion depends on the density distribution value of the quantized variable and the measure of the quantization error modified in accordance with (3.59).

In restoring a continuous signal from a quantized one, a digital-analogue converter is also used with uniform quantized output signals. To obtain the values of the quantization interval representatives that correspond to optimal quantization, a uniform quantized signal undergoes non-linear transformation (correction). It can be easily shown that the function describing the form of correction is close to the inverse function of nonlinear predistortion during quantization. For example, let $w_1(a)$ be the correction function, and $\hat{a}^r$ the representative of the rth interval (a^r, a^{r+1}) of uniform quantization of magnitude a. Then, the representation value of the rth interval must be located within this interval,

$$w^{-1}(a) < \hat{a}_k^r = w_1(\hat{a}) < w^{-1}(a^{r+1}) \quad , \tag{3.66}$$

from which it also follows that with small (a^{r+1}, a^r) the correction function is $w_1(a) \approx w^{-1}(a)$.

In the practical application of these results, the main question is the choice of the measure of the quantization errors, i.e., the criteria for quantization accuracy. They are determined by the nature of the problem to be solved after quantization and by the purpose of the quantized variable. Although in the above statement the criterion for quantization accuracy is the element-by-element criterion, when choosing it one must bear in mind the fact that the quantization error is a two-dimensional process which is, generally speaking, correlated with the quantized signal. Thus, if the brightness of a picture is subjected to quantization, the noise pattern of quantization usually becomes similar to that of the image and therefore can interfere with visual perception, even when the brightness distortions in the elements of the picture are insignificant. Large domains with smoothly changing brightness, separated from each other by edges where the brightness changes much more sharply, occur in many pictures. During quantization it can happen that one domain is quantized onto two neighbouring levels. As a result, during the restoration of the quantized picture there develops between these two parts of one domain an extended edge which is easily perceptible by the eye, since vision has a low threshold of contrast detection if there are large fields for comparison. This is the problem of so-called false contours encountered in quantizing pictures [3.1,10,22].

3.9 Examples of Optimum Quantization

Let us analyze a few examples of how the optimum distribution and number of quantization intervals are chosen. For simplicity we shall drop the subscript k in α_k, and the edges of the segment values of the quantized signal will be denoted α_{min} and α_{max}.

We shall begin with the definition of α_{min} and α_{max}. In picture-processing tasks, these values are usually either given, or determined on the basis of the prescribed probability that the quantized variable will exceed these edges values:

$$\int_{-\infty}^{\alpha_{min}} p(\alpha)d\alpha \leqslant \varepsilon_{1b} \quad ; \quad \int_{\alpha_{max}}^{\infty} p(\alpha)d\alpha \leqslant \varepsilon_{2b} \quad . \tag{3.67}$$

This corresponds to the prescribed criteria for restriction errors in (3.47a-48b) in the form

$$D_{1b}(\varepsilon) = D_{2b}(\varepsilon) = 1 \quad . \tag{3.68}$$

In the following examples we shall assume precisely such an approach to determining α_{min}, α_{max}, and we shall consider the choice of optimum quantization only within the prescribed range $(\alpha_{min},\alpha_{max})$. We shall number the quantization intervals from 0 to M - 1.

3.9.1 Example: The Threshold Metric

Let

$$D_{1b}(\varepsilon) = D_{2b}(\varepsilon) = 1 \quad ; \qquad \bar{D}_0(\Delta^r) = \begin{cases} 0, & |\Delta^r| \leqslant \Delta^r_{th} \\ 1, & |\Delta^r| > \Delta^r_{th} \end{cases} \quad , \tag{3.69}$$

Such a metric may be called a threshold metric and its corresponding accuracy criterion requires quantization errors to be imperceptible. For this criterion the optimal distribution of quantization intervals and the choice of the interval numbering are trivial, and reduce to solving

$$|\Delta^r| = \Delta^r_{th} \quad . \tag{3.70}$$

Thus the width of the rth quantization interval must be chosen to be equal to $2\Delta^r_{th}$, and the representative value of the rth level is the value of α in the middle of the rth interval. If Δ^r_{th} does not depend on r we obtain a uniform quantization scale. In this the number of quantization levels (intervals) is defined by

$$M_u = (\alpha_{max} - \alpha_{min}) / 2\Delta_{th} \quad . \tag{3.71}$$

There are important and frequently encountered cases in practice for which

$$\Delta_t^r = \delta_0 \hat{\alpha}^r \quad , \tag{3.72}$$

i.e., when the relative and not the absolute value of the quantization error is limited. Thus, in conformity with the Weber-Fechner psychophysical law, it is possible to describe to a first approximation the accuracy requirements imposed by the human visual system upon quantization of the picture brightness. Using (3.61) it is not difficult to show that in this case it is necessary to subject not the variable itself to uniform quantization, but its normalized logarithm

$$\frac{w(\alpha) - w(\alpha_{min})}{w(\alpha_{max}) - w(\alpha_{min})} = \frac{\ln(\alpha / \alpha_{min})}{\ln q} \quad , \tag{3.73}$$

where $q = \alpha_{max} / \alpha_{min}$.

The number of quantization levels in this logarithmic scale should be

$$M = (\ln q)/ \delta_0 \quad . \tag{3.74}$$

This turns out to be the case in practice: prior to uniform quantization the signal undergoes logarithmic predistortion (compression); the values for the representatives of the quantization interval during restoration are also chosen on the uniform scale; and then the synthesized signal is expanded.

Let us find the number of levels required during quantization of picture brightness. Psychophysical measurements show that in normal light the threshold of the relative contrast sensitivity of vision, δ_0, is of the order of 1.5 - 2% for a large test spot with long adaptation to the background [3.1]. Existing picture-reproducing devices give a dynamic brightness range of the order of $q \approx 100$. Substituting these values into (3.74), we obtain $M \approx 230$.

Direct experiments with quantized brightness using television picture-reproducing equipment have shown that 64 - 128 levels are sufficient [3.19]. The decrease in the number of levels required is partly associated with noise from the videosignal sensors and picture-producing equipment. At present 64 - 256 levels with logarithmic predistortion of the videosignal have been adopted in picture-quantization and picture-reproducing devices.

It is interesting to evaluate the gain g in the number of quantization levels which is achieved by logarithmic predistortion as opposed to uniform quantization, with the same accuracy. Clearly,

$$g = (q - 1)/\ln q \quad . \tag{3.75}$$

With high q the gain can be fairly large; with $q = 100$, $g \approx 20$. However, the real gains are usually not so great, since the error evaluation in the "worst" case, as in the procedure with threshold criteria, is excessively strict from a practical point of view.

3.9.2 Example: Power Criteria for the Absolute Value of the Quantization Error

By substituting the quantization error

$$\bar{D}_0(\Delta^r) = (\Delta_r)^{2n} \tag{3.76}$$

into (3.65) and solving the resulting differential equation, we obtain

$$\frac{w(\alpha) - w(\alpha_{min})}{w(\alpha_{max}) - w(\alpha_{min})} = \frac{\int_{\alpha_{min}}^{\alpha} [p(\alpha)]^{1/(2n+1)} d\alpha}{\int_{\alpha_{min}}^{\alpha_{max}} [p(\alpha)]^{1/(2n+1)} d\alpha} \quad . \tag{3.77}$$

Thus the necessary non-linear predistortion depends only on the probability distribution of the quantized value. The idea of this dependence is clear from

$$\Delta^r \approx \Delta / \left(\frac{dw(\alpha)}{d\alpha}\right) \sim |p(\alpha)|^{-1/2n+1} \quad ; \tag{3.78}$$

i.e., the quantization intervals of the values α are inversely proportional to the probability density of these values in the corresponding power.

In the widespread case where the mean-square quantization error is adopted ($n = 1$),

$$\frac{w(\alpha) - w(\alpha_{min})}{w(\alpha_{max}) - w(\alpha_{min})} = \frac{\int_{\alpha_{min}}^{\alpha} \sqrt[3]{p(\alpha)}\, d\alpha}{\int_{\alpha_{min}}^{\alpha_{max}} \sqrt[3]{p(\alpha)}\, d\alpha} \quad . \tag{3.79}$$

In some cases one can approximately consider the quantized coefficients of discrete signal representation to have a Gaussian-shaped probability distribution density in the interval $(\alpha_{min}, \alpha_{max})$:

$$p(\alpha) = c \exp\left(-\frac{(\alpha - \alpha_0)^2}{2\sigma_\alpha^2}\right) \quad , \tag{3.80}$$

where c is the normalizing constant.

Then with $n = 1$,

$$\frac{w(\alpha) - w(\alpha_{min})}{w(\alpha_{max}) - w(\alpha_{min})} = \frac{\Phi\left(\frac{\alpha - \alpha_0}{\sqrt{3}\,\sigma_\alpha}\right) - \Phi\left(\frac{\alpha_{min} - \alpha_0}{\sqrt{3}\,\sigma_\alpha}\right)}{\Phi\left(\frac{\alpha_{max} - \alpha_0}{\sqrt{3}\,\sigma_\alpha}\right) - \Phi\left(\frac{\alpha_{min} - \alpha_0}{\sqrt{3}\,\sigma_\alpha}\right)} \quad , \tag{3.81}$$

where

$$\Phi(x) = \frac{1}{\sqrt{2\pi}} \int_{-\infty}^{x} \exp\left(-\frac{t^2}{2}\right) dt \quad . \tag{3.82}$$

The gain in the number of quantization levels for a given accuracy depends in these cases on the parameter

$$q = (\alpha_{max} - \alpha_{min}) / \sigma_\alpha \quad , \tag{3.83}$$

and is

$$g = \frac{M_p}{M} = \frac{q}{\sqrt{2\pi}\,\sqrt[4]{27}} \, \frac{[\Phi(q/2) - \Phi(-q/2)]^{1/2}}{[\Phi(q/2\sqrt{3}) - \Phi(-q/2\sqrt{3})]^{3/2}} \quad . \tag{3.84}$$

With sufficiently large q this gain is roughly equal to $g/5.7$. Thus the gain will only be significant for very large q, when the "tails" substantially affect the distribution $p(\alpha)$.

3.9.3 Example: Power Criteria for the Relative Quantization Error

The relative quantization error is given by

$$\bar{D}_0(\Delta^r) = (\Delta/\alpha)^{2n} \quad . \tag{3.85}$$

From the Euler-Lagrange equation (3.65) for this case it is not difficult to obtain

$$\frac{w(\alpha) - w(\alpha_{min})}{w(\alpha_{max}) - w(\alpha_{min})} = \frac{\int_{\alpha_{min}}^{\alpha} \left(\frac{p(\alpha)}{\alpha}\right)^{1/2n+1} d\alpha}{\int_{\alpha_{min}}^{\alpha_{max}} \left(\frac{p(\alpha)}{\alpha}\right)^{1/2n+1} d\alpha} \quad . \tag{3.86}$$

If the variable α has a uniform distribution then

$$\frac{w(\alpha) - w(\alpha_{min})}{w(\alpha_{max}) - w(\alpha_{min})} = \frac{\alpha^{2n/2n+1} - \alpha_{min}^{2n/2n+1}}{\alpha_{max}^{2n/2n+1} - \alpha_{min}^{2n/2n+1}} \quad . \tag{3.87}$$

When the quantization accuracy is evaluated in terms of the average value of the modulus of the relative error ($n = 1/2$), we have

$$\frac{w(\alpha) - w(\alpha_{min})}{w(\alpha_{max}) - w(\alpha_{min})} = \frac{\sqrt{\alpha} - \sqrt{\alpha_{min}}}{\sqrt{\alpha_{max}} - \sqrt{\alpha_{min}}} \quad . \tag{3.88}$$

Let us find the gain in the number of quantization levels that is guaranteed by non-linear predistortion as in (3.88) as compared with uniform quantization. The quantization interval in this case is

$$\Delta^r = \Delta_0 / w'(\alpha) = \frac{2\Delta_0(\sqrt{\alpha_{max}} - \sqrt{\alpha_{min}})\sqrt{\alpha}}{[w(\alpha_{max}) - w(\alpha_{min})]} \quad , \tag{3.89}$$

where Δ_0 is the uniform quantization interval of $w(\alpha)$, and the average quantization error is

$$Q_{0,1} = \frac{2\Delta_0(\sqrt{\alpha_{max}} - \sqrt{\alpha_{min}})}{\alpha_{max} - \alpha_{min}} \int_{\alpha_{min}}^{\alpha_{max}} \frac{\sqrt{\alpha}}{\alpha} d\alpha = 4(\sqrt{q} - 1) / (\sqrt{q} + 1) M_1 \quad , \tag{3.90}$$

where M_1 is the number of quantization levels $w(\alpha)$, and $q = \alpha_{max} / \alpha_{min}$.

At the same time, with uniform quantization on M_2 levels,

$$Q_{0,2} = \frac{\Delta_0}{\alpha_{max} - \alpha_{min}} \int_{\alpha_{min}}^{\alpha_{max}} \frac{d\alpha}{\alpha} = \frac{1}{M_2} \ln q \quad . \tag{3.91}$$

By writing $\bar{Q}_{0,1} = \bar{Q}_{0,2}$, we obtain

$$g = \frac{M_2}{M_1} = \frac{\ln q}{4} \frac{\sqrt{q} + 1}{\sqrt{q} - 1} \quad . \tag{3.92}$$

Thus with a uniform distribution $p(\alpha)$ the gain is not large; with $q = 100$ it is 1.4. Comparing this result with the value $g = 20$ for the relative error

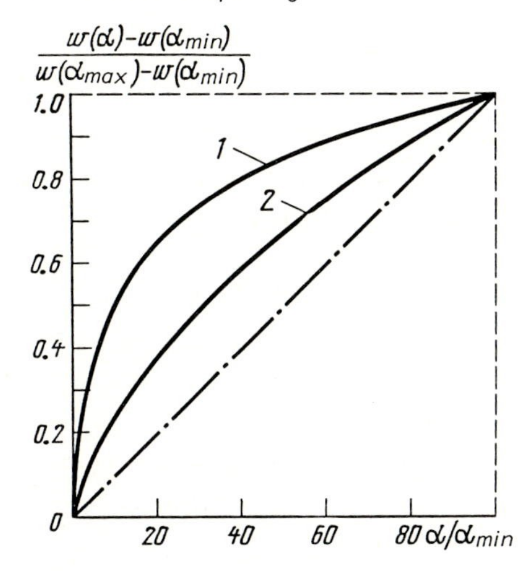

Fig. 3.6. Predistortion for optimal quantization: (*1*) logarithmic predistortion according to (3.73); (*2*) predistortion according to (3.88)

threshold criteria, we see that when the probability of various signal values is taken into account, the gain from the non-linear distribution of quantization levels is much more modest than with evaluation in the worst case. Figure 3.6 shows the predistortion characteristics obtained from these two related cases. Curve 1 is the logarithmic predistortion in (3.73), Curve 2 the predistortion in (3.88), and the broken line is the linear function corresponding to uniform quantization without predistortion.

3.10 Quantization in the Presence of Noise. Quantization and Representation of Numbers in Digital Processors

In determining the requirements placed on the continuous signal quantization method, it should be taken into account that the real sources of the continuous videosignal (such as photographic film, photomultipliers, television transmission tubes, fluorescence and heat- and X-ray-sensitive sensors), as well as the picture-reproducing devices (cathode-ray tubes, modulated light sources, and photosensitive materials) create their own noise. As a result of this noise the correspondence between the signal and the object of study is not absolutely accurate. In calculating the optimum nonlinear predistortion in quantization it can be assumed that these sources of noise, recalculated for the input of uniform quantizing equipment, will combine additively with quantization noise. Hence, the possible dependence of the noise intensity of the sensor and the synthesizer on the signal level can be established (for example, the noise variance of a photomultiplier is proportional to the magnitude of the signal). The criteria for quantization accuracy should also be formulated here, with account taken of the joint effect of sensor noise and quantization noise.

An important feature of the interaction between these two forms of distortion is the randomization of quantization noise, by which its correlation with the signal to be quantized is destroyed. As a result the admissible level of quantization noise can be somewhat increased. Thus, for example, the random noise of a video signal sensor destroys false contours in coarse quantization of picture brightness, thus decreasing their visibility. One method of reduced picture description is even based on this effect. The method consists in adding pseudorandom noise with independent samples to the video signal prior to coarse quantization, and in subtracting this noise from the quantized video signal during picture restoration [3.22].

An analoguous method can also be used in picture contrast correction in digital processing (Sect.7.2). However, the influence which random noise from the signal sensor and synthesizer has upon quantization noise is beneficial only to a certain degree. There is an optimum relation between these types of noise, depending on the properties of the quantized signal and the nature of the problem being solved. In the majority of cases, random noise from sensor and synthesizer must have approximately the same intensity (variance) as the quantization noise. Thus, on the basis of calculations on three different criteria, it is recommended in [3.8] that the mean-square value of random noise be approximately three times less than the width of the quantization interval (i.e., that the variance relation be 4/3, if quantization noise is considered to be uniformly distributed throughout the quantization interval).

The quantization procedure is used not only for the conversion of continuous signals to digital signals, but also in the transformation of digital signals in digital processors, for example when it is necessary to store the results of a calculation in digital memory of limited capacity. The above remarks on the optimum choice of quantization scales are also essentially applicable here.

It is interesting to evaluate the various methods of representing numbers in digital processors from this standpoint. Numbers are represented in either fixed- or floating-point format. In fixed-point representation uniform number quantization takes place, with the quantization step corresponding to the lowest significant digit of the computer word. Corresponding to floating-point representation is two-step quantization: a logarithmic quantization with quantization levels being coded by the exponent of the number and uniform quantization within the logarithmically ordered levels with indexing by the mantissa of the number. With this kind of mixed quantization the non-trivial question arises of how the given finite number of digits of the computer word should be distributed between the codes of the mantissa and the computer word should be distributed between the codes of the mantissa and of the exponent. In solving this problem it is useful to bear in mind that logarithmic quantization is optimal for the threshold criterion of relative error, and uniform quantization for the threshold criterion of absolute error. Moreover, one must take into account not only the necessity of reducing the quantity of digits used to represent numbers, but also the requirement for speed and simplicity of the arithmetic operations on numbers.

It should be mentioned in conclusion that at different stages in processing different methods of representing numbers may prove useful. Various floating-point formats can be used to represent numbers in the central processor and its memory unit.

Thus, for example, in [3.23] there is a description of a reduced representation of a number with a floating-point format (7 bits for the mantissa, 4 bits for the exponent, 2 bits for the two signs). This format was used to store the results of processing on magnetic tape in experiments on digital holography and interferometry [3.24].

3.11 Review of Picture-Coding Methods

The conversion of a picture into a digital signal for storage in computer memory and transmission through digital communication channels is called *picture coding*. A large variety of picture-coding methods currently exists, and has been fairly widely discussed in the literature. Digital television systems are usually considered the main field of application for picture coding. Recently, however, the problem of coding for picture storage and for display processors has gained in importance [3.25,26]. Almost all of the known coding methods can be fitted into the scheme shown in Fig.3.7.

The majority of these methods involve a three-stage procedure: separate discretization and quantization of samples of discrete representation, and subsequent statistical coding of the digital signal. Discretization is usually based on the sampling theorem and linear transformations, which help in finding discrete signal representations on various bases, or linearly predistort the signal so that sampling function bases can be used. The entire picture frame or a region of the picture can undergo linear transformation. In practice linear transformation for coding takes place in digital processors after preliminary oversampling according to the sampling theorem. The optimum size of the regions in two-dimensional transformations varies from a few tens of picture elements to a few hundred (the typical size is 16×16 elements) [3.27,28]. Of the recommended linear coding transforms, the so-called cosine transform [3.29] (see also Sect.4.7) and the slant transform [3.30] are noteworthy. Both correspond more closely to picture structure than the Walsh-Hadamard, Haar, and Fourier transforms, but are equally simple to implement in digital processors.

Sliding transforms are the most convenient for use in specialized television coding devices which work in real time at video rates. These transformations are effected by the convolution operation (in other words transformation by the basis, which depends on the difference between the arguments). Linear transformation consists here in decorrelation of the video signal by subtracting from its current value the value found by linear prediction from the previous ones. The prediction can be both one-dimensional and two-dimensional.

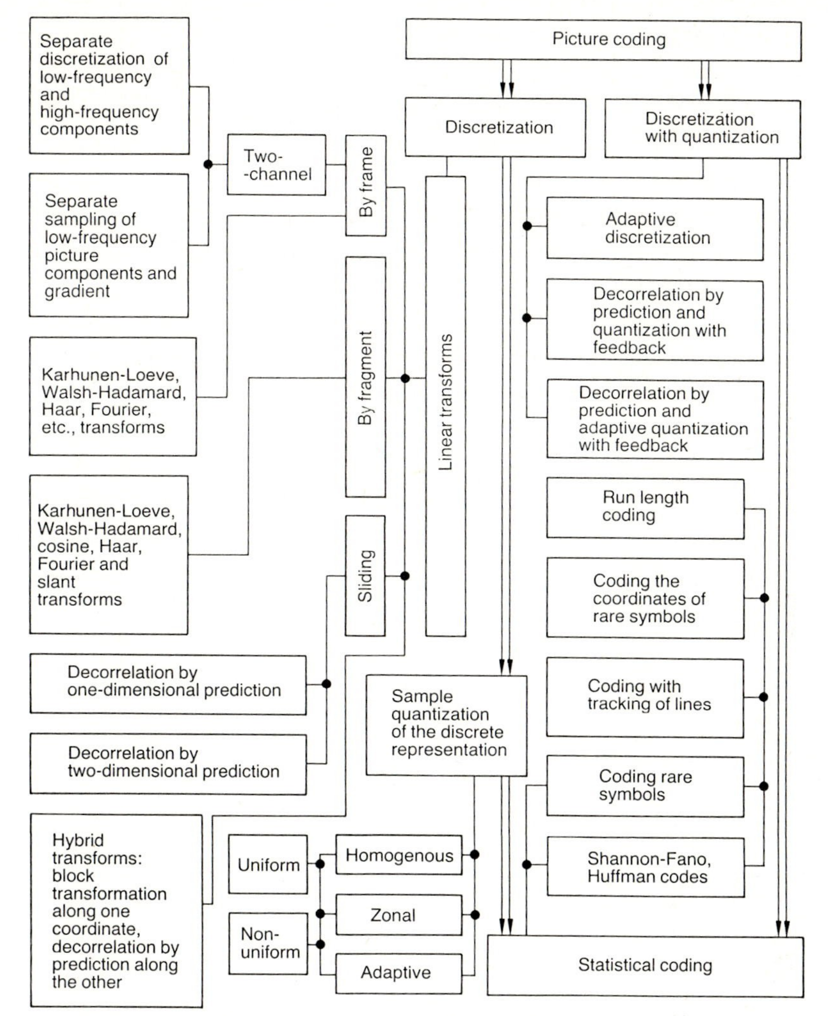

Fig. 3.7. Classification of the methods of picture coding

This coding method (in combination with the corresponding quantization of the decorrelated signal) is called differential pulse code modulation (DPCM) [3.4]. In picture readout and transmission systems with parallel and sequential-type sensors (parallel on one coordinate and scanning along the other), a promising method of transformation is a hybrid one. It consists in a block transformation along one coordinate and decorrelation along the other [3.31]. Hybrid transformations are also useful for interframe coding of frame sequence in the transmission of pictures of moving objects [3.32].

The simplest method of sample quantization of discrete representation is homogeneous quantization, in which the quantization rules (number and distribution of levels) are the same for all samples. It is usually employed when discretization is performed by measuring samples of the original continuous signal or by using sliding transformations. In other cases, zonal and adaptive non-homogeneous quantization is used.

In zonal quantization, samples are divided into groups (zones), each of which has its own quantization rule. The zonal quantization method may also include sample quantization, when two-channel discretization is used.

In adaptive quantization the division into groups is carried out adaptively. Various approaches are possible. For example, one may drop all transformation coefficients (samples) the sum of whose squares does not exceed a certain fraction of the sum of the squares of all the coefficients; or one may drop coefficients not exceeding the prescribed threshold in absolute value. The groups may also be divided according to the magnitude of the sum of the squared coefficients or by other methods.

Homogeneous and zonal quantization can be uniform or, for more economical quantization, non-uniform.

In certain cases it is possible and convenient (from the standpoint of feasibility) to develop a combined discretization and quantization procedure (Fig.3.7). Thus in adaptive discretization the arrangement of signal samples is determined by the results of measurements of the restoration error. In predictive decorrelation and in quantization with feedback, the predicted signal values are formed from the quantized values of the previous discrete signal values.

The final phase of coding is statistical coding of the sample quantization values. With statistical coding a significant saving in the volume of digital picture description can be achieved by using non-uniformity in the frequency with which the separate values of the quantized signal appear. To do this the non-uniform codes of Shannon-Fano and Huffman [3.1,33] are used, as well as various methods of coding rare symbols [3.1] in combination with the Shannon-Fano and Huffman codes. The latter are mainly used in systems with DPCM and adaptive discretization, and in two-channel systems (in coding the signal gradient). Of the methods for coding rare symbols, those using line tracing (formed, for example, by the signal gradient) may be especially mentioned as being among the most effective [3.34].

Detailed picture coding methods and an extensive bibliography are given in [3.4,33,35-38].

4 Discrete Representations of Linear Transforms

Along with digital methods of representing signals, a discrete representation is also necessary for signal transformation in digital processing. This chapter is devoted to different aspects of this problem.

The discrete representation of linear transforms and of the most commonly used shift-invariant transformations is discussed in Sects.4.1,2. Section 4.3 continues this discussion and introduces transversal, recursive, 2-D-separable, sequential, and parallel multistage digital filters. In Sect.4.4 digital filters are characterized using the concepts of impulse response and frequency transfer function, and new concepts are introduced to adapt this characterization to the description of pictures as continuous signals. The important practical question of boundary effects is discussed in Sect.4.5.

Section 4.6 opens the second part of the chapter, which is devoted to orthogonal transforms. The discrete Fourier transform (DFT) is treated in Sect.4.6. In Sect.4.7 the natural generalization of the shifted DFT (SDFT) is introduced and explained, and Sect.4.8 presents the main applications of DFT and SDFT in digital picture processing. In Sect.4.9, Walsh and similar transforms are introduced in matrix form. The description of the Haar transform given in Sect.4.10 is then used to construct the general matrix description of many other orthogonal transforms in Sect.4.11.

4.1 Problem Formulation and General Approach

This chapter is devoted to the description of discrete linear transforms (operators) that can be implemented in digital processors, and their relation to continuous representations. The way in which operators are described is determined entirely by the discrete representation of the processed signals.

Insofar as it is digital signals that are being processed, i.e., signals that have been sampled and quantized, one can, in choosing a method for discrete description of the operator, consider the domain of its definition to be finite dimensional. Let A_N be the linear space of the processed signals $\{a_N(t)\}$; $\{\varphi_k(t)\}$ the basis in this space; and L the transform operator. The space of the signals $\{b_N(\tau)\}$ after transformation is also finite dimensional (see Sect.2.6). Let us take $\{\theta_k(\tau)\}$ as the basis in this space. It then follows from (2.80 - 82) that

$$b_N(\tau) = \sum_{k=0}^{N-1} \beta_k \theta_k(\tau) \quad , \quad \text{where} \tag{4.1}$$

$$\beta_k = \sum_{n=0}^{N-1} \alpha_n \lambda_{n,k} \quad ; \tag{4.2}$$

$\alpha_n = (a_N, \boldsymbol{\psi}_n)$ are the representation coefficients $a_N(t)$ on the basis $\{\varphi_k(t)\}$; $\{\psi_n(t)\}$ is the basis reciprocal to $\{\varphi_k(t)\}$;

$$\lambda_{nk} = (L\, \boldsymbol{\varphi}_n, \boldsymbol{\eta}_k) \tag{4.3}$$

is the discrete representation of the operator L on the basis $\{\theta_k(\tau)\}$; and $\{\eta_k(\tau)\}$ is the basis reciprocal to $\{\theta_k(\tau)\}$.

Since we are concerned with the discrete representation of continuous operators, we can express the impulse response $H_N(t,\tau)$ of operator L (see Sect.2.7) in terms of $\{\lambda_{n,k}\}$ by substituting (4.2,3) into (4.1) and describing α_n by (2.23,28):

$$\begin{aligned} b_N(\tau) &= La_N(t) = \sum_{k=0}^{N-1} \sum_{n=0}^{N-1} \lambda_{n,k} \alpha_n \theta_k(\tau) = \sum_{k=0}^{N-1} \sum_{n=0}^{N-1} \left[\int_T a(t)\psi_n^*(t)dt\right] \lambda_{n,k} \theta_k(\tau) \\ &= \int_T a(t) \left[\sum_{k=0}^{N-1} \sum_{n=0}^{N-1} \lambda_{n,k} \psi_n^*(t) \theta_k(\tau)\right] dt \quad , \end{aligned} \tag{4.4}$$

which implies that the operator kernel is

$$H_N(t,\tau) = \sum_{k=0}^{N-1} \sum_{n=0}^{N-1} \lambda_{n,k} \psi_n^*(t) \theta_k(\tau) \quad . \tag{4.5}$$

Signals $\{a_N(t)\}$, considered as objects for digital processing, are approximations of continuous signals $\{a(t)\}$, their projection on N-dimensional space (Sect.3.2). Therefore signals $\{b_N(\tau)\}$ are the approximate results of the transformation of continuous signals by a continuous operator. In this sense $H_N(t,\tau)$ can be considered a finite-dimensional approximation of the kernel of the continuous transformation L.

4.2 Discrete Representation of Shift-Invariant Filters for Band-Limited Signals

Discrete signal representation in digital processing is usually based on sampling theory, and signals are considered to be functions with limited Fourier spectra. As already noted, a representation of this kind is very convenient in view of the simplicity with which it is obtained as a set of signal samples. The sampling basis functions $\mathrm{sinc}(2\pi Ft)$ used in this case are also convenient for representing shift-invariant linear filters.

To clarify the physical meaning of (4.1-5) in the given case we shall repeat their derivations for the sampling basis functions.

Let $(-F,F)$ be the frequency interval of the Fourier spectra of signals at the input of shift-invariant linear filters, so that they can be written in the form

$$a(t) = \sum_{k=-\infty}^{\infty} a_k \,\mathrm{sinc}\left[2\pi F\left(t - \frac{k}{2F}\right)\right] \quad . \tag{4.6}$$

Also, let $h(t-\tau)$ be the operator (filter) impulse response (Sect.2.8), i.e.,

$$b(t) = \int_{-\infty}^{\infty} a(\tau)\, h(t-\tau)d\tau = \int_{-\infty}^{\infty} a(t-\tau)\, h(\tau)d\tau \quad . \tag{4.7}$$

Since $a(t)$ is a band-limited signal,

$$a(t) = 2F \int_{-\infty}^{\infty} a(\tau)\,\mathrm{sinc}[2\pi F(t-\tau)]d\tau \quad ; \tag{4.8}$$

$$\begin{aligned} 2F \int_{-\infty}^{\infty} b(\tau)\,\mathrm{sinc}[2\pi F(t-\tau)]d\tau &= 2F \int_{-\infty}^{\infty}\int_{-\infty}^{\infty} a(\tau-\tau_1)\, h(\tau_1) \\ \times \mathrm{sinc}[2\pi F(t-\tau)]d\tau d\tau_1 &= \int_{-\infty}^{\infty} h(\tau_1)d\tau_1\left\{2F\int_{-\infty}^{\infty} a(\tau-\tau_1)\right. \\ \left.\times \mathrm{sinc}\left[2\pi F(t-\tau)\right]d\tau\right\} &= \int_{-\infty}^{\infty} a(t-\tau_1)\, h(\tau_1)d\tau_1 = b(t) \quad , \end{aligned} \tag{4.9}$$

i.e., $b(t)$ is also a band-limited function which can be restored from its samples $\{b_k\}$:

$$b(t) = \sum_{k=-\infty}^{\infty} b_k \,\mathrm{sinc}\left[2\pi F\left(t - \frac{k}{2F}\right)\right] \quad , \qquad \text{where} \tag{4.10}$$

$$b_k = b\left(\frac{k}{2F}\right) = 2F \int_{-\infty}^{\infty} b(t)\,\mathrm{sinc}\left[2\pi F\left(t - \frac{k}{2F}\right)\right]dt \quad . \tag{4.11}$$

Let us now rewrite (4.7) in the form

$$\begin{aligned} b(t) &= \int_{-\infty}^{\infty} a(\tau)\, h(t-\tau)d\tau = \int_{-\infty}^{\infty} \left\{\sum_{k=-\infty}^{\infty} a_k \,\mathrm{sinc}\left[2\pi F\left(t - \frac{k}{2F}\right)\right]\right\} h(t-\tau)d\tau \\ &= \sum_{k=-\infty}^{\infty} a_k \int_{-\infty}^{\infty} h(t-\tau)\,\mathrm{sinc}\left[2\pi F\left(\tau - \frac{k}{2F}\right)\right]d\tau = \frac{1}{2F}\sum_{k=-\infty}^{\infty} a_k\, \hat{h}\left(t - \frac{k}{2F}\right) \quad , \end{aligned} \tag{4.12}$$

where

$$\hat{h}(t) = 2F \int_{-\infty}^{\infty} h(\tau)\,\mathrm{sinc}\left[2\pi F(t-\tau)\right]d\tau \tag{4.13}$$

is a band-limited function derived by limiting the bandwidth of the impulse response $h(t)$ of the continuous filter.

Then the samples $\{b_k\}$ of signal $b(t)$ are expressed in terms of the samples $\{a_k\}$ of signal $a(t)$ and of the samples $\{\hat{h}_k\}$ filter impulse response $\hat{h}(t)$ with truncated spectrum:

$$b_k = \sum_{n=-\infty}^{\infty} a_n \hat{h}_{k-n} = \sum_{n=-\infty}^{\infty} a_{k-n} \hat{h}_n \quad , \quad \text{where} \tag{4.14}$$

$$\hat{h}_n = 2F \int_{-\infty}^{\infty} h(t) \operatorname{sinc}\left[2\pi F\left(t - \frac{n}{2F}\right)\right] dt \quad . \tag{4.15}$$

A comparison of (4.14) with (4.7) reveals that (4.14) is none other than the known rectangle formula for calculating the convolution integral (4.7) when the discretization step on the argument is equal to $(2F)^{-1}$. Equation (4.14), which is obviously a variant of (4.2), is one of the basic relations used in the digital implementation of linear shift-invariant filters. The finite-dimensional approximation of the continuous filter is based on it, by discarding in (4.14) terms with sufficiently small values of $\hat{h}_n$:

$$b_k \approx \sum_{n=-N_1}^{N_2} a_{k-n} \hat{h}_n \quad . \tag{4.16}$$

It is practically impossible to synthesize ideal sampling functions for signal discretization (Sect.3.5). However, if during signal discretization and continuous signal restoration only the basis functions with shift are used, as in (3.27,29), then the basic digital filtering formula remains valid. Only the definition of the $\hat{h}_n$ changes. In this case

$$\hat{h}_n = \int_{-\infty}^{\infty} h(\tau)d\tau \int_{-\infty}^{\infty} \varphi_d(t)\varphi_s(t - \tau + n\Delta t)dt \quad . \tag{4.17}$$

All the relations in this section can be simply extended to the two-dimensional case. Then (4.16) can be rewritten in the form

$$b_{k,\ell} \quad \sum_{m=-N_{11}}^{N_{21}} \sum_{n=-N_{12}}^{N_{22}} a_{k-m,\ell-n} \hat{h}_{m,n} \quad , \tag{4.18}$$

where $\{b_{k,\ell}\}$ $\{a_{k,\ell}\}$ $\{h_{m,n}\}$ are samples of signals and the filter impulse response the spectrum of which has been truncated with a rectangular window; the samples are taken on a rectangular raster.

4.3 Digital Filters

Both one-dimensional and two-dimensional linear transforms are used in picture processing. The digital implementation of these transformations is called *digital filtering*. Methods of two-dimensional digital filtering based on the representation of signals and filter impulse response according to the sampling theorem, as in (4.16-18), are called *filtering in the space domain*. It will be shown below that signals can also be filtered by transforming their spectra (Sect.4.8); this is called *filtering in the frequency domain* (or in the spatial frequency domain). In this section various approaches to digital filtering in the space domain are analyzed.

Let us begin by looking at one-dimensional filters. A digital filter implemented in a digital processor directly by (4.16) is termed transverse or non-recursive [4.1-3]. To perform computer calculations on the basis of this formula, it is necessary to execute (N_2+N_1+1) multiplications and (N_2+N_1) additions for each sample of transformed signals.

There is a class of operators whose kernels are such that the transformation (4.16) can be cast into the following recursive form;

$$b_k = \hat{h}_0 a_k + \sum_{n=1}^{N} \hat{g}_n b_{k-n} \quad . \tag{4.19}$$

It is not difficult to verify that the impulse response samples $\{h_n\}$ of such filters must satisfy the relation

$$\hat{h}_n = \sum_{k=1}^{n} \hat{g}_k \hat{h}_{n-k} \quad , \quad n = 1, 2, \ldots \quad . \tag{4.20}$$

Calculating the sequence $\{b_k\}$ by (4.19) requires fewer operations per sample than by (4.16), because use is made of the results of the previous calculations. For example, when $g_1 \neq 0$, $g_2 = g_3 = \ldots = 0$, and

$$b_k = h_0 a_k + g_1 b_{k-1} \quad , \tag{4.21}$$

i.e., every value b_k can be found by only two multiplications and one addition, while in the direct sum (4.16), the number of members is, generally speaking, infinitely large, since

$$\hat{h}_n = h_0 g_1^{n-1} \quad , \tag{4.22}$$

and g_1^{n-1} never vanishes. This advantage in the speed of calculation always forces us to seek an approximation to the required filter by a recursive one. In principle, such an approximation can be constructed on the basis of (4.20), but it is more convenient to use the frequency transfer functions, which will be considered in the following section.

Equation (4.19) describes the simplest digital recursive filter. In the general case, the digital recursive filter is defined by

$$b_k = \sum_{n=0}^{N_T-1} \hat{h}_n a_{k-n} + \sum_{n=1}^{N_R} \hat{g}_n b_{k-n} \quad , \tag{4.23}$$

where N_T and N_R are the number of terms in the transverse and recursive parts of the formula, respectively.

An example of a recursive representation is a filter commonly used in digital picture processing for calculating the current average signal over a given interval:

$$\bar{a}_k = \frac{1}{2N+1} \sum_{n=-N}^{N} a_{k-n} \quad . \tag{4.24}$$

In this case the samples of filter impulse response, $\hat{h}_n = 1/(2N+1)$, $n = -N, \ldots, 0, \ldots, N$, and (4.24) can be transformed into the recursive form:

$$\bar{a}_k = (a_{k+N} - a_{k-N-1})/(2N+1) + \bar{a}_{k-1} \quad . \tag{4.25}$$

It is easily seen that (4.25) is a particular case of (4.23). Hence, calculating by (4.25), it is possible to establish the current average signal not by 2N additions for one sample average, as in (4.24), but by only three. Remarkably, the number of operations here does not depend on the number of samples over which averaging takes place.

Equation (4.18) defines a two-dimensional non-recursive filter. Recursive filters can also be constructed in the two-dimensional case, but the direction of recursion must be assigned. If we consider the "past" signal values to be above and to the left of the given sample, and index the samples from left to right and top to bottom, then (4.23) for the one-dimensional recursive filter can be generalized to the two-dimensional case in the following way:

$$b_{k,\ell} = \sum_{m=0}^{N_{T_1}-1} \sum_{n=0}^{N_{T_2}-1} a_{k-m,\ell-n} \hat{h}_{m,n} + \sum_{m=1}^{N_{R_1}} \sum_{n=-N_{R_2}}^{N_{R2}} b_{k-m,\ell-n} \hat{g}_{m,n}$$
$$+ \sum_{n=1}^{N_{p_2}} b_{k,\ell-n} g_{0,n} \quad . \tag{4.26}$$

There is yet another class of two-dimensional digital filters which are of special interest from the standpoint of saving computer time. They are *two-dimensional separable filters*. The impulse response of these filters, $\hat{h}_{m,n}$, can be written as a product of one-dimensional functions: $\hat{h}_{m,n} = \hat{h}_m^{(1)} h_n^{(2)}$.

For such filters (4.17) becomes

$$b_{k,\ell} = \sum_{m=-N_{11}}^{N_{12}} \hat{h}_m^{(1)} \sum_{n=-N_{21}}^{N_{22}} \hat{h}_n^{(2)} a_{k-m,\ell-n} \quad , \tag{4.27}$$

which can be calculated recursively. Indeed, denoting the inner sum in (4.27) by $c_{k-m,\ell}$, we obtain

$$b_{k,\ell} = \sum_{m=-(N_{11}-1)}^{N_{12}} \hat{h}_m^{(1)} c_{k-1-(m-1),\ell} + \hat{h}_{-N_{11}}^{(1)} c_{k+N_{11},\ell} \quad , \tag{4.28}$$

where the variables $c_{k-1-(m-1),\ell}$ under the summation sign are the same as those used for the calculation of $b_{k-1,\ell}$ in the preceding step, apart from one, $c_{k-1-N_{12},\ell}$. Therefore to calculate a sample $b_{k,}$ by means of (4.28) it is necessary to perform $(N_{22}+N_{21}+N_{12}+N_{11}+2)$ multiplications and $(N_{22}+N_{21}+N_{12}+N_{11})$ additions and not $(N_{22}+N_{21}+1)(N_{12}+N_{11}+1)$ and $(N_{22}+N_{21}+1)(N_{12}+N_{11}+1)-1$, respectively, as calculation by (4.17) requires.

An example of a two-dimensional separable filter is the one used to obtain the current average signal value over a rectangular area:

$$\bar{a}_{k,\ell} = \frac{1}{(2N_1 + 1)(2N_2 + 1)} \sum_{m=-N_1}^{N_1} \sum_{n=-N_2}^{N_2} a_{k-m,\ell-n} \quad . \tag{4.29}$$

A separable representation (4.27) of the required two-dimensional filter impulse response is not always possible. One can attempt to approximate the required function $\hat{h}_{m,n}$ by a sum of the separable functions:

$$\hat{h}_{m,n} = \sum_{u=0}^{U-1} h_m^{(u)} g_n^{(u)} \quad . \tag{4.30}$$

If the number of terms U in this sum is not large, this exchange of one filter for several is also more efficient from the computing standpoint than filtering by (4.18). The problem of finding the best representation (4.30) is related to that of finding the best approximation of the finite-dimensional signal [4.4,5].

The possibility of representing two-dimensional filter impulse responses in a separable form depends to a large extent on the choice of coordinate system. For example, the impulse response of isotropic filters, which are often encountered in processing pictures obtained by optical imaging systems, is, in the Cartesian corrdinate system (t_1, t_2), a function of the squared coordinate sum: $(t_1^2 + t_2^2)$, and generally cannot be represented in a separable form. In the polar system of coordinates the very same impulse response is not only separable, but is also simply a function of one variable.

The choice of coordinate system in picture processing is also important as regards the possibility of considering filters to be shift invariant or spatially homogeneous if a special system of coordinates is chosen with nonlinear scales along the coordinate axes [4.6,7].

Representation (4.30) of a filter impulse response in the form of sums of impulse responses of a simpler type corresponds to parallel filtering by several filters and summation of the results. Such a representation can thus be called parallel. Also possible are sequential digital filter representations. With them the required impulse response appears in the form of a convolution of impulse responses of a simpler form:

$$\hat{h}_m = \sum_{m_{U-1}} h^{(U-1)}_{m_{U-1}} \sum_{m_{U-2}} h^{(U-2)}_{m_{U-2}} \dots \sum_{m_0} h^{(0)}_{m-\sum_{u=0}^{U-1} m_u} h^{(1)}_{m_0} \quad , \tag{4.31}$$

which corresponds to passing the signal successively through several filters.

In the one-dimensional case the sequential filter representation offers no advantages in the number of operations if the cascade filters cannot be constructed recursively, and is even inferior to the single-cascade representation. Thus, let N be the number of samples (length) of the filter impulse response of every member of the cascade in (4.31). Then the spread of pulse responses h_m with a cascade having U members is $[U(N-1)+1]$ samples. This means that if filtering is carried out by means of a single-member filter according to (4.16), then it is necessary to perform $[U(N-1)+1]$ operations for every signal sample. With sequential filtering, U filters require UN operations, i.e., more. But two-dimensional sequential filters can be significantly more convenient than single-filters, since in the two-dimensional case the latter require $[U(N-1)+1] \times [U(M-1)+1]$ operations instead of UNM for the equivalent sequential filter with an impulse response spread of $N \times M$ samples per member of the cascade.

Therefore sequential representations of two-dimensional filters are better able to save computer time in signal processing on computers and the corresponding hardware in special-purpose processors. An analysis of the possibility of filter representation in the sequence form is easier using the frequency transfer function, which is considered in the next section. For more details on the synthesis of 2-D digital filters, see [4.8].

4.4 Transfer Functions and Impulse Responses of Digital Filters

In digital picture processing it is important to remember that a picture as an object of processing and a picture obtained as a result of processing are continuous signals. Thus the following are continuous signals: the illuminance distribution of the photocathode in a television transmitting tube, the brightness distribution of a picture tube phosphor, the density distribution of a photographic film, etc. Therefore, in order to choose the correct parameters for digital processing, and in particular digital filters, it is necessray to have a description of them which is adequate for the description of continuous signals.

As noted in Sects.2.7,8, the most important characteristics of linear filters are their impulse responses and frequency transfer functions. Let us define these characteristics for digital filters and establish their relation to the corresponding characteristics of the continuous filters approximating them.

We shall define the discrete impulse response (DIR) of the digital filter described by (4.16) as

$$h_d(t) = \sum_{n=-N_1}^{N_2} \hat{h}_n \delta(t - k\Delta t) \quad , \tag{4.32a}$$

where Δt is the discretization interval corresponding to (3.27,29). DIR is an artificial concept which is convenient in describing digital filter, irrespective of how the samples of the signals they transform are obtained and of how continuous signals are restored from the result of digital filtering.

In order to take the effect of discretization and continuous signal restoration into account, we shall introduce the concept of digital filter continuous impulse response (CIR). CIR is the impulse response of a continuous filter, such that for a given continuous input signal, the same continuous output signal is obtained as the signal restored from the samples on the output of the digital filter. According to this definition, the CIR of a digital filter (4.16) is written as

$$h_c(t, \tau) = \sum_{k=0}^{K-1} \sum_{n=-N_1}^{N_2} \hat{h}_n \varphi_d[\tau - (k - n)\Delta t]\, \varphi_s(t - k\Delta t) \quad , \tag{4.32b}$$

where $\varphi_d \tau$ and $\varphi_s(t)$ are the weight functions describing the aperture (impulse response) of the devices for discretization and continuous signal restoration as in (3.27,29). K is the number of samples in the sum (3.29) over which signal restoration takes place after digital filtering.

Frequency transfer functions are defined as the Fourier transforms of pulse responses. Therefore in accordance with (2.98,112), we shall call the following the discrete frequency transfer function (DFTF) of the digital filter (4.16):

$$H_d(f) = \int_{-\infty}^{\infty} h_d(t) \exp(i2\pi ft)dt = \sum_{n=-N_1}^{N_2} \hat{h}_n \exp(i2\pi nf\Delta t) \quad . \tag{4.33a}$$

Note that it is in precisely this way that the digital filter frequency transfer function is defined in the literature on digital signal filtering [4.1-3].

We shall call the function

$$H_c(f) = \int_{-\infty}^{\infty}\int_{-\infty}^{\infty} h_c(t,\tau)\exp[i2\pi(ft - p\tau)]dt\,d\tau = \left[\sum_{n=-N_1}^{N_2} \hat{h}_n \exp(i2\pi pn\Delta t)\right]$$

$$\times \left\{\sum_{k=0}^{K-1} \exp[i2\pi k(f-p)\Delta t]\right\} \phi_s(f)\,\phi_d(-p) = H_d(p)\,\phi_s(f)\,\phi_d(-p)$$

$$\times \frac{\sin[\pi K(f-p)/2F]}{\sin[\pi(f-p)/2F]} \exp\left[\frac{i\pi(K-1)(f-p)}{2F}\right] \quad , \qquad (4.33b)$$

the continuous frequency transfer function (CFTF) of the digital filter. Here $2F = 1/\Delta t$ and $\phi_d(\ )$, $\phi_s(\ )$ are the Fourier transforms of $\varphi_d(\tau)$ and $\varphi_s(t)$, or frequency transfer functions of discretization devices and continuous signal restoration equipment.

CFTF is distinguished from DFTF by three factors: the frequency transfer function of the discretization device, the frequency transfer function of the restoration device and the window function

$$W(f) = \sin[\pi K(f-p)/2F]/\sin[\pi(f-p)/2F] \quad ,$$

which takes into account the influence of the restricted lengths of the signal sequences being processed. A fourth exponential factor is irrelevant, since it depends only on the method by which the members of the series are indexed.

Equation (4.33b) shows that as a result of the truncation of the signal sample series in the digital filter, its corresponding continuous filter turns out to be shift variant. Nevertheless, the main "energy" of the function $W(f)$ is concentrated on the interval $(-F/K, F/K)$, so that with large K, $W(f)$ can be considered a delta function, and the spatial variance of the filter can be ignored, considering that

$$H_c(f) \approx H_d(f)\,\phi_s(f)\,\phi_d(-f) \quad . \qquad (4.34)$$

In Sect.4.2 it was shown how values $\{h_n\}$, which define the digital filter, are linked to the impulse response of the corresponding linear filter (4.17). However, the restriction on the number of members was not taken into account there in the transition from the exact fc.mula (4.14) to the basic digital filtering formula (4.16). This restriction can be more easily analyzed in terms of frequency transfer functions.

Let $H(f)$ be the frequency transfer function of a continuous filter with impulse response $h(\tau)$:

$$h(\tau) = \int_{-\infty}^{\infty} H(f)\exp(-i2\pi f\tau)df \quad . \qquad (4.35)$$

By substituting (4.35) into (4.17) we obtain

$$h_n = \int_{-\infty}^{\infty} H(f)\, \phi_s(f)\, \phi_d(-f)\, \exp(-i2\pi fn/2F)df \quad ,$$

where $\phi_s(f)$ and $\phi_d(f)$ are the same as in (4.34). It can be easily seen that according to this formula, h_n are the Fourier coefficients of periodic functions

$$\hat{H}(f) = 2F \sum_{n=-\infty}^{\infty} H_t(f + 2nF) \quad , \quad \text{where}$$

$$H_t(f) = H(f)\, \phi_s(f)\, \phi_d(-f)$$

is the filter frequency transfer function truncated by the function $\phi_s(f)$ and $\phi_d(-f)$. Consequently,

$$\hat{H}(f) = \sum_{n=-\infty}^{\infty} \hat{h}_n \exp(i2\pi fn/2F) \quad .$$

Restricting the number of coefficients h_n as in (4.16) distorts this frequency transfer function. It is possible to take this distortion into account by writing

$$H_r(f) = \int_{-\infty}^{\infty} \left[\sum_{n=-N_1}^{N_2} \hat{h}_n \delta\left(t - \frac{n}{2F}\right)\right] \exp(i2\pi ft)dt = \int_{-\infty}^{\infty} \operatorname{rect} 2F \frac{t + N_1/2F}{N_1 + N_2}$$
$$\times \left[\sum_{n=-\infty}^{\infty} \hat{h}_n \delta\left(t - \frac{n}{2F}\right)\right] \exp(i2\pi ft)dt \quad . \tag{4.36}$$

Using the theorem on the spectrum of products (Table 2.2, line 16), we obtain

$$H_r(f) = \left[\frac{N_2 + N_1}{2F} \operatorname{sinc}\left(\pi \frac{N_2 + N_1}{2F} f\right) \exp\left(i\pi \frac{N_2 - N_1}{2F} f\right)\right]$$
$$\circledast \left[\sum_{n=-\infty}^{\infty} \hat{h}_n \exp\left(i2\pi n \frac{f}{2F}\right)\right] \quad , \tag{4.37}$$

where $\circledast$ denotes the convolution operation (2.109). The first of the convolved functions describes the influence which restricting the number of terms in (4.16) has on the digital filter frequency transfer function. The second is the frequency transfer function of a filter with an unlimited number of terms $\hat{H}(f)$ in (4.34). Substituting $\hat{H}(f)$ in place of the sum in (4.37), we obtain, finally,

$$\hat{H}_r(f) = (N_2 + N_1) \sum_{n=-\infty}^{\infty} \int_{-\infty}^{\infty} H\,(f_1 + 2nF)$$
$$\times \operatorname{sinc}\left[\pi \frac{N_2 + N_1}{2F}(f - f_1)\right] \exp\left[i\pi \frac{N_2 - N_1}{2F}(f - f_1)\right] df_1 \quad . \tag{4.38}$$

Thus, the discrete frequency transfer function $\hat{H}_r(f)$ of the digital filter, being an approximation of the continuous filter with frequency transfer function H(f), is a periodic repetition with step 2F of the truncated transfer function $H_t(f) = H(f)\,\phi_s(f)\,\phi_d(-f)$, provided the latter is smoothed by a function

$$(N_2 + N_1)\ \mathrm{sinc}[\pi(N_1 + N_2)\Delta f/2F)]\ \exp[i\pi(N_2 - N_1)\Delta f/2F] \quad .$$

It should be noted that such smoothing leads to the superposition of smoothed functions H(f) on frequencies close to the interval edges (-F, F).

The effects associated with the transfer from the continuous to the discrete filter with a limited number of impulse response samples are shown in Fig.4.1.

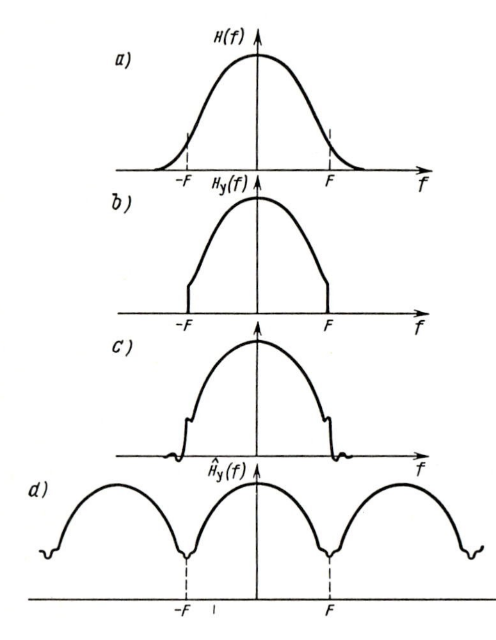

Fig. 4.1 a-d. Effects associated with the transfer from the continuous to the discrete filter: (a) continuous filter frequency transfer function, (b) the same function after spectrum truncation, (c) the frequency transfer function of the continuous filter with spectrum truncation and a response of limited length, (d) discrete transfer function of the discrete filter determined by the filter pulse response samples in (c)

As an example let us look at the discrete frequency transfer functions of filter (4.24), which calculates the current average of a series of samples, i.e., a filter with sampled impulse response

$$\hat{h}_n = 1/(2N + 1),\ n = -N, \ldots, 0, \ldots, N \quad . \tag{4.39}$$

Substituting (4.39) into (4.32), we find

$$\hat{H}_r(f) = \frac{1}{2N + 1} \sum_{n=-N}^{N} \exp\left(i2\pi f \frac{n}{2F}\right) = \frac{1}{2N + 1} \frac{\sin[\pi(2N + 1)f/2F]}{\sin(\pi f/2F)}$$

$$= \frac{\mathrm{sinc}[\pi(2N + 1)f/2F]}{\mathrm{sinc}(\pi f/2F)} \quad . \tag{4.40}$$

The discrete frequency transfer functions, constructed for N = 2 and N = 128 in (4.40), are presented in Fig.4.2. For purposes of comparison the broken line

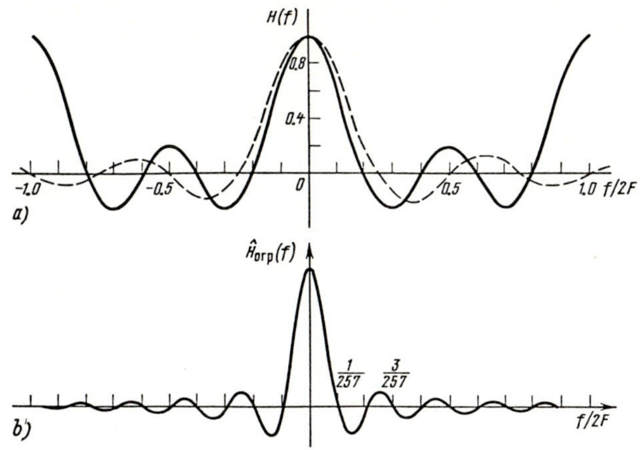

Fig. 4.2a,b. Discrete frequency transfer function (4.40) for (a) $N=2$ and (b) $N=128$

in Fig.4.2a shows the frequency transfer function of the corresponding continuous filter with impulse response

$$h(\tau) = N/F, \quad -N/2F \leqslant \tau \leqslant N/2F \quad . \tag{4.41}$$

It is not difficult to see that for low N there is an obvious difference between these transfer functions[1].

Let us now look at the discrete frequency transfer functions of the recursive filters which are defined generally by (4.23). To do this, we transform (4.23) into

$$b_k - \sum_{n=1}^{N_R} \hat{g}_n \, b_{k-n} = \sum_{n=0}^{N_T-1} \hat{h}_n \, a_{k-n} \quad , \tag{4.42}$$

from which the formula for spectra and frequency transfer functions is derived:

$$[1 - \hat{G}_r(f)]\beta(f) = \hat{H}_r(f)\,\alpha(f) \quad , \quad \text{where} \tag{4.43}$$

$$\hat{G}_r(f) = \sum_{n=1}^{N_R} \hat{g}_n \exp(i2\pi fn/2F) \quad ;$$

$$H_r(f) = \sum_{n=0}^{N_T-1} \hat{h}_n \exp(i2\pi fn/2F) \quad . \tag{4.44}$$

Consequently, the DFTF of the digital recursive filter can be written

[1] These frequency transfer functions coincide better if the summation interval for the continuous filter extends from $-(2N+1)/4F$ to $(2N+1)/4F$, i.e., if it somewhat differs from the discrete filter summation interval.

$$\hat{H}_d(f) = \frac{\hat{H}_r(f)}{1 - \hat{G}_r(f)} = \frac{\sum_{n=0}^{N_T-1} \hat{h}_n \exp(i2\pi fn / 2F)}{1 - \sum_{n=1}^{N_R} \hat{g}_n \exp(i2\pi fn / 2F)} \quad . \tag{4.45}$$

All these results can also be applied to the two-dimensional case. Thus the discrete frequency transfer function of the two-dimensional non-separable filter is defined by

$$\hat{H}_d(f_1, f_2) = \sum_{m=-N_{11}}^{N_{21}} \sum_{n=-N_{12}}^{N_{22}} \hat{h}_{m,n} \exp[i2\pi(f_1 m / F_1 + f_2 n / F_2)] \quad . \tag{4.46}$$

The DFTF of a two-dimensional separable filter is the separable function

$$\hat{H}_d(f_1, f_2) = \hat{H}_d^{(1)}(f_2)(\hat{H})_d^{(2)}(f_2) \quad , \tag{4.47}$$

where $\hat{H}_d^{(1)}(f_1)$ and $\hat{H}_d^{(2)}(f_2)$ are defined by an expression of type (4.3).

The DFTF of the two-dimensional recursive filter (4.26) is written as

$$\hat{H}_R(f_1, f_2) = \frac{\sum_{m=0}^{N_{T1}-1} \sum_{n=0}^{N_{T2}-1} \hat{h}_{m,n} \exp\left[i2\pi\left(\frac{mf_1}{F_1} + \frac{nf_2}{F_2}\right)\right]}{1 - \sum_{m=1}^{N_{R1}} \sum_{n=-N_{p2}}^{N_{R2}} \hat{g}_{m,n} \exp\left[i2\pi\left(\frac{mf_1}{F_1} + \frac{nf_2}{F_2}\right)\right] - \sum_{n=1}^{N_{R2}} \hat{g}_{0,n} \exp\left(i2\pi\frac{nf_2}{F_2}\right)} \quad . \tag{4.48}$$

Clearly, the frequency transfer function of the parallel-type filter is

$$H_p(f) = \sum_{u=0}^{U-1} H^{(u)}(f) \quad , \tag{4.49}$$

and the frequency transfer function of the sequence type filter is

$$H_s(f) = \prod_{u=0}^{U-1} H^{(u)}(f) \quad . \tag{4.50}$$

The corresponding continuous frequency transfer functions are determined by (4.33b).

4.5 Boundary Effects in Digital Filtering

Let us return to the basic formula (4.16) of digital filtering in the spatial domain:

$$b_k = \sum_{\ell=-N_1}^{N_2} a_{k-\ell}\hat{h}_\ell \quad . \tag{4.51}$$

The number of samples in the initial sequence of the input signal $\{a_k\}$ is always finite. Let them be numbered from 0 to N-1, $k = 0, 1, \ldots, N-1$. Then (4.51) determines only the samples of the output signal with indices from $k = N_2$ to $k = N-N_1 - 1$. Samples corresponding to the range from $k = 0$ to $k = N_2 - 1$ and from $k = N-N_1$ to $k = N - 1$ are not determined, because the samples of initial sequences with $k < 0$ and $k > N - 1$ are not given. Thus the length of the sequence in linear digital filtering by (4.51) is decreased.

However, it is usually inconvenient for the number of samples of the output signal to differ from the number of samples of the input signal. It is for this reason that a method of predetermining the initial sequence is adopted. There are two ways of doing this, one of which can be called the statistical and the other the algebraic approach.

The statistical approach consists in considering the given sequence of input samples to be the realization of some random process, and in predicting the values of the missing end terms of this series from the available members. In the simplest case, one attributes to the missing members the value of the mathematical expectation of the input signal (e.g., zero if the input signal has zero mean). Better results can be achieved in this regard by replacing the missing samples with the weighted sum of a specified number of the end samples. In particular cases of such predetermination, the values of the closest known samples are attributed to the unknown samples.

However, the missing samples are more frequently determined from considerations of another kind. Shift-invariant linear filters can also be realized by operations on the Fourier spectra of signals (Sect.4.8). It appears that filtering in the "frequency" domain is equivalent to filtering in the "spatial" domain by (4.51) if the initial signal is considered to be a periodic extension for index values $k < 0$ and $k > N - 1$. There are several possible methods for the periodic extension of signals: The simple extension

$$\tilde{a}_k = a_{(k)\mathrm{mod}N} = \begin{cases} a_{k+N} & , -N \leq k \leq -1 \quad ; \\ a_k & , 0 \leq k \leq N-1 \quad ; \\ a_{k-N} & , N \leq k \leq 2N-1 \quad ; \end{cases} \tag{4.52}$$

and two types of even extension

$$\tilde{a}_k = \begin{cases} a_{|k|} & , -(N-1) \leq k \leq -1 \quad ; \\ a_k & , 0 \leq k \leq N-1 \quad ; \\ a_{2N-k-2} & , N \leq k \leq 2N-2 \quad ; \end{cases} \tag{4.53a}$$

$$\tilde{a}_k = \begin{cases} a_{|k+1|} & , -N \leqslant k \leqslant -1 \quad ; \\ a_k & , 0 \leqslant k \leqslant N-1 \quad ; \\ a_{2N-k-1} & , N \leqslant k \leqslant 2N-1 \quad . \end{cases} \tag{4.53b}$$

The simple extension, which includes operations on spectra obtained by the discrete Fourier transform (Sect.4.6), distorts the signal structure unnaturally because it makes samples at different ends of the initial sequence become neighbors in the extended sequence. Digital filtering in the spatial and frequency domains therefore makes use of even methods of sequence predetermination. These methods of finite sequences extension are illustrated in Fig.4.3.

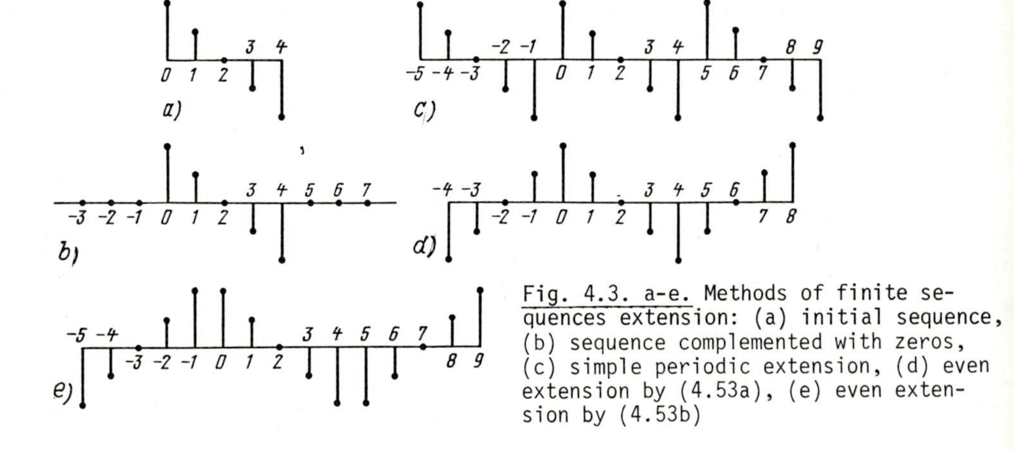

Fig. 4.3. a-e. Methods of finite sequences extension: (a) initial sequence, (b) sequence complemented with zeros, (c) simple periodic extension, (d) even extension by (4.53a), (e) even extension by (4.53b)

Two-dimensional signals are determined similarly: the simple extension

$$\tilde{a}_{k,\ell} = \begin{cases} a_{k-N_1,\ell+N_2} & , N_1 \leqslant k \leqslant 2N_1 - 1 \quad ; -N_2 \leqslant \ell \leqslant -1 \quad ; \\ a_{k+N_1,\ell+N_2} & , -N_1 \leqslant k \leqslant -1 \quad ; -N_2 \leqslant \ell \leqslant -1 \quad ; \\ a_{k,\ell+N_2} & , 0 \leqslant k \leqslant N_1 - 1 \quad ; -N_2 \leqslant \ell \leqslant -1 \quad ; \\ a_{k+N_1,\ell} & , -N_1 \leqslant k \leqslant 1 \quad ; 0 \leqslant \ell \leqslant N_2 - 1 \quad ; \\ a_{k,\ell} & , 0 \leqslant k \leqslant N_1 - 1 \quad ; 0 \leqslant \ell \leqslant N_2 - 1 \quad ; \\ a_{k-N_1,\ell} & , N_1 \leqslant k \leqslant 2N_1 - 1 \quad ; 0 \leqslant \ell \leqslant N_2 - 1 \quad ; \\ a_{k,\ell-N_2} & , 0 \leqslant k \leqslant N_1 + 1 \quad ; N_2 \leqslant \ell \leqslant 2N_2 - 1 \quad ; \\ a_{k+N_1,\ell-N_2} & , -N_1 \leqslant k \leqslant -1 \quad ; N_2 \leqslant \ell \leqslant 2N_2 - 1 \quad ; \\ a_{k-N_1,\ell-N_2} & , N_1 \leqslant k \leqslant 2N_1 - 1 \quad ; N_2 \leqslant \ell \leqslant 2N_2 - 1 \quad ; \end{cases} \tag{4.54}$$

$$\tilde{a}_{k,\ell} = \begin{cases} a_{2N_1-k-2,|\ell|}, & N_1 \leqslant k \leqslant 2N_1 - 2 \quad ; -(N_2-1) \leqslant \ell \leqslant -1 \quad ; \\ a_{|k|,|\ell|}, & -(N_1-1) \leqslant k \leqslant -1 \quad ; -(N_2-1) \leqslant \ell \leqslant -1 \quad ; \\ a_{k,|\ell|}, & 0 \leqslant k \leqslant N_1 - 1 \quad ; -(N_2-1) \leqslant \ell \leqslant -1 \quad ; \\ a_{|k|,\ell}, & -(N_1-1) \leqslant k \leqslant -1 \quad ; 0 \leqslant \ell \leqslant N_2 - 1 \quad ; \\ a_{k,\ell}, & 0 \leqslant k \leqslant N_1 - 1 \quad ; 0 \leqslant \ell \leqslant N_2 - 1 \quad ; \\ a_{2N_1-k-2,\ell}, & N_1 \leqslant k \leqslant 2N_1 - 2 \quad ; 0 \leqslant \ell \leqslant N_2 - 1 \quad ; \\ a_{k,2N_2-\ell-2}, & 0 \leqslant k \leqslant N_1 - 1 \quad ; N_2 \leqslant \ell \leqslant 2N_2 - 2 \quad ; \\ a_{|k|,2N_2-\ell-2}, & -(N_1-1) \leqslant k \leqslant -1 \quad ; N_2 \leqslant \ell \leqslant 2N_2 - 2 \quad ; \\ a_{2N_1-k-2,2N_2-\ell-2}, & N_1 \leqslant k \leqslant 2N_1 - 2 \quad ; N_2 \leqslant \ell \leqslant 2N_2 - 2 \quad ; \end{cases} \tag{4.55a}$$

$$\tilde{a}_{k,\ell} = \begin{cases} a_{2N_1-k-\ell,|\ell+1|}, & N_1 \leqslant k \leqslant 2N_1 - 1 \quad ; -N_2 \leqslant 1 \leqslant -1 \quad ; \\ a_{|k+1|,|\ell+1|}, & -N_1 \leqslant k \leqslant -1 \quad ; -N_2 \leqslant \ell \leqslant -1 \quad ; \\ a_{k,\quad +1}, & 0 \leqslant k \leqslant N_1 - 1 \quad ; -N_2 \leqslant \ell \leqslant -1 \quad ; \\ a_{|k+1|,\ell}, & -N_1 \leqslant k \leqslant -1 \quad , 0 \leqslant \ell \leqslant N_2 - 1 \quad ; \\ a_{k,\ell}, & 0 \leqslant k \leqslant N_1 - 1 \quad ; 0 \leqslant \ell \leqslant N_2 - 1 \quad ; \\ a_{2N_1-k-1,\ell}, & N_1 \leqslant k \leqslant 2N_1 - 1 \quad ; 0 \leqslant \ell \leqslant N_2 - 1 \quad ; \\ a_{k,2N_2-\ell-1}, & 0 \leqslant k \leqslant N_1 - 1 \quad ; N_2 \leqslant \ell \leqslant 2N_2 - 1 \quad ; \\ a_{|k+1|,2N_2-\ell-1}, & -N_1 \leqslant k \leqslant -1 \quad ; N_2 \leqslant \ell \leqslant 2N_2 - 1 \quad ; \\ a_{2N_1-k-1,2N_2-\ell-1}, & -N_1 \leqslant k \leqslant 2N_1 - 1 \quad ; N_2 \leqslant \ell \leqslant 2N_2 - 1 \quad . \end{cases} \tag{4.55b}$$

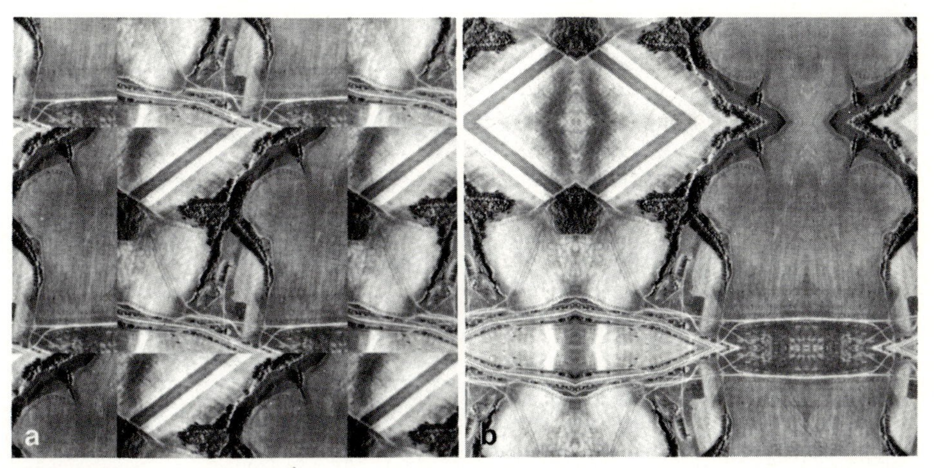

Fig. 4.4 a,b. Two-dimensional extension methods: (a) simple periodic extension, (b) even extension by (4.55b)

A visual idea of these two-dimensional extension methods can be obtained from Fig.4.4.

4.6 The Discrete Fourier Transform (DFT)

The Fourier transform (Sect.2.5) can be regarded as a linear transform with the kernel

$$H(f, t) = \exp(-i2\pi ft) \quad . \tag{4.56}$$

Let us find its discrete representation on the basis

$$\{\varphi_k(t) = \mathrm{sinc}[2\pi F(t - k/2F)]\} \tag{4.57}$$

for signals with spectrum limited to the interval $(-F, F)$ for which the following representation is valid:

$$a(t) = \sum_{k=-\infty}^{\infty} a(k/2F)\, \mathrm{sinc}[2\pi F(t - k/2F)] \quad . \tag{4.58}$$

The Fourier transform for such signals is

$$\alpha(f) = \int_{-\infty}^{\infty} a(t) \exp(i2\pi ft)dt = \sum_{k=-\infty}^{\infty} a(k/2F) \int_{-\infty}^{\infty} \mathrm{sinc}[2\pi F(t - k/2F)] \exp(i2\pi ft)dt = \frac{1}{2F} \sum_{k=-\infty}^{\infty} a(k/2F) \exp(i2\pi f\, k/2F) \quad . \tag{4.59}$$

Let us now consider the periodic signal

$$a_p(t) = \sum_{n=-\infty}^{\infty} a(t + nT) \quad . \tag{4.60}$$

Its spectrum is

$$\alpha_p(f) = 1/T \sum_{s=-\infty}^{\infty} \alpha_p(s/T)\, \delta(t - s/T) \quad , \tag{4.61}$$

where $\alpha_p(s/T)$ are spectrum samples of the signal $a_p(t)$ taken from the interval $(-T/2, T/2)$ (Table 2.2, line 19). If T is sufficiently large and the signal a(t) falls to zero sufficiently rapidly on that interval, so that distortions of it in the sum (4.60) resulting from aliasing can be ignored, then $\alpha_p(s/T) \approx \alpha(s/T)$. Then

$$\alpha_p(s/T) \approx 1/2F \sum_k a(k/2F) \exp(i2\pi\, ks/2TF) \quad , \tag{4.62}$$

providing summation over k takes place within the limits $|k| < TF$.

Values of T and F can always be chosen so that the variable 2TF is an integer. We shall denote it by N. Let us also write

$$\alpha_s = \sqrt{2F/T}\,\alpha(s/T)\exp(i2\pi\ ks/N) \quad , \quad a_k = a[(k-k_o)/2F] \ . \qquad (4.63)$$

Here k_o is chosen so that the summation in (4.62) runs from $k=0$ to $N-1$. We then obtain

$$\alpha_s = 1/\sqrt{N}\sum_{k=0}^{N-1} a_k \exp(i2\pi ks/N) \quad . \qquad (4.64)$$

This relation is called the discrete Fourier transform (DFT) [4.11].

The discrete Fourier transform is reversible:

$$a_k = 1/\sqrt{N}\sum_{s=0}^{N-1} {}_s \exp(-i2\ ks/N) \quad . \qquad (4.65)$$

Its kernel is the matrix

$$\mathbf{F}_N = \{1/\sqrt{N}\exp(i2\pi ks/N)\} \quad , \qquad (4.66)$$

which is a discrete representation of the kernel of the continuous Fourier transform.

Equation (4.66) is the analogue of (4.3). Note that it can be obtained directly from (4.3) for the basis $\{\varphi_k(t) = \mathrm{sinc}[2\pi F(t-k/2F)]\}$.

Coefficients $\{\alpha_s\}$ of the DFT sequence $\{a_k\}$ are approximately equal to spectrum samples taken at intervals $1/T$ of the signal $a(t)$ periodically extended with period T. This is the relation between the DFT and the continuous Fourier transform. From the proposed restricted signal length it, follows that the sampling theorem is valid for its spectrum, and, consequently, that it can be restored from the variables $\{\alpha_s\}$, which are the coefficients of the DFT signal samples.

The most commonly used properties of the one-dimensional DFT are presented in Table 4.1. For convenience in comparing them with the properties of the continuous Fourier transform, the right-hand column of Table 4.1 gives the number of the corresponding line in Table 2.2. The main difference between the DFT and the continuous Fourier transform is the cyclic or periodic nature: the number of sequence samples and its DFT are read modulo N, i.e., as in a circle; the number of points in the cycle is N (Table 4.1, line 2).

By analogy with the one-dimensional DFT, the two-dimensional DFT can be obtained by applying the 2-D sampling theorem to 2-D signals and spectra. Usually the only 2-D DFTs used are those derived from 2-D sampling theory in Cartesian coordinates:

$$\alpha_{r,s} = 1/\sqrt{N_1N_2}\sum_{k=0}^{N_1-1}\sum_{\ell=0}^{N_2-1} a_{k,\ell}\exp[i2\pi(kr/N_1+\ell s/N_2)] \ . \qquad (4.67)$$

These DFTs are convenient because they factorize to two 1-D DFTs, i.e., they are separable.

Table 4.1. Properties of the discrete Fourier transforms (DFT)

	Signal	DFT	No. of line in Table 2.2
1)	$a_k = \frac{1}{\sqrt{N}} \sum_{s=0}^{N-1} \alpha_s \times \exp\left(-\mathrm{i}2\pi \frac{ks}{N}\right)$	$\alpha_s = \frac{1}{\sqrt{N}} \sum_{k=0}^{N-1} a_k \times \exp\left(\mathrm{i}2\pi \frac{ks}{N}\right)$	1
2)	$a_k = a_{(k) \bmod N}$	$\alpha_s = \alpha_{(s) \bmod N}$	–
3)	$\{a_k^*\}$	$\{\alpha_{N-s}^*\}$	2
4)	$\{a_{N-k}\}$	$\{\alpha_{N-s}\}$	3
5)	$\{a_k = a_k^*\}$	$\{\alpha_s = \alpha_{N-s}^*\}$	4
6)	$\{a_k = -a_k^*\}$	$\{\alpha_s = -\alpha_{N-s}^*\}$	5
7)	$\{a_k = a_{N-k}\}$	$\{\alpha_s = \alpha_{N-s}\}$	6
8)	$\{a_k = -a_{N-k}\}$	$\{\alpha_s = -\alpha_{N-s}\}$	7
9)	$\{a_k = a_k^* = a_{N-k}\}$	$\{\alpha_s = \alpha_{N-s} = \alpha_{N-s}^* = \alpha_s^*\}$	8
10)	$\{a_k = -a_{N-k} = a_k^*\}$	$\{\alpha_s = -\alpha_{N-s} = \alpha_{N-s}^* = -\alpha_s^*\}$	9
11)	$\sum_{k=0}^{N-1} a_k b_k^* = \sum_{k=0}^{N-1} \alpha_s \beta_s^*$		17
12)	$\cos\left(2\pi \frac{s_0 k}{N} + \varphi\right)$	$\frac{1}{2}\left\{\frac{(-1)^{s+[s_0]}\sin \pi\sigma}{\sin \pi(s+s_0)/N} \times \exp\left[\mathrm{i}\pi \frac{N-1}{N}(s+s_0)+\varphi\right] + \frac{(-1)^{s-[s_0]}\sin \pi\sigma}{\sin \pi(s-s_0)/N} \times \exp\left[\mathrm{i}\pi \frac{(N-1)}{N} \times (s-s_0)-\sigma\right]\right\}$, $[s_0]$ is the integer part of s_0; σ is the remainder	15

Table 4.1 (continued)

Signal	DFT	No. of line in Table 2.2
13) $\{a_{(k+n) \bmod N}\}$	$\left\{\alpha_s \exp\left(-\mathrm{i}2\pi \frac{sn}{N}\right)\right\}$	10
14) $\frac{1}{\sqrt{N}} \sum_{n=0}^{N-1} a_n b_{(k-n) \bmod n}$	$\alpha_s \beta_s$	16
15) $\delta_{(k-n) \bmod N}$ $= \begin{cases} 1, & (k-n) \bmod N = 0; \\ 0, & (k-n) \bmod N \neq 0 \end{cases}$	$\frac{1}{\sqrt{N}} \exp\left(\mathrm{i}2\pi \frac{ns}{N}\right)$	12, 13
16) $\mathrm{rect}_N \frac{k-k_1}{k_2-k_1}$ $= \begin{cases} 0, & 0 \leqslant k < k_1; \\ 1, & k_1 \leqslant k \leqslant k_2; \\ 0, & k_2 < k \leqslant N-1 \end{cases}$	$\frac{1}{\sqrt{N}} \frac{\sin[\pi(k_2-k_1+1)s/N]}{\sin \pi s/N}$ $\times \exp[\mathrm{i}\pi(k_1+k_2)s/N]$	23
17) $\{a_{k=(l) \bmod N}\}$, $l = 0, 1, \ldots, LN-1$	$\sqrt{L}\, \alpha_{s_1} \delta_{s_2}$; $s = s_1 L + s_2$; $s_1 = 0, 1, \ldots, N-1$; $s_2 = 0, 1, \ldots, L-1$	19
18a) $\left\{\left[1 - \mathrm{rect}_{LN} \frac{l-N/2}{LN-N}\right]\right.$ $\left. \times\, a_{(l) \bmod N}\right\}$, $= 0, 1, \ldots, LN-1$; N is even	$\frac{1}{\sqrt{L}} \sum_{r=0}^{N-1} \alpha_s$ $\times \frac{\sin \pi \frac{N-1}{LN}(s-rL)}{N \sin[\pi(s-rL)/Ln]}$, $s = 0, 1, \ldots, LN-1$	21
18b) $\left[1 - \mathrm{rect}_{LN} \frac{l-(N+1)/2}{LN-(N+1)}\right]$ $\times\, a_{(l) \bmod N}$, $l = 0, 1, \ldots, LN-1$; N is odd	$\frac{1}{\sqrt{L}} \sum_{r=0}^{N-1} \alpha_r$ $\times \frac{\sin[\pi(s-rL)/L]}{N \sin[\pi(s-rL)/LN]}$, $s = 0, 1, \ldots, LN-1$	21

Table 4.1 (continued)

	Signal	DFT	No. of line in Table 2.2
19)	$\{a_{l_1}\delta_{l_2}\}$, $l=l_2N+l_1$; $l_1=0,1,\ldots,N-1$; $l_2=0,1,\ldots,L-1$	$\frac{1}{\sqrt{L}}\sum_{r=0}^{N-1}\alpha_r \times \frac{\sin[\pi(s-rL)/L]}{N\sin[\pi(s-rL)/LN]} \times \exp\left[\mathrm{i}\pi\frac{N-1}{LN}(s-rL)\right]$, $s=0,1,\ldots,LN-1$	21
20)	$\{a_{l_2}\delta_{l_1}\}$, $l=l_2L+l_1$; $l_1=0,1,\ldots,L-1$; $l_2=0,1,\ldots,N-1$;	$\frac{1}{\sqrt{L}}\{\alpha_{(s)\bmod N}\}$; $s=0,1,\ldots,LN-1$	22
21)	$\{a_{(pk)\bmod N}\}$, p and N do not have common divisors	$\alpha_{(qs)\bmod N}$, where $(pq)\bmod N=1$	24
22)	$\frac{1}{N}\sum_{k=0}^{N-1}a_k=\frac{1}{\sqrt{N}}\alpha_0$	$\frac{1}{N}\sum_{s=0}^{N-1}\alpha_s=\frac{1}{\sqrt{N}}a_0$	27
23)	$\frac{1}{N}\sum_{k=0}^{N-1}(-1)^k a_k=\frac{1}{\sqrt{N}}\alpha_{N/2}$	$\frac{1}{N}\sum_{s=0}^{N-1}(-1)^s\alpha_s=\frac{1}{\sqrt{N}}a_{N/2}$	–
24)	$\frac{1}{N}\sum_{k=0}^{(N/2)-1}a_{2k}=\frac{\alpha_0+\alpha_{N/2}}{2\sqrt{N}}$, $\frac{1}{N}\sum_{k=0}^{(N/2)-1}a_{2k+1}=\frac{\alpha_0-\alpha_{N/2}}{2\sqrt{N}}$	$\frac{1}{N}\sum_{s=0}^{(N/2)-1}\alpha_{2s}=\frac{a_0+a_{N/2}}{2\sqrt{N}}$, $\frac{1}{N}\sum_{s=0}^{(N/2)-1}\alpha_{2s+1}=\frac{a_0+a_{N/2}}{2\sqrt{N}}$,	–
25)	$\sum_{n=0}^{k}a_n$	$\left[\frac{1}{\sqrt{N}}\sum_{r=0}^{N-1}(N-r)\alpha_r\right]\delta_s + \frac{\alpha_s}{1-\exp(\mathrm{i}2\pi s/N)}(1-\delta_s)$	28
26)	a_k-a_{k-1}	$\alpha_s(1-\exp[(\mathrm{i}2\pi s)/N]$	29

The inverse 2-D DFT is written

$$a_{k,\ell} = 1/\sqrt{N_1 N_2} \sum_{k=0}^{N_1-1} \sum_{\ell=0}^{N_2-1} \alpha_{r,s} \exp[-i2\pi(kr/N_1 + is/N_2)] \quad . \tag{4.68}$$

Some properties of the 2-D DFT are presented in Table 4.2. Two-dimensional periodicity is characteristic of the 2-D DFT. The coefficients of the 2-D DFT can be regarded as samples of 2-D continuous signal spectra, periodically extended over a plane in the Cartesian coordinate system, as in Fig.4.4a.

Table 4.2. Properties of the two-dimensional DFT

Signal	2-D DFT
1) $a_{k,l} = \frac{1}{\sqrt{N_1 N_2}} \sum_{r=0}^{N_1-1} \sum_{s=0}^{N_2-1} \alpha_{r,s} \times \exp\left[-i2\pi\left(\frac{kr}{N_1} + \frac{ls}{N_2}\right)\right]$	$\alpha_{r,s} = \frac{1}{\sqrt{N_1 N_2}} \sum_{k=0}^{N_1-1} \sum_{l=0}^{N_2-1} a_{k,l} \times \exp\left[i2\pi\left(\frac{kr}{N_1} + \frac{ls}{N_2}\right)\right]$
2) $a_{k,l} = a_{(k) \bmod N_1, (l) \bmod N_2}$	$\alpha_{r,s} = \alpha_{(r) \bmod N_1, (s) \bmod N_2}$
3) $\{a^*_{k,l}\}$	$\{\alpha^*_{N_1-r, N_2-s}\}$
4) $\{a_{k,l} = a^*_{k,l}\}$	$\{\alpha_{r,s} = \alpha^*_{N_1-r, N_2-s}\}$
5) $\{a^*_{N_1-k, N_2-l}\}$	$\{\alpha^*_{r,s}\}$
6) $\{a_{k,l} = a^*_{N_1-k, N_2-l}\}$	$\{\alpha_{r,s} = \alpha^*_{r,s}\}$
7) $\{a_{N_1-k, N_2-l}\}$	$\{\alpha_{N_1-r, N_2-s}\}$
8) $\{a_{N_1-k, l}\}$	$\{\alpha_{N_1-r, s}\}$
9) $\{a_{k,l} = a_{N_1-k, N_2-l}\}$	$\{\alpha_{r,s} = \alpha_{N_1-r, N_2-s}\}$
10) $\{a_{k,l} = a_{N_1-k, l}\}$	$\{\alpha_{r,s} = \alpha_{N_1-r, s}\}$
11) $\{a_{k,l} = -a_{N_1-k, l}\}$	$\{\alpha_{r,s} = -\alpha_{N_1-r, s}\}$
12) $\{a_{k,l} = a_{N_1-k, N_2-l} = a^*_{k,l}\}$	$\{\alpha_{r,s} = \alpha_{N_1-r, N_2-s} = \alpha^*_{r,s}\}$
13) $\sum_{k=0}^{N_1-1} \sum_{l=0}^{N_2-1} a_{k,l} b^*_{k,l} = \sum_{r=0}^{N_1-1} \sum_{s=0}^{N_2-1} \alpha_{r,s} \beta^*_{r,s}$	

Table 4.2 (continued)

Signal	2-D DFT
14) $\frac{1}{\sqrt{N_1 N_2}} \sum_{n=0}^{N_1-1} \sum_{m=0}^{N_2-1} a_{n,m} \times b_{(k-n) \bmod N_1, (l-m) \bmod N_2}$	$\alpha_{r,s} \beta_{r,s}$
15) $a_{(k+n) \bmod N_1, (l+m) \bmod N_2}$	$\alpha_{r,s} \exp\left[-i2\pi\left(\frac{nr}{N_1}+\frac{ms}{N_2}\right)\right]$
16) $\{a_{(n) \bmod N_1, (m) \bmod N_2}\}$ $n = 0, 1, \ldots, L_1 N_1 - 1;$ $m = 0, 1, \ldots, L_2 N_2 - 1$	$\sqrt{L_1 L_2}\, \alpha_{p_1, q_1} \delta_{p_2, q_2},$ $r = p_1 L_1 + p_2; s = q_1 L_2 + q_2;$ $p_1 = 0, 1, \ldots, N_1 - 1;$ $p_2 = 0, 1, \ldots, L_1 - 1;$ $q_1 = 0, 1, \ldots, N_2 - 1;$ $q_2 = 0, 1, \vdots, L_2 - 1$
17) $\frac{1}{N_1 N_2} \sum_{k=0}^{N_1-1} \sum_{l=0}^{N_2-1} a_{k,l} = \frac{\alpha_{0,0}}{\sqrt{N_1 N_2}}$	$\frac{1}{N_1 N_2} \sum_{r=0}^{N_1-1} \sum_{s=0}^{N_2-1} \alpha_{r,s} = \frac{a_{0,0}}{\sqrt{N_1 N_2}}$
18) $a_{k,0} = \frac{1}{\sqrt{N_1 N_2}} \sum_{r=0}^{N_1-1} \left(\sum_{s=0}^{N_2-1} \alpha_{r,s}\right) \times \exp\left(-i2\pi \frac{kr}{N_1}\right)$	$\alpha_{0,s} = \frac{1}{\sqrt{N_1 N_2}} \sum_{k=0}^{N_1-1} \left(\sum_{l=0}^{N_2-1} a_{k,l}\right) \times \exp\left(i2\pi \frac{ls}{N_2}\right)$
19) $a_{N_1/2, N_2/2} = \frac{1}{\sqrt{N_1 N_2}} \sum_{r=0}^{N_1-1} \sum_{s=0}^{N_2-1} (-1)^{r+s} \alpha_{r,s},$ N_1, N_2 being even	$\alpha_{N_1/2, N_2/2} = \frac{1}{\sqrt{N_1 N_2}} \sum_{k=0}^{N_1-1} \sum_{l=0}^{N_2-1} (-1)^{k+l} a_{k,l},$

Sometimes instead of representing the DFT in the form of the sum of (4.64, 65), it is more convenient to use matrix notation. The 1-D DFT is then be written

$$\boldsymbol{\alpha} = \mathbf{F}_N \mathbf{a} \quad , \tag{4.69}$$

where $\mathbf{a}$ is a column matrix the elements of which are the samples of the transformed sequence $\{a_k\}$, $\boldsymbol{\alpha}$ is the column matrix composed of the DFT coefficients $\{\alpha_s\}$, and $\mathbf{F}_N$ is the square DFT matrix of order N (4.66).

The DFT matrix belongs to the class of so-called unitary matrices, whose inverse obtained by transposing the original matrix and replacing the elements by their complex conjugates.

The two-dimensional DFT can be written analogously to (4.69) as

$$\boldsymbol{\alpha} = F_{N_2} a F_{N_1} \quad , \tag{4.70}$$

where **a** and $\boldsymbol{\alpha}$, the initial and transformed signals are $N_1 \times N_2$ matrices, and $\mathbf{F}_{N_1}$ and $\mathbf{F}_{N_2}$ are square DFT matrices (4.66) $N_1 \times N_1$ and $N_2 \times N_2$ respectively.

4.7 Shifted, Odd and Even DFTs

In addition to the basic version of the DFT (4.64), a whole class of transforms can be introduced which are obtained by displacing the position of the signal and its spectrum samples relative to the origin of the respective coordinate system during signal and spectrum discretization [4.9-11].

Let us first consider the 1-D case. If the signal samples are so positioned that the origin of coordinates falls midway between the 0 and -1 samples, we obtain the following version of the DFT:

$$\alpha_s = 1/\sqrt{N} \sum_{k=0}^{N-1} a_k \exp[i2\pi(k + 1/2)s/N] \quad . \tag{4.71a}$$

Its inverse transformation is expressed by

$$a_k = 1/\sqrt{N} \sum_{s=0}^{N-1} \alpha_s \exp[-i2\pi(k + 1/2)s/N] \quad . \tag{4.71b}$$

If the samples are similarly shifted in the frequency plane, we obtain a transformation of the form (introduced in [4.12]):

$$\alpha_s = 1/\sqrt{N} \sum_{k=0}^{N-1} a_k \exp[i2\pi(k + 1/2)(s + 1/2)/N] \quad . \tag{4.72a}$$

Its inverse transformation is

$$a_k = 1/\overline{N} \sum_{s=0}^{N-1} \alpha_s \exp[-i2\ (k + 1/2)(s + 1/2)/N] \quad . \tag{4.72b}$$

When the shift is made in the frequency domain only, the corresponding pair of transformations (introduced in [4.13]) are:

$$\alpha_s = 1/\sqrt{N} \sum_{k=0}^{N-1} a_k \exp[i2\pi k(s + 1/2)/N] \quad ; \tag{4.73a}$$

$$a_k = 1/\sqrt{N} \sum_{s=0}^{N-1} \alpha_s \exp[-i2\pi k(s + 1/2)/N] \quad , \qquad (4.73b)$$

which are dual to the transformation pair (4.71a,b).

In general, a shift in the signal samples by u and in the spectrum by v gives the corresponding pair of transforms:

$$\alpha_s^{u,v} = 1/\sqrt{N} \sum_{k=0}^{N-1} a_k \exp[i2\pi(k + u)(s + v)/N] \quad ; \qquad (4.74a)$$

$$a_k^{u,v} = 1/\sqrt{N} \sum_{s=0}^{N-1} \alpha_s^{u,v} \exp[-i2\pi(k + u)(s + v)/N] \quad . \qquad (4.74b)$$

The superscript u and v of a_k in (4.74b) are so placed in order to emphasize that in contrast to the initial sequence $\{a_k\}$, determined for $k = 0, 1, \ldots, N-1$, $\{a_k^{u,v}\}$ are determined for any k. With $k = 0, 1, \ldots, N-1$, the sequence $\{a_k^{u,v}\}$ coincides with the original $\{a_k\}$.

Let us call the transformation pair (4.74a,b) the direct and inverse shifted discrete Fourier transforms, and denote them SDFT(u,v) and ISDFT(u,v), respectively. The conventional discrete Fourier transform discussed in the previous section is obviously SDFT(0,0).

SDFT(u,v) may be expressed in terms of DFT:

$$\alpha_s^{u,v} = 1/\sqrt{N} \sum_{k=0}^{N-1} a_k \exp[i2\pi(k + u)(s + v)/N] = \Big\{1/\sqrt{N} \sum_{k=0}^{N-1} [a_k \exp(i2\pi v(k + u)/N)] \exp(i2\pi ks/N)\Big\} \times \exp(i2\pi us/N) \qquad (4.75)$$

$$= \langle \mathrm{DFT}\; a_k \exp(i2\pi v(k + u)/N)]\rangle \exp(i2\pi us/N) \quad .$$

It is therefore possible to calculate SDFT(u,v) from DFT. However, for reasons of simplicity in calculation it is convenient to modify SDFT(u,v) somewhat by removing from the transform kernel the factor $\exp(i2\pi uv/N)$, which is independent of k and s, and by writing the transform pair as

$$\alpha_s^{u,v} = 1/\sqrt{N} \left[\sum_{k=0}^{N-1} a_k \exp[i2\pi k(s + v)/N]\right] \exp(i2\pi us/N) \quad ; \qquad (4.76a)$$

$$a_k^{u,v} = 1/\sqrt{N} \left[\sum_{s=0}^{N-1} \alpha_s^{u,v} \exp[-i2\pi s(k + u)/N]\right] \exp(-i2\pi kv/N) \quad . \qquad (4.76b)$$

Let us call this pair of transforms the modified shifted Fourier transforms [MSDFT(u,v)]. The properties of SDFTs and MSDFTs are similar.

If u and v are integers, SDFT(u,v) reduces to the DFT of cyclically shifted sequences, and the properties of SDFT(u,v) therefore coincide with the properties of DFT or SDFT(0,0). If u and v are fractions, this is not the case.

The main properties of SDFT(u,v) are presented in Table 4.3. For convenience in comparing them with the properties of the conventional DFT, the right-hand column of Table 4.3 gives the numbers of the corresponding lines in Table 4.1.

Table 4.3. Properties of the SDFT

Signal	SDFT (u, v)	No. of line in Table 4.1
1) $a_k^{u,v} = \frac{1}{\sqrt{N}} \sum_{s=0}^{N-1} \alpha_s^{u,v} \times \exp\left(-i2\pi \frac{(k+u)(s+v)}{N}\right)$	$\alpha_s^{u,v} = \frac{1}{\sqrt{N}} \sum_{k=0}^{N-1} a_k \times \exp\left(-i2\pi \frac{(k+u)(s+v)}{N}\right)$	1
2) $a_{k+hN}^{u,v} = a_k^{u,v} \exp(-i2\pi hv)$	$\alpha_{s+gN}^{u,v} = \alpha_s^{u,v} \exp(i2\pi gu)$	2
3a) $a_k^* \exp\left[-i2\pi \frac{2v(k+u)}{N}\right]$	$(\alpha_{N-s}^{u,v})^* \exp(i2\pi u)$	3
3b) a_k^*	If $2v$ is an integer, $(\alpha_{N-2v-s}^{u,v})^* \exp(i2\pi u)$	
4a) $(a_{N-k}^{u,v})^* \exp(-i2\pi v)$	$\alpha_s^* \exp\left[i2\pi \frac{2u(s+v)}{N}\right]$	3
4b) If $2u$ is an integer, $(a_{N-2u-k}^{u,v})^* \exp(-i2\pi v)$	α_s^*	
5a) $a_{N-k-1}^{u,v} \times \exp\left[-i2\pi \frac{(k+u)(2v-1)}{N}\right]$	$\alpha_{N-s-1}^{u,v} \times \exp\left[i2\pi \frac{(s+v)(2u-1)}{N}\right] \times \exp\left\{-i2\pi\left[(u+v) + \frac{(2u-1)(2v-1)}{N}\right]\right\}$	
5b) If $2u$ and $2v$ are integers, $(-1)^{2u} a_{N-2u-k}$	$(-1)^{2v} \alpha_{N-2v-s}$	
6a) $a_k = \pm a_k^* \times \exp\left[-i2\pi \frac{2v(k+u)}{N}\right]$	$\alpha_s^{u,v} = \pm(\alpha_{N-s}^{u,v}) \exp(i2\pi u)$	5, 6

Table 4.3 (continued)

Signal	SDFT (u, v)	No. of line in Table 4.1
6b) $a_k = \pm a_k^*$	If $2v$ is an integer, $\alpha_s^{u,v} = \pm(\alpha_{N-2v-s}^{u,v})^*(-1)^{2u}$	5, 6
7) If $2u$, $2v$ are integers, $a_k^{u,v} = \pm a_{N-2u-k}^{u,v}$	$\alpha_s^{u,v} = \pm(-1)^{2(u+v)}\alpha_{N-2v-s}^{u,v}$	7, 8
8) If $2u$, $2v$ are integers, $a_k^{u,v} = (a_k^{u,v})^* = \pm a_{N-2u-k}^{u,v}$	$\alpha_s^{u,v} = (-1)^{2u}(\alpha_{N-2v-s}^{u,v})^* = \pm(-1)^{2(u+v)}\alpha_{N-2v-s}^{u,v}$	9, 10
9) $a_{k+k_0}^{u,v}$	$\alpha_s^{u,v}\exp\left[-i2\pi\dfrac{k_0(s+v)}{N}\right]$	13
10) $a_k^{u,v}\exp\left[i2\pi\dfrac{s_0(k+u)}{N}\right]$	$\alpha_{s+s_0}^{u,v}$	13
11a) ${}^{s_0}a_k^{p,q/u,v} = \dfrac{1}{\sqrt{N}}\sum_{s=0}^{M-1}\alpha_{s+s_0}^{u,v}\exp\left[-i2\pi\dfrac{(k+p)(s+q)}{N}\right] = \sum_{n=0}^{N-1}\left(a_n\exp\left[i2\pi\dfrac{(n+u)(s_0+v-q)}{N}\right]\right) \times \dfrac{\sin[\pi M(k-n-u+p)/N]}{N\sin[\pi(k-n-u+p)/N]} \times \exp\left[-i2\pi\dfrac{(k-n-u+p)(q+(M-1)/2)}{N}\right]$		
11b) With $q = -(M-1)/2$; $s_0 + v = q$ ${}^{s_0}a^{p,q/u,v} = \left\{\dfrac{1}{\sqrt{N}}\sum_{n=0}^{N-1}a_n\dfrac{\sin[\pi M(n-k+u-p)/N]}{N\sin[\pi(n-k+u-p)/N]}\right\}$		–

A comparison of Table 4.1 with Table 4.3 reveals that SDFT(u,v) has a number of peculiarities in comparison with the standard DFT. SDFT(u,v) has a more general rule of sequence extension for the numbers k and s that differ from (0, 1, ..., N-1) (see Table 4.3, line 2); a more general definition of odd and even sequences (Table 4.3, line 7); a more general formula for signal restoration from a spectrum, allowing signal interpolation by the trans-

Table 4.3 (continued)

Signal	SDFT (u, v)	No. of line in Table 4.1
12a) $^{s_a,s_b}c_k^{u_c,v_c} = \frac{1}{\sqrt{N}} \sum_{s=0}^{M-1} \alpha_{s+s_a}^{u_a,v_a} \beta_{s+s_b}^{u_b,v_b} \exp\left[-i2\pi \frac{(k+u_c)(s+v_c)}{N}\right]$ $= \frac{1}{\sqrt{N}} \sum_{n=0}^{N-1} a_n \exp\left[i2\pi \frac{(n+u_a)(s_a+v_a-v_c)}{N}\right] b_{k-n+[u_c-(u_a+u_b)]}^{\text{int}};$ $b_k^{\text{int}} = \sum_{m=0}^{N-1} \left\{ b_m^{u_b,v_b} \exp\left[i2\pi \frac{(m+u_b)(s_b+v_b-v_c)}{N}\right]\right\}$ $\times \frac{\sin[\pi M(m-k)/N]}{N \sin[\pi(m-k)/N]} \exp\left[i2\pi \frac{(m-k)(v_c+(M-1)/2)}{N}\right]$		14
12b) If $[u_c-(u_a+u_b)]$ is an integer, $^{s_a,s_b}c_k^{u_c,v_c} = \frac{1}{\sqrt{N}} \sum_{n=0}^{N-1} \left\{ a_n \exp\left[i2\pi \frac{(n+u_a)(s_a+v_a-v_c)}{N}\right]\right\}$ $\times b_{k-n+[u_c-(u_a+u_b)]}^{u_b,v_b} \exp\left[i2\pi \frac{(k-n+u_c-u_a)(s_b+v_b-v_c)}{N}\right]$		14

form pair SDFT(u,v) and ISDFT(p,q) with the correspondingly chosen u,v,p,q (Table 4.3, lines 11a,b); a more general convolution theorem (Table 4.3, line 12a,b), etc. Owing to these peculiarities, some varieties of SDFT are more convenient in signal processing than DFT (see below, and also Sect.4.8,5.6).

Let us consider in more detail some SDFT(u,v)'s of even signals. Corresponding to the conventional DFT (Table 4.3, line 7) is the following definition of an even signal composed of 2N samples:

$$\tilde{a}_k = \begin{cases} a_k & , k = 0, 1, \ldots, N-1 \quad ; \\ a_n & , k = N \quad ; \\ a_{2N-k} & , k = N+1, \ldots, 2N-1 \quad , \end{cases} \tag{4.77a}$$

and the following definition of a_k for $k < 0$, $k \geqslant 2N$:

$$\tilde{a}_{k+2hN} = \tilde{a}_k \quad . \tag{4.77b}$$

Hence it follows that

$$\tilde{a}_k = \tilde{a}_{2N-k} \quad . \tag{4.77c}$$

By substituting (4.77) into (4.64) we find that the discrete spectrum of such a signal is defined by

$$\tilde{\alpha}_s^{0,0} = 2/\sqrt{2N}\left[\sum_{k=0}^{N-1} a_k \cos(\pi ks/N) - a_0/2 + (-1)^s a_N/2\right]$$

$$s = 0, 1, \ldots, N \tag{4.78}$$

and has the same symmetry as the signal

$$\tilde{\alpha}_s^{0,0} = \tilde{\alpha}_{2N-s}^{0,0} \quad , \quad s = 0, 1, \ldots, 2N-1 \quad ; \quad \tilde{\alpha}_{s+2gN}^{0,0} = \tilde{\alpha}_s^{0,0} \quad . \tag{4.79}$$

The inverse transform therefore reduces to an expression analogous to (4.78):

$$\alpha_k = 2/\sqrt{2N}\left[\sum_{s-0}^{N-1} \alpha_s \cos(\pi ks/N) - \alpha_0/2 + (-1)^k \alpha_N/2\right] \quad . \tag{4.80}$$

For SDFT(0,1/2) the signal is even

$$\tilde{\alpha}_k = \begin{cases} a_k & , 0 \leqslant k \leqslant N-1 \quad ; \\ a_N & , k = N \quad ; \\ a_{2N-k} & , N+1 \leqslant k \leqslant 2N-1 \quad , \end{cases} \tag{4.81}$$

i.e., the same as for DFT or SDFT(0,0).

By substituting (4.81) into (4.73a) we obtain

$$\tilde{\alpha}_s^{0,1/2} = 1/\sqrt{2N}\left[2i \sum_{k=0}^{N-1} a_k \sin[2\pi k(s+1/2)/2N] + i(-1)^s a_N + a_0\right] \quad . \tag{4.82}$$

Since $a_0 \neq 0$, this spectrum is not generally speaking fully symmetrical:

$$\tilde{\alpha}_{2N-s-1}^{0,1/2} = -\tilde{\alpha}_s^{0,1/2} + 2\, a_0/\sqrt{2N} \quad . \tag{4.83}$$

By substituting this relation into (4.73b) we obtain the inverse transform

$$\tilde{a}_k = -2i/\sqrt{2N} \sum_{s=0}^{N-1} \alpha_s \sin[2\pi k(s+1/2)/2N] + ia_0\left[(-1)^k - 1\right]/2N \sin(\pi k/2N) \tag{4.84}$$

When $a_0 = 0$, the signal (4.81) for SDFT(0,1/2) becomes symmetrical (Table 4.3, lines 2,7):

$$\tilde{a}_k^{0,1/2} = \tilde{a}_{2N-k} \quad ; \quad \tilde{a}_{k+2hN}^{0,1/2} = (-1)^h \tilde{a}_k \quad ,$$

$$\tilde{\alpha}_s^{0,1/2} = -\tilde{\alpha}_{2N-s-1}^{0,1/2} \quad , \quad \tilde{\alpha}_{s+2gN}^{0,1/2} = \tilde{\alpha}_s^{0,1/2} \quad . \tag{4.85}$$

Such a signal is even relative to points $k = N$ and odd relative to points $k = 0$. Its spectrum $\tilde{\alpha}_s^{0,1/2}$ is an odd sequence.

For SDFT(1/2,1/2) the signal

$$\tilde{a}_k^{1/2,1/2} = \tilde{a}_{2N-1-k}^{1/2,1/2} \quad , \quad \tilde{a}_{k+2hN}^{1/2,1/2} = (-1)^h \tilde{a}_k \quad , \tag{4.86}$$

is even relative to points $k = N$ and odd relative to conditional points $k = -1/2$. It is easy to show that its transform takes the form

$$\tilde{\alpha}_s^{1/2,1/2} = 2i / \sqrt{2N} \sum_{k=0}^{N-1} \tilde{a}_k \sin[\pi(k + 1/2)(s + 1/2) / N] \quad , \tag{4.87}$$

as a result of which the spectrum $\alpha_s^{1/2,1/2}$ has the same type of symmetry as the signal in (4.81):

$$\tilde{\alpha}_s^{1/2,1/2} = \tilde{\alpha}_{2N-1-s}^{1/2,1/2} \quad , \quad \tilde{\alpha}_{s+2gN}^{1/2,1/2} = (-1)^g \alpha_s^{1/2,1/2} \quad . \tag{4.88}$$

The inverse transform in this case is also analogous to the direct one (4.87):

$$\tilde{\alpha}_k^{1/2,1/2} = -\, 2i / \sqrt{2N} \sum_{k=0}^{N-1} \alpha_s \sin[\pi(k + 1/2)(s + 1/2) / N] \quad . \tag{4.89}$$

Finally, SDFT(1/2,0) has a corresponding even signal (Table 4.3, lines 2,7):

$$\tilde{a}_k^{1/2,0} = \tilde{a}_{2N-1-k}^{1/2,0} \quad , \quad \tilde{\alpha}_{k+2hN}^{1/2,0} = \tilde{a}_k^{1/2,0} \quad . \tag{4.90}$$

By substituting (4.90) into (4.71a) we find that for an even signal such as this, SDFT(1/2,0) reduces to

$$\tilde{\alpha}_s^{1/2,0} = 2 / \sqrt{2N} \sum_{k=0}^{N-1} a_k \cos[\pi(k + 1/2)s / N] \quad , \; s = 1, 2, \ldots, N-1 \quad .$$

$$\tilde{\alpha}_0^{1/2,0} = 1 / \sqrt{2N} \sum_{k=0}^{N-1} a_k \quad . \tag{4.91}$$

The discrete spectrum derived from this formula is odd relative to $s = N$ and even relative to $s = 0$:

$$\tilde{\alpha}_s^{1/2,0} = -\,\tilde{\alpha}_{2N-s}^{1/2,0} \quad , \quad \tilde{\alpha}_N^{1/2,0} = 0 \quad , \quad \tilde{\alpha}_{s+2gN}^{1/2,0} = (-1)^g \tilde{a}_s^{1/2,0} \quad . \tag{4.92}$$

Transform (4.91) has a corresponding inverse transform

$$\tilde{a}_k = 2 / \sqrt{2N} \sum_{s=0}^{N-1} \alpha_s \cos[\pi(k + 1/2)s / N] \quad , \; k = 0, 1, \ldots, N-1 \quad . \tag{4.93}$$

The pair of transforms (4.91,93) were suggested in [4.14] (see also [4.15]), where they are called discrete cosine transforms.

The basis $\{\cos[\pi(k + 1/2)s / N]\}$ closely approximate the eigenfunctions of the integral equation (3.40) for exponential correlation functions, and thus gives better results in picture coding than any other known bases [4.14,15][2].

---

[2] The good results achieved with cosine transforms can best be explained by the rapid convergence at the edges of the sequence, since the even signal extension corresponds to the cosine transform, eliminating discontinuities at the edges.

A useful property of SDFT(1/2,0) is that when the symmetry of the signal is even, its spectrum has odd symmetry. Therefore SDFT(1/2,0) is convenient for calculating spectra and signal convolutions when signals are evenly extended (Sects.4.8,5.6).

The SDFT of odd signals can be considered in the same way. From them one can derive the SDFT(1,1) of a signal of the form

$$\tilde{a}_k^{1,1} = a_{2N-2-k} \quad , \; k = 0,\, 1,\, \dots,\, N-2,\, N \quad , \; N+1,\, \dots,\, 2N-2 \quad ;$$
$$\tilde{a}_{N-1}^{1,1} = \tilde{a}_{2N-1}^{1,1} = 0 \quad . \tag{4.94}$$

It is not difficult to show that the transform of such a signal reduces to a transform of the form

$$\alpha_s^{1,1} = 2 / \sqrt{2N} \sum_{k=0}^{N-2} a_k \sin[\pi(k+1)(s+1) / N] \quad , \tag{4.95}$$

which was suggested in [4.16] for picture coding and is known as the sine transform.

Let us now turn to 2-D transforms. Two-dimensional SDFT(u,v;w,z)'s can, like 2-D DFTs, be defined as separating into two 1-D transforms:

$$\alpha_{z,s}^{u,v;w,z} = 1 / \sqrt{N_1 N_2} \sum_{k=0}^{N_1-1} \sum_{=0}^{N_2-1} a_{k,\ell} \times \exp\{i2\pi[(k+u)(r+v) / N_1$$
$$+ (\ell + w)(s + z) / N_2]\} \quad . \tag{4.96a}$$

They have corresponding inverse transforms

$$a_{k,\ell}^{u,v;w,z} = 1 / \sqrt{N_1 N_2} \sum_{r=0}^{N_1-1} \sum_{s=0}^{N_2-1} \alpha_{r,s}^{u,v;w,z}$$
$$\times \exp[-i2\pi(k+u)(r+v) / N_1 + (\ell + w)(s + z) / N_2)] \quad . \tag{4.96b}$$

Some properties of 2-D SDFT(u,v;w,z)'s are presented in Table 4.4.

Two-dimensional odd and even SDFTs are defined similarly. Thus, the 2-D SDFT(1/2,0;1/2,0) generates the 2-D discrete cosine transform:

$$\alpha_{r,s}^{1/2,0;1/2,0} = 2 / \sqrt{N_1 N_2} \times \sum_{k=0}^{N_1-1} \sum_{\ell=0}^{N_2-1} a_{k,\ell} \cos[\pi(k + 1/2) r / N_1]$$
$$\times \cos[\pi(\ell + 1/2) s / N_2] \quad , \quad r,\, s \neq 0 \quad ; \tag{4.97a}$$

$$\alpha_{0,s}^{1/2,0;1/2,0} = 1 / \sqrt{N_1 N_2} \sum_{k=0}^{N_1-1} \sum_{\ell=0}^{N_2-1} a_{k,\ell} \cos[\pi(\ell + 1/2) s / N_2] \quad ;$$

$$\alpha_{r,0}^{1/2,0;1/2,0} = 1 / \sqrt{N_1N_2} \sum_{k=0}^{N_1-1} \sum_{\ell=0}^{N_2-1} a_{k,\ell} \cos[\pi(k + 1/2)r / N_1]$$

$$\alpha_{0,0}^{1/2,0;1/2,0} = 1 / 2\sqrt{N_1N_2} \sum_{k=0}^{N_1-1} \sum_{\ell=0}^{N_2-1} a_{k,\ell} \quad ; \tag{4.97b}$$

$$a_{k,\ell}^{1/2,0;1/2,0} = 1 / \sqrt{N_1N_2} \sum_{r=0}^{N_1-1} \sum_{s=0}^{N_2-1} \alpha_{r,s}^{1/2,0;1/2,0} \cos[\pi(k + 1/2)r / N_1] \times \cos[\pi(\ell + 1/2)s / N_2] \quad .$$

Corresponding to them are the following types of signal symmetry:

$$a_{k,\ell}^{1/2,0;1/2,0} = a_{2N_1-k-1,\ell}^{1/2,0;1/2,0} = a_{k,2N_2-1-\ell}^{1/2,0;1/2,0} = a_{2N_1-1-k,2N_2-1-\ell}^{1/2,0;1/2,0} \quad ,$$

$$a_{k+2hN,\ell+2gN_2}^{1/2,0;1/2,0} = a_{k,\ell} \tag{4.98a}$$

and the following types of spectrum symmetry:

Table 4.4. Properties of the two-dimensional SDFTs

Signal	SDFT
1) $a_{k,l}^{u,v;w,z} = \frac{1}{\sqrt{N_1N_2}} \times \sum_{r=0}^{N_1-1} \sum_{s=0}^{N_2-1} \alpha_{r,s}^{u,v;w,z} \times \exp\left\{-i2\pi\left[\frac{(k+u)(r+v)}{N_1} + \frac{(l+w)(s+z)}{N_2}\right]\right\}$	$\alpha_{r,s}^{u,v;w,z} = \frac{1}{\sqrt{N_1N_2}} \sum_{k=0}^{N_1-1} \sum_{l=0}^{N_2-1} a_{k,l} \times \exp\left\{i2\pi\left[\frac{(k+u)(r+v)}{N_1} + \frac{(l+w)(s+z)}{N_2}\right]\right\}$
2) $a_{k+hN_1,l+gN_2}^{u,v;w,z} = a_{k,l}^{u,v;w,z} \times \exp[-i2\pi(hv+gz)]$	$\alpha_{r+hN_1,s+gN_2}^{u,v;w,z} = \alpha_{r,s}^{u,v;w,z} \times \exp[i2\pi(hu+gw)]$
3a) $(a_{k,l}^{u,v;w,z})^* \times \exp\left[-i2\pi\left(2v\frac{k+u}{N_1} + 2z\frac{l+w}{N_2}\right)\right]$	$(\alpha_{N_1-r,N_2-s}^{u,v;w,z})^* \exp[i2\pi(u+w)]$
3b) $(a_{k,l}^{u,v;w,z})^*$	If $2u$, $2z$ are integers, $(\alpha_{N_1-2v-r,N_2-2z-s}^{u,v;w,z})^*(-1)^{2(u+w)}$

Table 4.4 (continued)

Signal	SDFT
4a) $a^{u,v;w,z}_{N_1-1-k,N_2-1-l} \times \exp\left\{-\mathrm{i}2\pi\left[\frac{(2v-1)(k+u)}{N_1} + \frac{(2z-1)(l+w)}{N_2}\right]\right\}$	$\alpha^{u,v;w,z}_{N_1-1-r,N_2-1-s} \times \left\{\exp \mathrm{i}2\pi\left[\frac{(2u-1)(r+v)}{N_1} + \frac{(2w-1)(s+z)}{N_2}\right]\right\} \exp\left\{-\mathrm{i}2\pi \times \left[(u+w+v+z) + \frac{(2u-1)(2v-1)}{N_1} + \frac{(2w-1)(2z-1)}{N_2}\right]\right\}$
4b) If $2u$, $2v$, $2w$, $2z$ are integers, $a^{u,v;w,z}_{N_1-k-2u,N_2-l-2w} \times (-1)^{2(u+w)}$	$\alpha^{u,v;w,z}_{N_1-r-2v,N_2-s-2z} \times (-1)^{2(v+z)}$
5) $a^{\mathrm{int}}_{k,l} = \frac{1}{\sqrt{N_1 N_2}} \sum_{r=0}^{M_1-1} \sum_{s=0}^{M_2-1} \alpha^{u,v;w,z}_{r+r_0,s+s_0} \exp\left\{-\mathrm{i}2\pi\left[\frac{(k+u_0)(r+v_0)}{N_2} + \frac{(l+w_0)(s+z_0)}{N_2}\right]\right\} = \sum_{n=0}^{N_1-1} \sum_{m=0}^{N_2-1} \left\{a_{n,m} \exp\left(\mathrm{i}2\pi\left[\frac{(n+u)(r_0+v+v_0)}{N_1} + \frac{(m+w)(s_0+z-z_0)}{N_2}\right]\right)\right\} \frac{\sin[\pi M_1(n-k+u-u_0)/N_1]}{N_1 \sin[\pi(n-k+u-u_0)/N_1)]} \times \frac{\sin[\pi M_2(m-l+w-w_0)/N_2]}{N_2 \sin[\pi(m-l+w-w_0)/N_2]} \exp\left\{\mathrm{i}2\pi\left[\frac{(n-k+u-u_0)(v_0-(M_1-1)/2)}{N_1} + \frac{(m-l+w-w_0)(z_0-(M_2-1)/2)}{N_2}\right]\right\}$	

$$\alpha^{1/2,0;1/2,0}_{r,s} = -\alpha^{1/2,0;1/2,0}_{2N_1-r,s} = -\alpha^{1/2,0;1/2,0}_{r,2N_2-s} = \alpha^{1/2,0;1/2,0}_{2N_1-r,2N_2-s} \quad ,$$
$$\alpha^{1/2,0;1/2,0}_{N_1,s} = \alpha^{1/2,0;1/2,0}_{r,N_2} = \alpha^{1/2,0;1/2,0}_{N_1,N_2} = 0 \quad . \qquad (4.98b)$$

Of special interest in the 2-D case is the SDFT constructed for the discretization of signals and/or their spectra on a hexagonal raster. With a hexagonal raster the set of samples along one coordinate is situated half the distance between samples on each neighboring line [4.17]. Therefore 2-D

Table 4.4 (continued)

6) $$c_{k,l}=\frac{1}{\sqrt{N_1N_2}}\sum_{r=0}^{N_1-1}\sum_{s=0}^{N_2-1}\alpha_{r+r_a,\,s+s_a}^{u_a,\,v_a;\,w_a,\,z_a}\,\beta_{r+r_b,\,s+s_b}^{u_b,\,v_b;\,w_b,\,z_b}$$
$$\times\exp\left\{-i2\pi\left[\frac{(k+u_c)(r+v_c)}{N_1}+\frac{(l+w_c)(s+z_c)}{N_2}\right]\right\}$$
$$=\frac{1}{\sqrt{N_1N_2}}\left\{\sum_{n=0}^{N_1-1}\sum_{m=0}^{N_2-1}a_{n,m}\exp\left(i2\pi\left[\frac{(n+u_a)(r_a+v_a-v_c)}{N_1}\right.\right.\right.$$
$$\left.\left.+\frac{(m+w_a)(s_a+z_a-z_c)}{N_2}\right]\right)b_{k-n+[u_c-(u_a+u_b)],\,l-m+[w_c-(w_a+w_b)]}^{u_b,\,v_b;\,w_b,\,z_b}$$
$$\times\exp\left\{i2\pi\left[\frac{(k+u_c)(r_b+v_b-v_c)}{N_1}+\frac{(l+w_c)(s_b+z_b-z_c)}{N_2}\right]\right\};$$

$[u_c-(u_a+u_b)]$; $[w_c-(w_a+w_b)]$ are integers

SDFTs for signal and spectrum samples distributed on a hexagonal raster are defined as

$$\alpha_{r,s}^{h,h}=1/\sqrt{NM}\sum_{k=0}^{N-1}\sum_{=0}^{M-1}a_{k,\ell}\exp[i2\pi([4k+1+(-1)^{\ell}][4r+1+(-1)^{s}]/4N$$
$$+\ell s/N)] \quad .$$

With a hexagonal raster only for signals or only for spectra, they are defined repectively as

$$\alpha_{r,s}^{h,r}=1/\sqrt{NM}\sum_{k=0}^{N-1}\sum_{=0}^{M-1}a_{k,\ell}\exp[i2\pi([4k+1+(-1)^{s}]r/4N+\ell s/M)] \quad ;$$

$$\alpha_{r,s}^{r,h}=1/\sqrt{NM}\sum_{k=0}^{N-1}\sum_{=0}^{M-1}a_{k,\ell}\exp[i2\pi([4s+1+(-1)^{\ell}]k/4N+\ell s/M)] \quad .$$

From these formulae it is clear that SDFTs on a hexagonal raster differ from the standard 2-D DFT in that for every even line, one of the 1-D transforms into which the SDFT separates is SDFT(1/2,1/2) for a hexagonal signal and spectrum raster; SDFT(1/2,0) for a hexagonal signal raster and a rectangular spectrum raster; or SDFT(0,1/2) for a rectangular signal raster and a hexagonal spectrum raster.

The "hexagonal" discrete spectra and signals are related to the "rectangular" ones by interpolation formulae similar to those presented in Table 4.4, line 5. Thus for example, the samples of the "hexagonal" spectrum of a signal set on a rectangular raster are related to the samples of its "rectangular" spectrum by

$$\alpha_{r,s}^{r,h} = \sum_{p=0}^{N-1} \alpha_{r,s}^{r,r} \sin[\pi[4r-4p+1+(-1)^s]/4)]/N\sin[\pi[4r-4p+1+(-1)^s]/4N)]$$

$$\times \exp[i\pi(N-1)/4N)] \quad .$$

Corresponding to the above-discussed "hexagonal" SDFTs are discrete signals whose continuous originals are defined over parts of a plane restricted by rectangles. Hexagonal DFTs for continuous signals defined over part of a plane restricted by perfect hexagons were introduced in [4.18].

4.8 Using Discrete Fourier Transforms

4.8.1 Calculating Convolutions

One of the most important applications of the discrete Fourier transform is in the calculation of the digital convolution of signals. As shown in Sect.4.2, the digital form of the convolution integral is defined by (4.16):

$$b_k = \sum_{\ell=-N_1}^{N_2} \hat{h}_\ell a_{k-\ell} \quad , \tag{4.99}$$

where $\{a_k\}$ are the signal samples, and h_ℓ are samples of the impulse response of a truncated linear shift-invariant filter with a truncated frequency transfer function.

By substituting the variables and renumbering the h_ℓ samples, we can rewrite (4.99) as

$$b_k = \sum_{n=0}^{N-1} \hat{h}_n \, a_{k+N_0-n} \quad , \tag{4.100}$$

where N_0 is the value of n corresponding to point $\ell = 0$ in (4.99); $N \geq N_1 + N_2 + 1$.

Let us compare this expression with that for cyclic convolution derived from the inverse SDFT of the product of the cyclically shifted SDFTs $\{\xi_{s+s_x}^{u_x,v_x}\}$ and $\{\eta_{s+s_y}^{u_y,v_y}\}$ of sequences $\{x_k\}$ and $\{y_k\}$ (Table 4.3, lines 12a,b):

$$\begin{aligned} {}^{s_x,s_y}C_k^{u_c,v_c} &= 1/\sqrt{N} \sum_{s=0}^{M-1} \xi_{s+s_x}^{u_x,v_x} \, \eta_{s-s_y}^{u_y,v_y} \times \exp[-i2\pi(k+u_c)(s+v_c)/N] \\ &= 1/\sqrt{N} \sum_{n=0}^{N-1} y_n \exp[i2\pi(n+u_y)(s_y+v_y-v_c)/N] \times x_{k-n+[u_c-(u_x+v_y)]}^{\mathrm{int}} \quad , \end{aligned} \tag{4.101}$$

where

$$x_k^{int} = \sum_{m=0}^{N-1} \left[x_m^{u_x,v_x} \exp[i2\pi(m + u_x)(s_x + v_x - v_c) / N] \right] \sin[\pi M(m - k) / N]$$
$$/ N \sin[\pi(m - k) / N] \exp[i2\pi(m - k)(v_c + (M - 1) / 2) / N] \quad , \tag{4.102}$$

$$\xi_s^{u_x,v_x} = 1 / \sqrt{N} \sum_{k=0}^{N-1} x_k \exp[i2\pi(k + u_x)(s + v_x) / N] \quad , \tag{4.103a}$$

$$\eta_s^{u_y,v_y} = 1 / \sqrt{N} \sum_{k=0}^{N-1} y_k \exp[i2\pi(k + u_y)(s + v_y) / N] \quad . \tag{4.103b}$$

It is easy to see that (4.101) reduces to the equation for discrete convolution (4.100) if the following substitutions are made:

$$b_k = {}^{s_x,s_y}C_k^{u_c,v_c} \quad ; \quad a_k = x_k^{int} \quad ;$$
$$\hat{h}_n = y_n \exp[i2\pi(n + u_y)(s_y + v_y - v_c) / N] \quad ; \tag{4.104}$$
$$N_0 = u_c - (u_x + u_y) \quad ,$$

and $a_k = x_k^{int}$ is considered to be continued to values of k that are below zero and above N-1, according to the following rule derived from (4.102):

$$a_{k+gN} = x_{k+gN}^{int} = a_k \exp(-i2\pi g v_c) = x_k^{int} \exp(-i2\pi g v_c) \quad . \tag{4.105}$$

Thus, the same problem of edge effects that was considered in Sect.4.5 is also involved in calculating convolution with DFTs, but here the method of extending the signal is determined by the choice of the parameter for the inverse SDFT.

In using discrete Fourier transforms it must be borne in mind that it is most efficient to calculate them with fast-Fourier-transform algorithms [4.1,2,15] (Chap.5). This imposes certain constraints on the number of signal samples N. One usually chosses a number equal to an integer power of two, which is necessary for the most efficient FFT algorithms. If the actual number of samples is not an integer power of two, then the signal is padded, with account taken of the law of periodic extension (4.105), to the closest such number of samples, so that there are no signal gaps on the boundaries of the new extended sequence. Good results are obtained when the absent samples are determined by linear interpolation between the end samples of the periodically extended sequence. This method of supplementing sequences is shown in Fig.4.5 for simple periodic extension (as in Fig.4.3c), which corresponds to the standard DFT.

The number of supplementary samples should not be lower than the number of non-zero samples of the impulse response $\hat{h}_\ell$. If it is lower, the signal se-

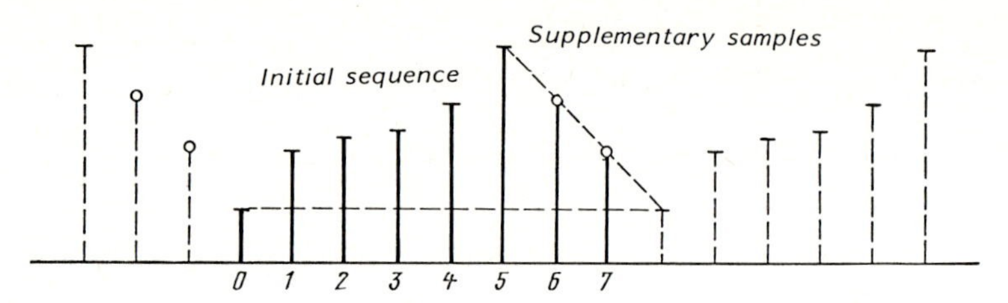

Fig. 4.5. Method of supplementing sequences for simple periodic extension

quence with the supplementary samples that have been found must be extended evenly to double the length; then this double sequence must be worked with. The sequence $\hat{h}_\ell$ must be padded with zeros to the same length. It should be noted that even if the length of the initial signal sequence is an integer power of two, it is always prudent during filtering to supplement the sequence evenly to double the length. This is done in order to avoid boundary effects resulting from the periodic extension of sequences when they are being convolved by the DFT. In using SDFT(1/2,0), an additional loss of computer time due to the doubling of the sequence length can be fairly easily avoided by applying combined transform algorithms (Sect.5.6).

If the number of non-zero samples of the impulse response of filter $\hat{h}_\ell$ is much smaller than the number of signal samples, signal convolution with such an impulse response can be calculated more rapidly not for the whole signal at once, but by parts, breaking down the initial signal sequence into sub-sequences. Here the question of signal predetermination only arises for the end subsequences – the first and the last. The other subsequences must be chosen to overlap some non-zero samples of the filter impulse response, and the surplus samples (midway along overlapping lengths on either side) must be discarded after joining up the convolution results; alternatively, the sequence must be supplemented with zeros on the edges so that the length of the impulse response is doubled and the overlapping segments are then added when joining up the convolution results [4.1].

When the number of samples of the filter impulse response is small in absolute terms, or when the required filter can be represented recursively, the use of the DFT to calculate the convolution can be inefficient because of the time required. The precise quantitative relations determining the efficiency of the DFT depends on the type of processor used [4.1,19].

All of the above is also applicable to the 2-D case, including 2-D signal predetermination.

Let us note in conclusion that SDFT can be used, as follows from (4.102), to obtain convolution samples arbitrarily placed between the prescribed signal samples. This is associated with the possibility of signal interpolation using SDFT, which is considered below.

4.8.2 Signal Interpolation

Discrete Fourier transforms are a convenient way of finding unknown intermediate signal samples (i.e., samples distributed between known ones) from the known samples (signal interpolation). The optimal interpolation of continuous signals to which the sampling theorem applies is determined by the theorem:

$$a(k + x) = \sum_{n=-\infty}^{\infty} a_n \operatorname{sinc}[\pi(k - n + x)] \quad . \tag{4.106}$$

For sequences of finite length this relation can be approximated with the aid of the SDFT and ISDFT transforms. For the SDFT(u,v) and ISDFT(p,q) of sequence $\{a_k\}$, $k = 0, 1, \ldots, N-1$, we have (Table 4.3, line 11a):

$$\begin{aligned} {}^{s_0}a_k^{p,q/u,v} &= 1/\sqrt{N} \sum_{s=0}^{M-1} \alpha_{s+s_0}^{u,v} \exp[-i2\pi(k + p)(s + q)/N] \\ &= \sum_{n=0}^{N-1} \{a_n \exp[i2\pi(n + u)(s_0 + v - q)/N]\} \sin[\pi M(k - n - u + p) \\ &\quad / N]/N \sin[\pi(k - n - u + p)/N] \exp\{-i2\pi[k - n - u + p][q - (M - 1) \\ &\quad /2]/N\} \quad . \end{aligned} \tag{4.107}$$

If we choose

$$q = (M - 1)/2 \quad , \quad s_0 + v - q = 0 \quad , \tag{4.108}$$

then

$$\begin{aligned} {}^{s_0}a_k^{p,q/u,v} &= \sum_{n=0}^{N-1} a_n \sin\{\pi M[k - n + (p - u)]/N\}/N \\ &\quad \times \sin\{[\pi(k - n + (p - u)]/N\} \quad . \end{aligned} \tag{4.109}$$

This formula can be considered a discrete analogue of (4.106) with $x = (p - u)$. For odd N one has to set $M = N$. If, as is usual when using FFT algorithms, N is an even number, $M = N - 1$ must be chosen; i.e., in performing the inverse transformation for sequences of length N, the end $(N-1)^{th}$ Fourier coefficient $\alpha_{N-1}^{u,v}$ must be set to zero.

As in signal convolution with the discrete Fourier transform, signal interpolation by (4.109) leads to considerable error on the edges of the sequence. In order to decrease the error, it is prudent to extend the interpolated sequence evenly.

Sequence interpolation can also be performed with the standard DFT by symmetrically supplementing the signal spectrum with zeros to a length that exceeds the initial length by as many times as the number of additional samples required for one initial signal sample (Table 4.1, lines 18a,b). However, the standard DFT is less economical than the SDFT in terms of computer time and memory. To obtain, with the aid of the FFT algorithm, L additional samples for each sample of a sequence of N elements, approximately $NL \log_2(NL)$ operations and a memory capacity of NL are required when interpolating by means of DFT, and $NL \log_2(2N)$ operations and N memory cells when using SDFT. Moreover, the additional samples obtained by using the standard DFT are equidistant, whereas with SDFT it is possible, in principle, to obtain intermediate samples with random shift relative to the initial ones.

In addition to these applications for signal convolution and interpolation, discrete Fourier transforms are also used in coding (Sect.3.11) and in evaluating Fourier spectra and signal and picture correlation functions (Chap.6).

4.9 Walsh and Similar Transforms

The discrete Walsh transform, like the discrete Fourier transform, can be regarded as a discrete analogue of the continuous transformation of signals over a basis composed of Walsh functions. There are three versions of this transform: the Walsh-Hadamard, Paley, and Walsh transforms. They differ in the way in which the base functions are ordered. All are determined on sequences whose number of terms is equal to an integer power of 2.

Let us begin by considering the Walsh transform. Walsh transforms were described above in Sect.2.3. Like discrete Fourier transforms, the discrete Walsh transform is constructed as the transformation of a sequence of signal samples over a basis composed of Walsh function samples taken at a discrete sequence of points:

$$\alpha_s = \sum_{k=0}^{N-1} a_k \, \mathrm{wal}_s(k) \quad . \tag{4.110}$$

Let the number of signal samples N be 2^n, $k = 0, 1, \ldots, 2^n - 1$. The number of basis functions is the same: $s = 0, 1, \ldots, 2^n - 1$. Then using (2.46), we can write the values of the Walsh basis function $\mathrm{wal}_s(k)$ as[3]

[3] Note that the numbering of binary digits in (4.111) differs from that used in (2.49), in that in the former, k and s are integers, and the digits of their binary representation are numbered by value j from right to left. In (2.46) ξ, the argument of the Walsh function, was less than unity. In order to reverse the numbering order of the argument digits from (2.49) to (4.111), the numbering of the Gray code digits of the function number had to be reversed.

$$\mathrm{wal}_s(k) = 1/\sqrt{N}\,(-1)^{\sum_{j=0}^{n-1} s^G_{n-1-j} k_j} \tag{4.111}$$

where s^G_{n-1-j} are the Gray code digits of the function number s, taken in reverse order (i.e. reading left to right), and k are the binary code digits of the sample number.

By substituting (4.111) into (4.110) we obtain

$$\alpha_s = 1/\sqrt{2n} \sum_{k=0}^{2^n-1} a_k (-1)^{\sum_{j=0}^{n-1} s^G_{n-1-j} k_j} \quad . \tag{4.112}$$

Since

$$k = \sum_{j=0}^{n-1} k_j 2^j \quad , \tag{4.113}$$

the summation in (4.112) over k can be achieved as a multidimensional summation over k_j:

$$\alpha_s = 1/\sqrt{2n} \sum_{k_{n-1}=0}^{1} (-1)^{s^G_0 k_{n-1}} \dots \sum_{k_j=0}^{1} (-1)^{s^G_{n-1-j} k_j} \dots \sum_{k=0}^{1} a_{k_{n-1},\dots,k_0} (-1)^{s^G_{n-1} k_0} \quad . \tag{4.114}$$

A representation such as this is the basis of the fast Walsh transform algorithm (Chap.5).

The inverse Walsh transform formula is the same as the direct transform formula (4.106):

$$a_k = 1/\sqrt{2n} \sum_{s=0}^{2^n-1} \alpha_s (-1)^{\sum_{j=0}^{n-1} s^G_{n-j-1} k_j} \quad . \tag{4.115}$$

As follows from (4.111), the Walsh functions in the Walsh transform are reverse binary ordered Gray codes (i.e. they are read in reverse order) of their number. If it is recalled that the Walsh functions are generated by the Rademacher functions, then the meaning of such an ordering is that every successive Walsh function differs in number from the previous one Walsh by the addition or subtraction of only one Rademacher function.

These functions can be ordered simply by their binary indexes. Transformation by such functions:

$$\mathrm{had}_s(k) = 1/\sqrt{2n}\,(-1)^{\sum_{j=0}^{n-1} s_j k_j} \tag{4.116}$$

is called the Walsh-Hadamard transform (or in some sources, the Hadamard or BIFOR transform [4.15]):

$$\alpha_s = 1/\sqrt{2^n} \sum_{k=0}^{2^n-1} a_k (-1)^{\sum_{j=0}^{n-1} s_j k_j} \tag{4.117}$$

Yet another method of ordering was introduced by Paley [4.20]. In it the Walsh functions are distributed in the order determined by their bit reversed codes:

$$\mathrm{pal}_s(k) = 1/\sqrt{2^n}\, (-1)^{\sum_{j=0}^{n-1} s_{n-j-1} k_j} \quad . \tag{4.118}$$

The method by which basis functions are ordered is determined solely by considerations of convenience, since every transform coefficient corresponds to its basis function from one and the same set of functions, independently of numbering. It is usually convenient to link the order of the basis functions with the convergence of the signal representation so that the representation coefficient which is senior in number (and corresponds to a basis function which likewise has the senior numbered index) makes a lower contribution to the signal.

Sometimes (see Sect.6.5) another property is important, namely smoothness of the spectrum which is considered a function of the number of spectral components. It is then interesting to compare the behaviour of the spectrum of one and the same picture in Walsh basis functions with various ordering methods. The graphs in Fig.4.6b-d show how the squared modulus of the components of the 1-D spectrum of the line of the picture shown in Fig.4.6a depends on index number. Various orderings are shown. Part (b) shows ordering due to Paley, (c) due to Walsh, and (d) due to Hadamard (the component numbers increase from left to right). It can be seen that the smoothest spectrum is obtained by Walsh ordering. Hadamard ordering gives a spectrum of chaotic, unpredictable character.

These discrete transforms can be represented in matrix form as the product of the column matrix of the initial sequence and the transformation matrix. As a result the column matrix of transform $\boldsymbol{\alpha}$ is obtained:

$$\boldsymbol{\alpha} = \mathrm{HAD}_{2^n} \mathbf{a} \quad , \tag{4.119a}$$

$$\boldsymbol{\alpha} = \mathrm{WAL}_{2^n} \mathbf{a} \quad , \tag{4.119b}$$

$$\boldsymbol{\alpha} = \mathrm{PAL}_{2^n} \mathbf{a} \quad , \tag{4.119c}$$

where HAD_{2^n}, WAL_{2^n}, PAL_{2^n} are Walsh transform matrices in which the rows (basis functions) are determined by (4.116,111,118), respectively.

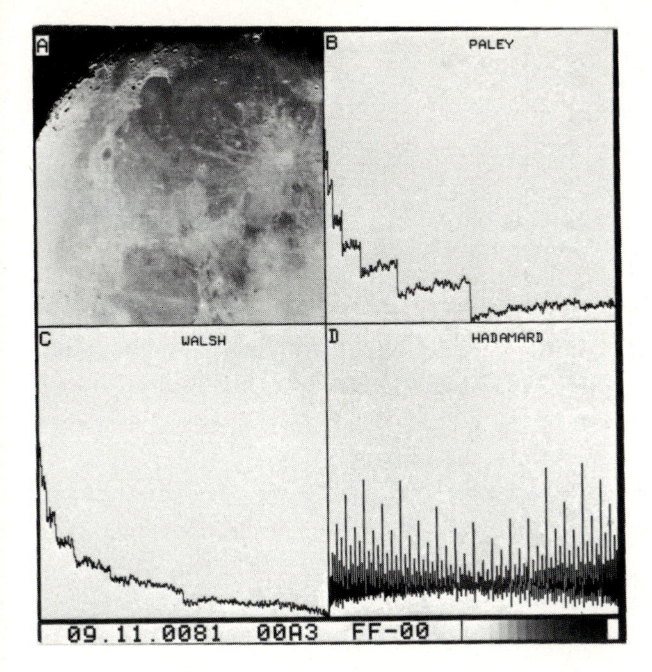

Fig. 4.6. Behavior of the picture spectrum in Walsh basis functions with various orderings: (a) test picture, (b) picture sepctrum in ordering due to Paley, (b) due to Walsh, (d) due to Hadamard (the spectrum component numbers increase from left to right)

For example, the HAD_{2^3}, WAL_{2^3} and PAL_{2^3} matrices are as follows:

$$HAD_{2^3} = \begin{bmatrix} 1 & 1 & 1 & 1 & 1 & 1 & 1 & 1 \\ 1 & -1 & 1 & -1 & 1 & -1 & 1 & -1 \\ 1 & 1 & -1 & -1 & 1 & 1 & -1 & -1 \\ 1 & -1 & -1 & 1 & 1 & -1 & -1 & 1 \\ 1 & 1 & 1 & 1 & -1 & -1 & -1 & -1 \\ 1 & -1 & 1 & -1 & -1 & 1 & -1 & 1 \\ 1 & 1 & -1 & -1 & -1 & -1 & 1 & 1 \\ 1 & -1 & -1 & 1 & -1 & 1 & 1 & -1 \end{bmatrix} \begin{matrix} 0 \\ 7 \\ 3 \\ 4 \\ 1 \\ 6 \\ 2 \\ 5 \end{matrix} \qquad (4.120a)$$

$$WAL_{2^3} = \begin{bmatrix} 1 & 1 & 1 & 1 & 1 & 1 & 1 & 1 \\ 1 & 1 & 1 & 1 & -1 & -1 & -1 & -1 \\ 1 & 1 & -1 & -1 & -1 & -1 & 1 & 1 \\ 1 & 1 & -1 & -1 & 1 & 1 & -1 & -1 \\ 1 & -1 & -1 & 1 & 1 & -1 & -1 & 1 \\ 1 & -1 & -1 & 1 & -1 & 1 & 1 & -1 \\ 1 & -1 & 1 & -1 & -1 & 1 & -1 & 1 \\ 1 & -1 & 1 & -1 & 1 & -1 & 1 & -1 \end{bmatrix} \begin{matrix} 0 \\ 1 \\ 2 \\ 3 \\ 4 \\ 5 \\ 6 \\ 7 \end{matrix} \qquad (4.120b)$$

$$\text{PAL}_2 3 = \begin{bmatrix} 1 & 1 & 1 & 1 & 1 & 1 & 1 & 1 \\ 1 & 1 & 1 & 1 & -1 & -1 & -1 & -1 \\ 1 & 1 & -1 & -1 & 1 & 1 & -1 & -1 \\ 1 & 1 & -1 & -1 & -1 & -1 & 1 & 1 \\ 1 & -1 & 1 & -1 & 1 & -1 & 1 & -1 \\ 1 & -1 & 1 & -1 & -1 & 1 & -1 & 1 \\ 1 & -1 & -1 & 1 & 1 & -1 & -1 & 1 \\ 1 & -1 & -1 & 1 & -1 & 1 & 1 & -1 \end{bmatrix} \begin{matrix} 0 \\ 1 \\ 3 \\ 2 \\ 7 \\ 6 \\ 4 \\ 5 \end{matrix} \qquad (4.120c)$$

The number of sign changes in the corresponding matrix line is indicated to the right of each matrix in (4.120). A comparison of these figures reveals that the lines of the Walsh matrix $\text{WAL}_2 n$ are ordered according to the number of sign changes, which is known as the sequency ([4.21]).

The Walsh transform matrices are symmetrical, i.e., they are not affected by transposition. They are also orthogonal, since they are inverse to themselves.

Thus the Walsh transform belongs, like the DFT, to the class of unitary transforms (for real matrices, the concepts of unitariness and orthogonality are the same).

The WAL and PAL matrices reduce to the Hadamard HAD matrix if they are multiplied by the appropriate permutation matrices. The latter has one s at that element of row s whose suffix k coincides with the number of the element of the sequence under permutation, which must transfer from the number k to s. Multiplication of the column matrix by such matrices rearranges its elements, while multiplication of the matrix by the permutation matrices rearranges its rows.

Let $M^{\text{rev}}_{2^n}$ be the bit reversal matrix, and $M^{G/D}_{2n}$ the matrix that converts Gray code into direct binary code. Then

$$\text{PAL}_2 n = M^{\text{rev}}_{2^n} \text{HAD}_2 n \quad , \qquad (4.121a)$$

$$\text{WAL}_2 n = M^{G/D}_{2^n} M^{\text{rev}}_{2^n} \text{HAD}_2 n \quad . \qquad (4.121b)$$

Since the Walsh-Paley matrix, the bit reversal matrix, and the Hadamard matrix are symmetrical, the following is also valid:

$$\text{PAL}_2 n = \text{HAD}_2 n \, M^{\text{rev}}_{2^n} \quad . \qquad (4.121c)$$

An important feature of the Hadamard matrix is that it belongs to the so-called Kronecker matrices [4.22,23], i.e., it can be represented in the form of the Kronecker direct product[4] of the second-order Hadamard matrix

$$h_2 = \begin{bmatrix} 1 & 1 \\ 1 & -1 \end{bmatrix} \quad :$$

$$\mathrm{HAD}_{2^n} = \bigotimes_{j=0}^{n-1} h_2 = h_2 \otimes h_2 \otimes \ldots \otimes h_2 = (h_2)^{[n]} \quad ,$$

where the signs $\bigotimes_{j=0}^{n-1}$ and $\otimes$ denote the Kronecker products of n and two matrices respectively, and [n] is the nth Kronecker power of the matrices.

A remarkable feature of Kronecker matrices is that they can be represented (factorized) as products of sparse matrices, i.e., matrices the majority of whose elements are zero. Because of this, considerably fewer operations have to be carried out during multiplication on Kronecker matrices than is usual during matrix multiplication (Chap.5).

Since 2-D Walsh functions are defined as the product of 1-D ones, it is not difficult to move from the latter to the former. It is simplest to denote them as matrix products, for example

$$\boldsymbol{\alpha}_{2^n} = \mathrm{HAD}_{2^n}\, \mathbf{a}_{2^n}\, \mathrm{HAD}_{2^n} \quad , \tag{4.123}$$

where $\boldsymbol{\alpha}_{2^n}$ and $\mathbf{a}_{2^n}$ are 2-D sequences (matrices of order $2^n \times 2^n$).

4.10 The Haar Transform. Additional Elements of Matrix Calculus

Discrete Haar transforms are transformations of the signal sample sequence $\{a_k\}$, $k = 0, 1, \ldots, N-1$, the number N being equal to an integer power of two, over a basis composed of Haar function samples $\mathrm{har}_s(k)$ (2.51):

$$\alpha_s = \sum_{k=0}^{N-1} a_k\, \mathrm{har}_s^{(n)}(k) \quad . \tag{4.124}$$

Let $N = 2^n$. Then according to (2.51),

$$\mathrm{har}_s^{(n)}(k) = 2^{(s_u-n)/2} (-1)^{k_{n-s_u-1}} \delta([k]_{n-s_u} \oplus s_{\mathrm{mod}2^{s_u}}) \quad , \tag{4.125}$$

where s_u is the number of the most significant digit equal to 1 in the binary representation of the number s; $[k]_{n-s_u}$ is a binary number composed of the s_u most significant digits of the binary number k; $\oplus$ denotes logical addition;

[4] The Kronecker product of two matrices is defined as a matrix composed of submatrices each of which is the product of one element of the first matrix times the entire second matrix. For more details on the properties of the Kronecker products of matrices, see [4.24] and Sect.5.1

$s_{\mathrm{mod}2^{s_u}}$ is a binary number composed of the s_u least significant digits of s; and $\delta(\ldots)$ is the Kronecker delta-function (symbol)[5].

In matrix form the Haar transform is described by Haar matrices whose rows are composed of the functions $\mathrm{har}_s^{(n)}(k)$:

$$\mathrm{HAR}_{2^n} = \{\mathrm{har}_s^{(n)}(k)\} \quad . \tag{4.126}$$

For example, the matrix HAR_{2^3} has the following form:

$$\mathrm{HAR}_{2^n} = \frac{1}{\sqrt{8}} \begin{bmatrix} 1 & 1 & 1 & 1 & 1 & 1 & 1 & 1 \\ \hline 1 & 1 & 1 & 1 & -1 & -1 & -1 & -1 \\ \hline 2^{1/2} & 2^{1/2} & -2^{1/2} & -2^{1/2} & 0 & 0 & 0 & 0 \\ 0 & 0 & 0 & 0 & 2^{1/2} & 2^{1/2} & -2^{1/2} & -2^{1/2} \\ \hline 2 & -2 & 0 & 0 & 0 & 0 & 0 & 0 \\ 0 & 0 & 2 & -2 & 0 & 0 & 0 & 0 \\ 0 & 0 & 0 & 0 & 2 & -2 & 0 & 0 \\ 0 & 0 & 0 & 0 & 0 & 0 & 2 & -2 \end{bmatrix} \quad . \tag{4.127}$$

Haar matrices are orthogonal but not symmetrical. Unlike Hadamard matrices, they are not Kronecker matrices, but can be represented as a sum of Kronecker matrices. Indeed, nth-order Haar matrices can be broken down into $(n+1)$ Kronecker-type submatrices according to the number of the most significant digit equal to 1 in the row number, as shown in (4.127) by dashed lines.

This representation also appears to be convenient for describing other orthogonal transforms, as will be shown later (Sect.4.11). To give a more formal description it is useful to introduce the following special definitions, notation and elementary matrices:

Definition 1. The vertical sum of two matrices $M^{(1)}_{r,s}$ and $M^{(2)}_{p,s}$ is called the matrix $M_{r+p,s}$, whose first r rows are the rows of the $M^{(1)}_{r,s}$ matrix, and whose last p rows are the rows of $M^{(2)}_{p,s}$ matrix.

Definition 2. The vertical sum of Kronecker matrices is called a layered Kronecker matrix.

---

[5] The numbering order of digits in the binary representation of number k in (4.125) is from right to left, the reverse of that adopted in (2.51).

Designations:

$M^{(1)}_{r,s} \boxminus M^{(2)}_{p,s}$ is the vertical sum of two matrices.

$\overset{n-1}{\underset{k=0}{\boxminus}} M^{(k)}$ is the vertical sum of n matrices.

$$\bar{\delta}(k) = 1 - 0^k = \begin{cases} 0, & k = 0 \quad ; \\ 1, & k \neq 0 \quad ; \end{cases}$$

$$M^{\bar{\delta}(k)} = \begin{cases} 1, & k = 0 \quad ; \\ M, & k \neq 0 \quad . \end{cases}$$

Elementary matrices:

$$E_2^0 = [1 \quad 1] \quad ; \quad E_2^1 = [1 \ -1] \quad ;$$

$$d_u = \begin{bmatrix} 1 & 0 \\ 0 & \exp(i2\pi u \ / \ _2n \end{bmatrix} \quad ; \tag{4.128}$$

$$I_2 = d_0 = \begin{bmatrix} 1 & 0 \\ 0 & 1 \end{bmatrix} \quad .$$

With the aid of the above, the Haar transform matrix of order 2^n can be written as a layered Kronecker matrix in the following way:

$$HAR_{2n} = \overset{n}{\underset{k=0}{\boxminus}} \left[\left(2^{[k-1]/2} \ d_0^{[k-1]} \otimes E_2^1 \right)^{\bar{\delta}(k)} \otimes (E_2^0)^{[n-k]} \right] \quad . \tag{4.129}$$

Recursive notation is also possible:

$$HAR_{2^n} = \left[HAR_{2^{n-1}} \otimes E_2^0 \right] \boxminus \left[2^{(n-1)/2} \ d_0^{[n-1]} \otimes E_2^1 \right] \tag{4.130}$$

Thus, for example,

$$HAR_{2^3} = [1\,1]^{[3]} \boxminus \left([1\ -1] \otimes [1\,1]^{[2]} \right) \boxminus \left(\begin{bmatrix} \sqrt{2} & 0 \\ 0 & \sqrt{2} \end{bmatrix} \otimes [1\ -1] \otimes [1\,1] \right)$$

$$\boxminus \begin{bmatrix} 2&0&0&0 \\ 0&2&0&0 \\ 0&0&2&0 \\ 0&0&0&2 \end{bmatrix} \otimes [1\ -1] \quad = [1\,1\,1\,1\,1\,1\,1\,1] \boxminus [1\,1\,1\,1\ \ -1\ -1\ -1\ -1]$$

$$\boxminus \begin{bmatrix} 2&2&-2&-2&0&0&0&0 \\ 0&0&0&0&2&2&-2&-2 \end{bmatrix} \boxminus \begin{bmatrix} 2&-2&0&0&0&0&0&0 \\ 0&0&2&-2&0&0&0&0 \\ 0&0&0&0&2&-2&0&0 \\ 0&0&0&0&0&0&2&-2 \end{bmatrix} \quad . \tag{4.131}$$

4.11 Other Orthogonal Transforms. General Representation. Review of Applications

On the basis of the DFT, Walsh-Hadamard and Haar transforms, many other orthogonal transforms can be constructed. In [4.25], for example, a hybrid

Hadamard-Haar transform is proposed whose the rth order matrix of dimensionality 2^n is defined by

$$\mathrm{HDHR}^{(r)}_{2^n} = \mathrm{HAD}_{2^r} \otimes \mathrm{HAR}_{2^{n-r}} \quad . \tag{4.132}$$

A recursive definition of the so-called modified Hadamard transform is given in [4.15]:

$$\mathrm{MHAD}_{2^n} = \begin{bmatrix} \mathrm{MHAD}_{2^{n-1}} & \mathrm{MHAD}_{2^{n-1}} \\ 2^{(n-1)/2}\, I_{2^{n-1}} & -2^{(n-1)/2}\, I_{2^{n-1}} \end{bmatrix} = \left[E_2^0 \otimes \mathrm{MHAD}_{2^{n-1}}\right]$$
$$\boxminus \left[2^{(n-1)/2}\, E_2^1 \otimes I_{2^{n-1}}\right] \tag{4.133}$$

and is described in its relation to the Haar transform.

The matrix of the so-called generalized Hadamard transform of rth order and dimensionality 2^{nr} (transformation by the Vilenkin-Chrestenson functions) was investigated in [4.20,22]. It is defined as the rth Kronecker power of the DFT matrices F_{2^n}:

$$G^{(r)}_{2^{nr}} = (F_{2^n})^{[r]} \quad . \tag{4.134}$$

In [4.26] there is a description of the so-called R transform, which is constructed on the basis of the Walsh-Hadamard transform by replacing every sum in (4.114) by its absolute value. This transform is not reversible.

Several transforms have been suggested for picture coding: the slant transform [4.27]; "slant-Haar" transforms [4.28]; the generalized Haar transform [4.29]; the complex Haar transform [4.30]; the S transform [4.31]; the generalized Fourier-Haar transform [4.32]; and transform by "discrete linear basis" [4.33].

The most important common property of these transforms is that they have so-called fast algorithms (Chap.5). This illustrates their profound kinship. This kinship can be revealed and the transforms described in a uniform way by using a relatively small set of elementary matrices. The structure of the transform matrices can be linked with the existence of fast algorithms by using the concept, introduced in Sect.4.10, of layered Kronecker matrices [4.34-36]. An integrated representation of the above transforms and a few others is presented in Table 4.5.

Table 4.5. Representation of orthogonal transform matrices as layered Kronecker matrices

Haar transform	$\mathrm{HAR}_{2^n} = \underset{k=0}{\overset{n}{\boxminus}} [(2^{(k-1)/2} d_0^{[k-1]} \otimes E_2^1)^{\bar{\delta}(k)} \otimes (E_2^0)^{[n-k]}]$
Complex Haar transform	$\mathrm{CHAR}_{2^n} = \underset{k=0}{\overset{n-1}{\boxminus}} \{[(h_2 d_2 \otimes 2^{k/2} d_0^k)^{\bar{\delta}(k-1)} \otimes E_2^1]^{\bar{\delta}(k)} \otimes (E_2^0)^{[n-k]}\}$
Generalized first-order Haar transform	$\mathrm{GHAR}_{2^n}^{(1)} = \underset{k=0}{\overset{n}{\boxminus}} \{[(\bar{d}_0 h_2 d_2 \otimes 2^{k/2} d_0^{[k]})^{\bar{\delta}(k-1)} \otimes E_2^1]^{\bar{\delta}(k)} \otimes (E_2^0)^{[n-k]}\}$
Generalized second-order Haar transform	$\mathrm{GHAR}_{2^n}^{(2)} = \underset{k=0}{\overset{n}{\boxminus}} (\{[har_{2^2}^{(2)}(d_6 \otimes d_4) M_{2^2}^{G/D} M_{2^2}^{\mathrm{rev}}]^{\bar{\delta}(k-2)}$ $\otimes \bar{d}_0 h_2 d_2\}^{\bar{\delta}(k-1)} \otimes E_2^1)^{\bar{\delta}(k)} \otimes (E_2^0)^{[n-k]}]$
Modified Hadamard transform	$\mathrm{MHAD}_{2^n} = [E_2^0 \otimes \mathrm{MHAD}_{2^{n-1}}] \square [E_2^1 \otimes (2^{(n-1)/2} d_0^{[n-1]})]$ $= \underset{k=0}{\overset{n}{\boxminus}} [(E_2^0)^{[n-k]} \otimes (E_2^1 \otimes 2^{(k-1)/2} d_0^{[k-1]})^{\bar{\delta}(k)}]$
Walsh-Hadamard transform	$\mathrm{HAD}_{2^n} = (h_2)^{[n]} = \underset{k=0}{\overset{n}{\boxminus}} [(E_2^0)^{[n-k]} \otimes (E_2^1 \otimes h_2^{[k-1]})^{\bar{\delta}(k)}]$
Walsh transform	$\mathrm{WAL}_{2^n} = (\mathrm{WAL}_{2^{n-1}} \otimes E_2^0) \boxminus [\bar{d}_0^{[n-1]} \mathrm{WAL}_{2^{n-1}} \otimes E_2^1]$ $= \underset{k=0}{\overset{n}{\boxminus}} \{[(\bar{d}_0^{[k-1]} \mathrm{WAL}_{2^{k-1}}) \otimes E_2^1]^{\bar{\delta}(k)} \otimes (E_2^0)^{[n-k]}\}$
Walsh-Paley transform	$\mathrm{PAL}_{2^n} = (\mathrm{PAL}_{2^{n-1}} \otimes E_2^0) \boxminus (\mathrm{PAL}_{2^{n-1}} \otimes E_2^1)$ $= \underset{k=0}{\overset{n}{\boxminus}} [(\mathrm{PAL}_{2^{k-1}} \otimes E_2^1)^{\bar{\delta}(k)} \otimes (E_2^0)^{[n-k]}]$
Hadamard-Haar transform	$\mathrm{HDHR}_{2^n}^{(m)} = \mathrm{HAD}_{2^m} \otimes \mathrm{HAR}_{2^{n-m}}$ $= h_2^{[m]} \otimes \underset{k=0}{\overset{n-m}{\boxminus}} [(2^{(k-1)/2} d_0^{[k-1]} \otimes E_2^1)^{\bar{\delta}(k)} \otimes (E_2^0)^{[n-m-k]}]$
S transforms	$S_{2^n} = \underset{k_1=0}{\overset{n/2-1}{\boxminus}} \left\{ \left[\left(\underset{k_2=0}{\overset{1}{\boxminus}} \left[\dfrac{\Delta_{k_2}^{[2(k_1-1)]} \otimes E_2^0}{2^{k_1-1} d_0^{[2(k_1-1)]} \otimes E_2^1} \right] \right) \otimes E_2^1 \right]^{\bar{\delta}(k_1)} \right.$ $\left. \otimes (E_2^0)^{[n-2k_1]} \right\}$

Table 4.5 (continued)

Discrete Fourier transform	$F_{2^n} = \boxminus_{k=0}^{2^n-1} \bigotimes_{j=0}^{n-1} E_2^0 d_{2^j k}$ $= M^{\text{rev}}_{2^n} \left[\frac{E_2^0 \otimes M^{\text{rev}}_{2^{n-1}} F_{2^{n-1}}}{E_2^1 \otimes M^{\text{rev}}_{2^{n-1}} F_{2^{n-1}} \left(\bigotimes_{m=0}^{n-2} d_{2^m}\right)}\right]$ $= M^{\text{rev}}_{2^n} \boxminus_{k=0}^{n} \left\{ (E_2^0)^{[n-k]} \otimes \left[E_2^1 \otimes (M^{\text{rev}}_{2^{k-1}} F_{2^{k-1}}) \left(\bigotimes_{m=0}^{k-2} d_{2^m}\right)\right]^{\bar{\delta}(k)} \right\}$
Generalized Hadamard transforms	$\text{VCF}^{(m)}_{2^{nm}} = (F_{2^n})^{[m]}$
Slant transform	SLANT_{2^n} $= M^{\text{rev}}_{2^n} \left[\frac{E_2^0 \otimes M^{\text{rev}}_{2^{n-1}} \text{SLANT}_{2^{n-1}} M^{D/G}_{2^{n-1}}}{E_2^1 \otimes \text{M}^{\text{rev}}_{2^{n-1}} \left(h_\varphi \boxplus \boxplus_{j=0}^{2^{n-2}-2} d_0\right) \text{SLANT}_{2^{n-1}} M^{D/G}}\right]$ $\times M^{D/G}_{2^n} = M^{\text{rev}}_{2^n} \left(\boxminus_{k=0}^{n} \left\{ (E_2^0)^{[n-k]} \otimes \left[E_2^1 \otimes M^{\text{rev}}_{2^{k-1}} \right.\right.\right.$ $\left.\left. \times \left(h_\varphi \boxplus \boxplus_{j=0}^{2^{k-2}-2} d_0\right) \text{SLANT}_{2^{k-1}} M^{D/G}_{2^{k-1}} \right]^{\bar{\delta}(k)} \right\} M^{D/G}_{2^n};$ $\varphi = \arccos \sqrt{(2^{2n-2}-1)/(2^{2n}-1)}$
Generalized Fourier-Haar transform	$\text{GFH}^{(k)}_{2^n} = \left[\frac{(E_2^0)^{[n-k]} \otimes \text{HAD}_{2^k}}{(R_n^k \text{GFH}_{2^n}) \otimes (d_0)^k}\right]$
Permutation transform by bit reversal	$M^{\text{rev}}_{2^n} = \boxminus_{k=0}^{n} [(M^{\text{rev}}_{2^{k-1}} \otimes G_2^1)^{\bar{\delta}(k)} \otimes (G_2^0)^{[n-k]}]$
Permutation transform from Gray code to direct binary code	$M^{G/D}_{2^n} = \boxminus_{k=0}^{n} [(G_2^0)^{[n-k]} \otimes (G_2^1 \otimes \bar{d}_0^{[k-1]} M^{G/D}_{2^{k-1}})^{\bar{\delta}(k)}]$

The following notation supplements that introduced in Sect.4.10:

$\boxplus$ is the direct sum of two matrices;

$\underset{k=0}{\overset{n-1}{\boxplus}}$ is the direct sum of n matrices[6];

$$G_2^0 = [1\,0] \quad ; \quad G_2^1 = [0\,1] \quad ; \quad h_2 = E_2^0 \boxminus E_2^1 = \begin{bmatrix} 1 & 1 \\ 1 & -1 \end{bmatrix} \quad ;$$

$$I_2 = \bar{d}_0 = G_2^1 \boxminus G_2^0 = \begin{bmatrix} 0 & 1 \\ 1 & 0 \end{bmatrix} \quad ;$$

$$\Delta_k = \begin{bmatrix} 0^k & 0 \\ 0 & 0^k \end{bmatrix} = \begin{cases} d_0, & k = 0 \quad ; \\ 0, & k \neq 0 \quad ; \end{cases}$$

$$har_{2^2}^{(2)} = (h_2 \otimes E_2^0) \boxminus (\bar{d}_0\, h_2\, d_0 \otimes E_2^1) \quad ;$$

$$h_\varphi = \begin{bmatrix} \cos\varphi & \sin\varphi \\ -\sin\varphi & \cos\varphi \end{bmatrix} \quad , \tag{4.135}$$

$M_{2^k}^{rev}$ is the bit reversal matrix, i.e., permutation of the elements of a vector in accordance with the inverse binary code of their numbers;

$M_{2^k}^{D/G}$ is the permutation matrix of the elements of a vector in accordance with the Gray code of their numbers;

$M_{2^k}^{G/D}$ is the inverse of the $M_{2^k}^{D/G}$ permutation matrix;

$R_{2^n}^k$ is the compression matrix of dimensionality $(2^k + 2^n)$, whose first 2^k rows are composed of zero vectors, and whose last 2^n rows form an identity matrix of size $2^n \times 2^n$.

For the sake of simplicity, the normalizing factor $1/2^{n/2}$ has been omitted in all of the transforms in Table 4.5.

This presentation creates a convenient basis for comparing and systematizing orthogonal transforms. Thus, it is easy to see that from HAR matrices MHAD matrices can be derived by changing the sequence order of the elementary matrices in Kronecker products; that the HAD matrix can be derived by replacing the elementary matrices d_0 with h_2; and finally that DFT matrices with binary row inversion (bit reversal) can be derived by introducing diagonal matrices d_u. A comparison of the matrices $M_{2^n}^{rev}$ and PAL_{2^n}, $M_{2^n}^{G/D}$ and WAL_{2^n}, and F_{2^n} and $SLANT_{2^n}$ is instructive, and vividly illustrates the role that elementary matrices play in the construction of transform matrices. Moreover, this representa-

---

6 For a definition of the direct matrix sum, see Sect.5.1 and [4.24].

tion, together with the factorization theorems introduced in Chap.5, allows easy construction of fast algorithms of these transforms in their most general forms using the technique of matrix algebra. It also offers a deeper understanding of how matrices of orthogonal transforms should in general be constructed, if fast algorithms are desired for them. It follows that new types of transforms having guaranteed fast algorithms can be built; this is done by introducing new types of elementary matrices or combining them in one transform, as well as by combining various layered Kronecker matrices with the aid of vertical sums and Kronecker products.

On the basis of the 1-D transforms described, the corresponding 2-D separable transforms can be constructed as double 1-D ones:

$$\boldsymbol{\alpha} = \mathbf{M}\mathbf{a}\mathbf{M} \quad , \tag{4.136}$$

where $\mathbf{M}$ is one of the transform matrices described above; $\mathbf{a}$ is a 2-D discrete signal; and $\boldsymbol{\alpha}$ is its transformation.

Note that all of the unitary picture transforms currently used in digital picture processing are separable, i.e., are calculated separately along columns and rows of the 2-D signal. Owing to this, the number of operations necessary for calculation decreases. Separable transforms can also be constructed by choosing various matrices for transformation over the rows and columns:

$$\boldsymbol{\alpha} = \mathbf{M}_1\mathbf{a}\mathbf{M}_2 \quad . \tag{4.137}$$

Mixed transforms used in specialized digital picture-coding devices are derived in this way [4.37].

The applications of unitary transforms in picture processing can be divided into three groups: picture coding; feature extraction for the preparation and recognition of pictures; and generalized filtering.

Picture coding is at present the main application of transforms (apart from the DFT). Moreover, the majority of transforms were specially introduced for use in coding.

The coefficients in the representation of a picture which result from a transformation can be regarded as features of it and used in picture preparation (Chap.8) and for recognition. An example of a transform specially developed for feature extraction in pattern recognition is the R transform. The applications of transforms for coding and recognition are linked with one another. As a rule, transforms that give better results for coding are also better for feature extraction.

The use of unitary transforms in signal filtering is based on a generalization of the concept of filtering in the frequency domain of the discrete Fourier transform. In signal filtering using DFT the following signal transformation takes place:

$$\hat{\mathbf{a}} = \mathbf{F}^{-1}\mathbf{H}_F\mathbf{Fa} \quad , \tag{4.138}$$

where $\mathbf{a}$ is the sequence of the initial signal; $\mathbf{F}$ is the DFT matrix; $\mathbf{H}_F$ is the diagonal matrix consisting of DFT coefficients of the impulse response of the required filter; $\mathbf{F}^{-1}$ is the inverse DFT matrix; and $\hat{\mathbf{a}}$ is the signal after filtering.

If another transform $\mathbf{T}$ is used instead of the DFT during signal transformation, then the matrix $\mathbf{H}_F$, which describe the filter, shold be replaced by another matrix that is generally speaking not diagonal:

$$\mathbf{a} = \mathbf{T}^{-1}\mathbf{H}_T\mathbf{Ta} \quad . \tag{4.139}$$

If the matrix $\mathbf{H}_T$ is chosen carefully, the same filtering result can be achieved as with the use of DFT by (4.138). For this the following conditions must be fulfilled:

$$H_F = FT^{-1}H_TTF^{-1} = M^{T-F}\mathbf{H}_T\mathbf{M}^{F-T} \qquad \text{or} \tag{4.140a}$$

$$\mathbf{H}_T = \mathbf{T}\mathbf{F}^{-1}\mathbf{H}_F\mathbf{F}\mathbf{T}^{-1} = (\mathbf{M}^{T-F})^{-1}\mathbf{H}_F(\mathbf{M}^{F-T})^{-1} \quad , \quad \text{where} \tag{4.140b}$$

$$\mathbf{M}^{T-F} = \mathbf{F}\mathbf{T}^{-1} \quad , \quad \mathbf{M}^{F-T} = \mathbf{T}\mathbf{F}^{-1} \tag{4.140c}$$

are the matrices of the conversion from transform $\mathbf{T}$ to DFT and back.

This approach was proposed in [4.38-40] for the generalization of optimal linear (Wiener) filtering (see also [4.15]).

Depending on the form of the transform T and the properties of the required filter, the complexity of performing filtering (4.139), measured in terms of the number of operations say, can change. A fast Walsh-Hadamard transform may prove more convenient to use than the DFT, despite the greater complexity of multiplication by the non-diagonal filter matrix (Sect.7.5).

5 Linear Transform Algorithms

The discrete representation of linear transformations discussed in the preceeding chapter requires efficient algorithmic implementation. The various aspects of this problem form the subject of this chapter.

In Sect.5.1 the principal concepts of so-called fast algorithms are explained, and several theorems of matrix factorization are formulated and proved as a basis for matrix derivation of the fast algorithms. In Sects. 5.2-4 they are used to derive the fast Haar transform, fast Walsh transform, and fast Fourier transform algorithms in their most general matrix representation. A review of the fast algorithms of some other transforms, as well as of the peculiarities of 2-D transforms, is presented in Sect.5.5. Section 5.6 describes combined algorithms based on the use of the two- to fourfold redundancy of real and real even sequences. In Sect.5.7 the recursive method of calculating local Fourier spectra is presented, in Sect.5.8 some of the newest fast algorithms for the DFT and digital convolution are briefly explained.

5.1 Fast Algorithms of Discrete Orthogonal Transforms

The orthogonal transforms considered in the previous chapter are described by the general formula

$$\alpha_s = \sum_{k=0}^{N-1} a_k \varphi_s(k) \quad , \tag{5.1}$$

where $\{a_k\}$, $k = 0, 1, \ldots, N-1$ is the signal sample sequence, and $\{\varphi_s(k)\}$ are the transform basis functions.

If calculations are performed directly according to this formula, then N^2 multiplications and $(N-1)N$ additions must be performed to find all the N coefficients α_s (in the DFT the multiplication and addition of complex numbers). Even with moderate values of N, the number of operations can be very large. Therefore in the digital processing of signals and, in particular of pictures, which are usually defined by an enormous number of samples, the DFT and other orthogonal transforms became practicable only after the development of so-called "fast" transform algorithms for which the number of operations grows no faster than $N\log_2 N$ as N is increased.

The idea underlying these fast algorithms becomes clear if we consider 2-D separable transforms whose basis functions are the products of 1-D basis functions:

$$\alpha_{r,s} = \sum_{k=0}^{N_1-1} \sum_{\ell=0}^{N_2-1} a_{k,\ell}\varphi_r(k)\varphi_s(\ell) \quad . \tag{5.2}$$

Since they reduce to two 1-D transforms

$$\alpha_{r,s} = \sum_{k=0}^{N_1-1} \varphi_r(k) \sum_{\ell=0}^{N_2-1} a_{k,\ell}\varphi_s(\ell) \quad , \tag{5.3}$$

the number of operations is $N_2^2N_1 + N_1^2N_2 = N_1N_2(N_1 + N_2)$ instead of $(N_1N_2)^2$, as would have been the case if the 2-D transform had not been separable. Thus fast algorithms of orthogonal transforms are based on the possibility of representing 1-D transforms as separable multidimentional summations. There exist now many methods of describing such algorithms for different transforms [5.1-11].

The sections that follow present a unified approach to the derivation of fast algorithms, which is based on the representation of transform matrices in the form of layered Kronecker matrices (Sect.4.11). It is shown that transform matrices of order 2^n can be factorized, i.e., represented in the form of the product of n sparse matrices containing at most two elements differing from zero in each row. Factorized representations can also be derived for permutation matrices and for certain matrices that link the matrices of various transforms. The following definitions and theorems of matrix algebra are used.

Definition 1. The direct sum $\mathbf{M}_1 \boxplus \mathbf{M}_2$ of matrices $\mathbf{M}_1$ and $\mathbf{M}_2$ is the block matrix M of the form

$$\mathbf{M} = \mathbf{M}_1 \boxplus \mathbf{M}_2 = \begin{bmatrix} M_1 & 0 \\ 0 & M_2 \end{bmatrix} \quad . \tag{5.4}$$

The direct sum has the following properties, which are essential for what follows:

$$\mathbf{A}_1\mathbf{A}_2 \dots \mathbf{A}_n \boxplus B_1B_2 \dots B_n = (\mathbf{A}_1 \boxplus B_1)(\mathbf{A}_2 \boxplus B_2) \dots (\mathbf{A}_n \boxplus B_n) \quad ; \tag{5.5}$$

$$(A_1 \boxplus A_2 \boxplus \dots \boxplus A_n)^T = \mathbf{A}_1^T \boxplus A_2^T \boxplus \dots \boxplus A_n^T \quad , \tag{5.6}$$

where the superscript T denotes the transpose.

Definition 2 [5.2,5]. The right-hand, direct (Kronecker) product of matrices $\mathbf{M}^{(1)}_{r,s}$ and $\mathbf{M}^{(2)}_{p,q}$ is denoted by

$$\mathbf{M} = \mathbf{M}^{(1)}_{r,s} \otimes \mathbf{M}^{(2)}_{p,q} \quad , \tag{5.7}$$

and is composed of $r \times s$ submatrices, each of which is the product of the corresponding element of matrix $\mathbf{M}^{(1)}_{r,s}$ and the entire matrix $\mathbf{M}^{(2)}_{p,q}$.

The properties of the direct matrix product are:

$$(\mathrm{A} + \mathrm{B}) \otimes \mathbf{C} = \mathbf{A} \otimes \mathbf{C} + \mathrm{B} \otimes \mathbf{C} \quad ; \tag{5.8}$$

$$\mathbf{A} \otimes (\mathrm{B} + \mathrm{C}) = \mathbf{A} \otimes \mathrm{B} + \mathbf{A} \otimes \mathbf{C} \quad ; \tag{5.9}$$

$$(\mathbf{A}_1\mathbf{A}_2 \ldots \mathbf{A}_n) \otimes (\mathrm{B}_1\mathbf{B}_2 \ldots \mathbf{B}_n) = (\mathbf{A}_1 \otimes \mathrm{B}_1)\,(\mathbf{A}_2 \otimes \mathrm{B}_2) \ldots (\mathrm{A}_n \otimes \mathrm{B}_n) \quad ; \tag{5.10a}$$

$$(\mathrm{A}_1 \otimes \mathbf{A}_2 \otimes \ldots \otimes \mathbf{A}_n)\,(\mathrm{B}_1 \otimes \mathrm{B}_2 \otimes \ldots \otimes \mathrm{B}_n) = (\mathbf{A}_1\mathrm{B}_1) \otimes (\mathbf{A}_2\mathrm{B}_2) \otimes \ldots \otimes (\mathbf{A}_n\mathrm{B}_n) \quad . \tag{5.10b}$$

Theorem 1.

$$\mathop{\boxminus}_{k=0}^{N-1} \left(\mathbf{M}^{(k)}_{r_k s_k}\, \mathbf{N}^{(k)}_{s_k,q}\right) = \left(\mathop{\boxplus}_{k=0}^{n-1} \mathbf{M}^{(k)}_{r_k,s_k}\right)\left(\mathop{\boxminus}_{k=0}^{n-1} \mathrm{N}_{s_k,q}\right) \quad . \tag{5.11a}$$

The proof of this theorem is evident from Fig.5.1a.

Corollary 1. If $\mathbf{N}_{s_k,q} = \mathbf{I}_{s_k}$, where $\mathbf{I}_{s_k}$ is the identity matrix of order s_k, then

$$\mathop{\boxminus}_{k=0}^{n-1} \mathbf{M}^{(k)}_{r_k,s_k} = \left(\mathop{\boxplus}_{k=0}^{n-1} \mathbf{M}^{(k)}_{r_k,s_k}\right)\left(\mathop{\boxminus}_{k=0}^{n-1} \mathrm{I}_{s_k}\right) \quad . \tag{5.11b}$$

Theorem 2 (formulated and proved for square matrices by GOOD [5.5]):

$$\mathbf{M}_{r,s} \otimes \mathbf{N}_{p,q} = (\mathbf{M}_{r,s} \otimes \mathrm{I}_p)(\mathrm{I}_s \otimes \mathbf{N}_{p,q}) \quad . \tag{5.12}$$

The proof of this theorem is illustrated by Fig.5.1b.

Corollary 2. With $\mathrm{N}^{(k)}_{s_k,q} = \mathrm{N}_{s,q}$,

$$\mathop{\boxminus}_{k=0}^{n-1} \left(\mathrm{M}^{(k)}_{r_k,s}\, \mathbf{N}_{s,q}\right) = \left(\mathop{\boxminus}_{k=0}^{n-1} \mathbf{M}^{(k)}_{r_k,s}\right) \mathbf{N}_{s,q} \quad . \tag{5.13}$$

Corollary 2 follows immediately from Theorems 1 and 2 and Corollary 1. This relation can also be checked directly.

Corollary 3.

$$\mathop{\boxminus}_{k=0}^{n-1} \left(\mathbf{M}^{(k)}_{r_k,s_k} \otimes \mathbf{N}^{(k)}_{p_k,q_k}\right) = \left[\mathop{\boxplus}_{k=0}^{n-1} \left(\mathbf{M}^{(k)}_{r_k,s_k} \otimes \mathbf{I}_{p_k}\right)\right] \mathop{\boxminus}_{k=0}^{n-1} \left(\mathbf{I}_{s_k} \otimes \mathbf{N}^{(k)}_{p_k,q_k}\right) \quad . \tag{5.14}$$

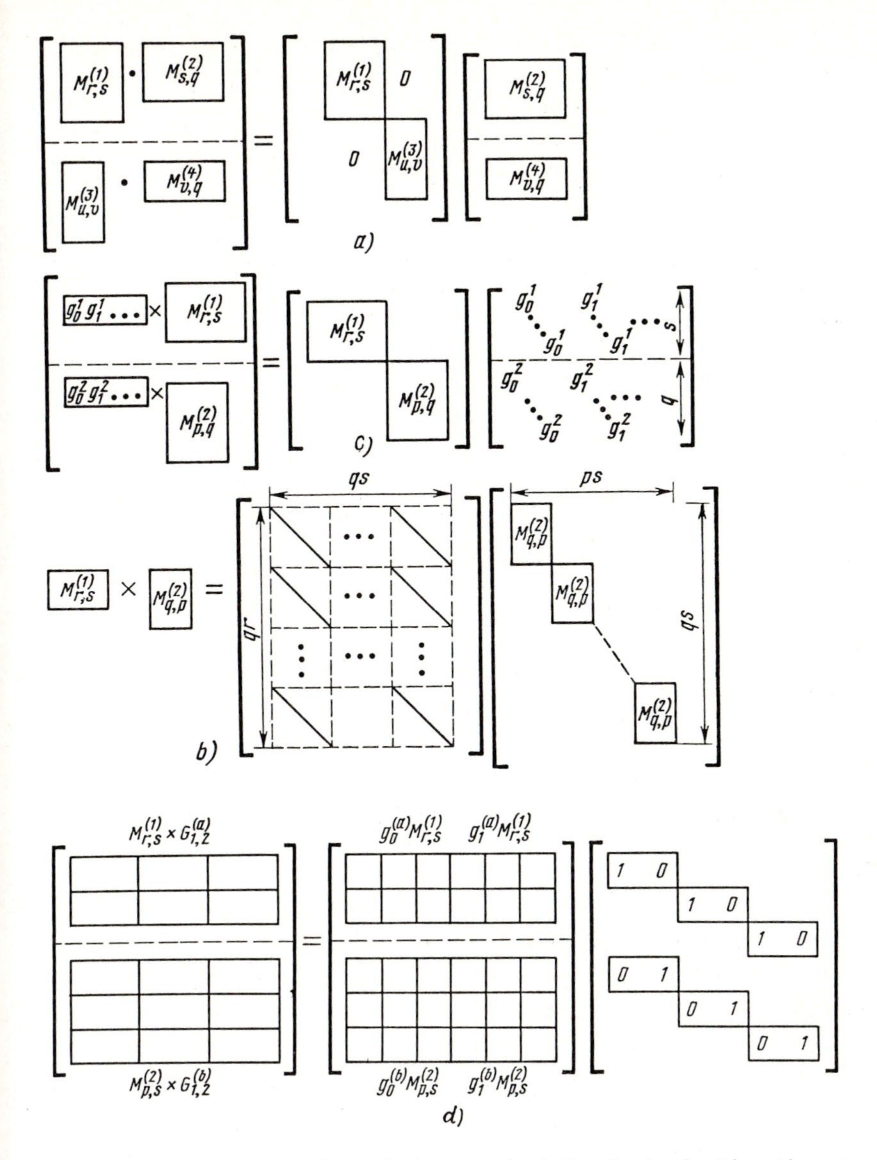

Fig. 5.1a-d. Illustration of the proof of the factorization theorem. (a) Theorem 1, (b) Theorem 2, (c) Theorem 3, (d) Theorem 4

Corollary 4. If $\mathbf{G}_{q_k}^{(k)}$ is a row matrix of q_k elements, then

$$\mathop{\boxminus}_{k=0}^{n-1} \left(\mathbf{M}_{r_k,s_k}^{(k)} \otimes \mathbf{G}_{q_k}^{(k)}\right) = \left[\mathop{\boxplus}_{k=0}^{n-1} \left(\mathbf{M}_{r_k,s_k}^{(k)}\right)\right]\left[\mathop{\boxminus}_{k=0}^{n-1} \left(\mathbf{I}_{s_k} \otimes \mathbf{G}_{q_k}^{(k)}\right]\right. \quad . \tag{5.15}$$

<u>Corollary 5.</u> If $\mathbf{G}_s^{(M),k}$ and $\mathbf{G}_q^{(N),\ell}$ are the kth and ℓth rows of matrices $\mathbf{M}_{r,s}$ and $\mathbf{N}_{p-q}$ respectively, then

$$\begin{aligned}
\mathbf{M}_{r,s} \otimes \mathbf{N}_{p,q} &= \left(\mathop{\boxminus}_{k=0}^{r-1} \mathbf{G}_s^{(M),k}\right) \otimes \left(\mathop{\boxminus}_{\ell=0}^{p-1} \mathbf{G}_q^{(N),\ell}\right) = \mathop{\boxminus}_{k=0}^{r-1} \left(\mathop{\boxminus}_{\ell=0}^{p-1} \mathbf{G}_s^{(M),k} \otimes \mathbf{G}_q^{(N),\ell}\right) \\
&= \mathop{\boxminus}_{k=0}^{r-1} \left(\mathop{\boxplus}_{\ell=0}^{p-1} \mathbf{G}_s^{(M),k}\right)\left[\mathop{\boxminus}_{\ell=0}^{p-1} \left(\mathbf{I}_s \otimes \mathbf{G}_q^{(N),\ell}\right)\right] = \left[\mathop{\boxminus}_{k=0}^{r-1} \left(\mathbf{I}_p \otimes \mathbf{G}_s^{(M),k}\right)\right] \\
&\times \left[\mathop{\boxminus}_{\ell=0}^{p-1} \left(\mathbf{I}_s \otimes \mathbf{G}_q^{(N),\ell}\right)\right] \quad .
\end{aligned} \tag{5.16}$$

In the transition here from the second to the third line, Theorem 1 is used for the vertical sum of matrices over ℓ. In the transition from the third to the fourth line, Corollary 2 for the vertical matrix sums over k is used, as well as the obvious relation for direct sums of similar matrices over ℓ.

For matrix rows the following theorems are also valid:

<u>Theorem 3.</u>

$$\mathop{\boxminus}_{k=0}^{n-1} \left(\mathbf{G}_{q_k}^{(k)} \otimes \mathbf{M}_{r_k,s_k}^{(k)}\right) = \left(\mathop{\boxplus}_{k=0}^{n-1} \mathbf{M}_{r_k,s_k}^{(k)}\right)\left[\mathop{\boxminus}_{k=0}^{n-1} \left(\mathbf{G}_{q_k}^{(k)} \otimes \mathbf{I}_{s_k}\right)\right] \quad . \tag{5.17}$$

<u>Theorem 4.</u>

$$\begin{aligned}
\left(\mathbf{M}_{r,s}^{(1)} \otimes \mathbf{G}_2^{(1)}\right) \boxminus \left(\mathbf{M}_{p,s}^{(2)} \otimes \mathbf{G}_2^{(2)}\right) &= \left[\left(\mathbf{G}_2^{(1)} \otimes \mathbf{M}_{r,s}^{(1)}\right) \boxminus \left(\mathbf{G}_2^{(2)} \otimes \mathbf{M}_{p,s}^{(2)}\right)\right] \\
&\times \left[\left(\mathbf{I}_s \otimes \mathbf{G}_2^{0}\right) \boxminus \left(\mathbf{I}_s \otimes \mathbf{G}_2^{1}\right)\right] \quad .
\end{aligned} \tag{5.18}$$

Figures 5.1c,d illustrate the proofs of Theorems 3 and 4.

Theorems 2 and 3, as well as Corollaries 3 - 5, are used to factorize matrices into products of sparse matrices, i.e., to construct fast algorithms for the multiplication of these matrices. Thus it is not difficult to show from Theorem 3 that the number of additions (subtractions) and multiplications for multiplication of the vector by the non-factorized matrix on the left-hand side of (5.17) is $\sum_{k=0}^{n-1} s_k(q_k r_k)$, whereas sequential multiplication of the vector by the matrix on the right-hand side of (5.17) requires $\sum_{k=0}^{n-1} q_k s_k + \sum_{k=0}^{n-1} r_k s_k = \sum_{k=0}^{n-1} s_k(q_k + r_k)$ operations; i.e., when $q_k r_k > 2$, the required number of operation decreases.

5.2 Fast Haar Transform (FHT) Algorithms

In accordance with Table 4.5, Haar transform matrices can be written as

$$\mathrm{HAR}_{2^n} = \mathop{\boxminus}_{k=0}^{n} \left[\left(2^{(k-1)/2}\, \mathbf{d}_0^{[k-1]} \otimes E_2^1\right)^{\bar{\delta}(k)} \otimes (E_2^0)^{[n-k]}\right] \quad . \tag{5.19}$$

Let us transform this matrix into a product of sparse matrices using the theorems of Sect.5.1. Note firstly that according to Theorem 1 one can extract from the matrix sum (5.19) the diagonal matrix

$$D_{2^n}^{\mathrm{HAR}} = 1 \boxplus \mathop{\boxplus}_{k=0}^{n-1} 2^{k/2}\, \mathbf{d}_0^{[k]} \quad , \tag{5.20}$$

so that

$$\mathbf{HAR}_{2^n} = 1/\, 2^{n/2}\, D_{2^n}^{\mathrm{HAR}} \mathop{\boxminus}_{k=0}^{n} \left[\left(\mathbf{d}_0^{[k-1]} \otimes E_2^1\right)^{\bar{\delta}(k)} \otimes (\mathbf{E}_2^0)^{[n-k]}\right] \quad . \tag{5.21}$$

To simplify future calculations, let us denote the vertical sum on the right-hand side of (5.21) as $\overline{\mathbf{HAR}}_{2^n}$. Let us separate in that sum the last term, for which $k = n$, and transform the remaining sum:

$$\begin{aligned}\overline{\mathbf{HAR}}_{2^n} &= \left\{\mathop{\boxminus}_{k=0}^{n-1} \left[\left(\mathbf{d}_0^{[k-1]} \otimes E_2^1\right)^{\bar{\delta}(k)} \otimes (E_2^0)^{[n-1-k]}\right]\right\} \otimes E_2^0 \boxminus \left(\mathbf{d}_0^{[n-1]} \otimes E_2^1\right) \\ &= \begin{bmatrix} \overline{\mathbf{HAR}}_{2^{n-1}} \otimes E_2^0 \\ \mathbf{d}_0^{[n-1]} \otimes E_2^1 \end{bmatrix} \quad .\end{aligned} \tag{5.22}$$

By applying Corollary 4 of Sect.5.1 for Kronecker submatrices in (5.22) and substituting

$$\mathbf{d}_0^{[n-1]} = \mathbf{I}_{2^{n-1}} \quad , \tag{5.23}$$

we obtain

$$\mathbf{HAR}_{2^n} = \left(\overline{\mathbf{HAR}}_{2^{n-1}} \boxplus \mathbf{I}_{2^{n-1}}\right) \begin{bmatrix} \mathbf{I}_{2^{n-1}} \otimes E_2^0 \\ \mathbf{I}_{2^{n-1}} \otimes E_2^1 \end{bmatrix} \quad . \tag{5.24}$$

Equation (5.24) is the recursive expression for Haar matrices. Using it to express $\mathbf{HAR}_{2^{n-2}}$, we obtain

$$\begin{aligned}\overline{\mathbf{HAR}}_{2^n} &= \left\{\left(\overline{\mathbf{HAR}}_{2^{n-2}} \boxplus \mathbf{I}_{2^{n-2}}\right) \begin{bmatrix} \mathbf{I}_{2^{n-2}} \otimes E_2^0 \\ \mathbf{I}_{2^{n-2}} \otimes E_2^1 \end{bmatrix} \boxplus \mathbf{I}_{2^{n-1}}\right\} \begin{bmatrix} \mathbf{I}_{2^{n-1}} \otimes E_2^0 \\ \mathbf{I}_{2^{n-1}} \otimes E_2^1 \end{bmatrix} \\ &= \left\{\left(\overline{\mathbf{HAR}}_{2^{n-2}} \boxplus \mathbf{I}_{2^{n-2}}\right) \begin{bmatrix} \mathbf{I}_{2^{n-2}} \otimes E_2^0 \\ \mathbf{I}_{2^{n-2}} \otimes E_2^1 \end{bmatrix} \boxplus \mathbf{I}_{2^{n-1}}\mathbf{I}_{2^{n-1}}\right\} \begin{bmatrix} \mathbf{I}_{2^{n-1}} \otimes E_2^0 \\ \mathbf{I}_{2^{n-1}} \otimes E_2^1 \end{bmatrix} \quad .\end{aligned} \tag{5.25}$$

For further transformation let us use the property (5.5) of the direct matrix sum, which in the present case tells us that

$$\overline{\mathbf{HAR}}_{2^n} = \left(\overline{\mathbf{HAR}}_{2^{n-2}} \boxplus \mathbf{I}_{2^{n-2}} \boxplus \mathbf{I}_{2^{n-1}}\right)\left(\begin{bmatrix} \mathbf{I}_{2^{n-2}} \otimes \mathbf{E}_2^0 \\ \mathbf{I}_{2^{n-2}} \otimes \mathbf{E}_2^1 \end{bmatrix} \boxplus \mathbf{I}_{2^{n-1}}\right)\begin{bmatrix} \mathbf{I}_{2^{n-1}} \otimes \mathbf{E}_2^0 \\ \mathbf{I}_{2^{n-1}} \otimes \mathbf{E}_2^1 \end{bmatrix} , \tag{5.26}$$

or, since

$$\mathbf{I}_{2^{n-2}} \boxplus \mathbf{I}_{2^{n-1}} = \mathbf{I}_{3\cdot 2^{n-2}} , \tag{5.27}$$

$$\overline{\mathbf{HAR}}_{2^n} = \left(\overline{\mathbf{HAR}}_{2^{n-2}} \boxplus \mathbf{I}_{3\cdot 2^{n-2}}\right)\left(\begin{bmatrix} \mathbf{I}_{2^{n-2}} \otimes \mathbf{E}_2^0 \\ \mathbf{I}_{2^{n-2}} \otimes \mathbf{E}_2^1 \end{bmatrix} \boxplus \mathbf{I}_{2^{n-1}}\right)\begin{bmatrix} \mathbf{I}_{2^{n-1}} \otimes \mathbf{E}_2^0 \\ \mathbf{I}_{2^{n-1}} \otimes \mathbf{E}_2^1 \end{bmatrix} . \tag{5.28}$$

Continuing in the same way we finally obtain

$$\overline{\mathbf{HAR}}_{2^n} = 1/2^{n/2}\, \mathbf{D}_{2^n}^{\mathrm{HAR}} \prod_{i=0}^{n-1} \left(\begin{bmatrix} \mathbf{I}_{2^i} \oplus \mathbf{E}_2^0 \\ \mathbf{I}_{2^i} \oplus \mathbf{E}_2^1 \end{bmatrix} \boxplus \mathbf{I}_{2^n - 2^{i+1}}\right) . \tag{5.29}$$

In this way the Haar matrix is represented as a product of n sparse matrices. In every i-th matrix of such a product there are 2^{i+1} rows with only two elements differing from zero, and $2^n - 2^{i+1}$ rows with only one element differing from zero. Therefore multiplication of the column matrix by such a matrix requires 2^{i+1} additions or subtractions. The total number of such operations is

$$N_{as} = \sum_{i=0}^{n-1} 2^{i+1} = 2(2^n - 1) . \tag{5.30}$$

Moreover, multiplication by the diagonal matrix $(1/2^{n/2})D_{2^n}^{\mathrm{HAR}}$ requires 2^n multiplications[1]. Thus the number of operations required for the fast Haar transform is proportional to the length of the series being transformed. The fast Haar transform is the fastest of all transforms used.

For the sake of visual clarity it is convenient to represent fast transform algorithms in the form of graphs, at whose nodes the initial signal samples and calculation results are distributed and whose edges link the summed values and the results of summation. The numbers along the edges indicate the coefficient by which the value in the node is multiplied; from this node the

---

[1] Half of these operations are multiplications by an integer power of 2, and can be replaced by shifts, which are faster operations than multiplication in the arithmetic registers of digital processors.

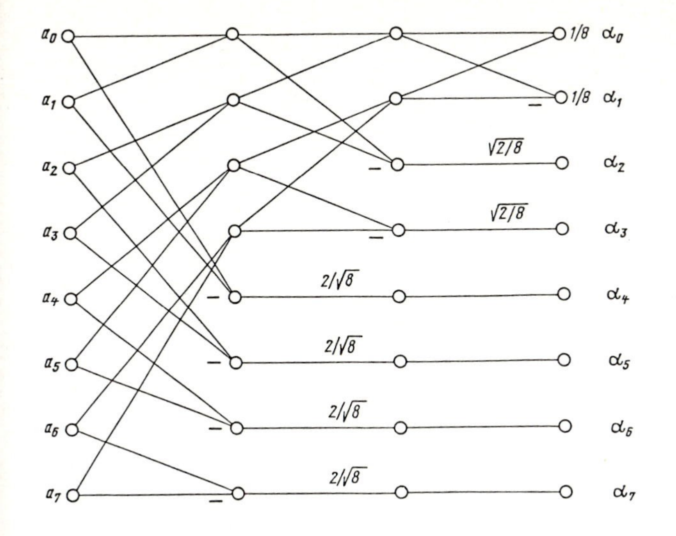

Fig. 5.2. Graph of the fast Haar transform algorithm

given edge is derived (units are not indicated). A fast Haar transform graph corresponding to (5.29) is shown in Fig.5.2 for $n = 3$.

The inverse Haar transform can be represented as the product of a transposed Haar matrix and the column matrix:

$$\mathrm{HAR}_{2^n}^{-1} = (\mathbf{HAR}_{2^n})^{T} \quad . \tag{5.31}$$

Since the relation

$$(\mathbf{M}_1\mathbf{M}_2 \ \dots \ \mathbf{M}_n)^{T} = \mathbf{M}_n^T\mathbf{M}_{n-1}^T \ \dots \ \mathbf{M}_2^T\mathbf{M}_1^T \tag{5.32}$$

is fulfilled for the transposition of the matrix product, and since (5.6) is fulfilled for the direct sum,

$$\mathbf{HAR}_{2^n}^{-1} = 1\,/\,2^{n/2}\ \mathbf{D}_{2^n}^{\mathrm{HAR}} \prod_{i=0}^{n-1} \left(\begin{bmatrix} \mathbf{I}_{2^{n-1-i}} \otimes \mathrm{E}_2^0 \\ \mathbf{I}_{2^{n-1-i}} \otimes \mathrm{E}_2^1 \end{bmatrix} \boxplus \mathbf{I}_{2^n-2^{n-1-i}} \right) \quad . \tag{5.33}$$

5.3 Fast Walsh Transform (FWT) Algorithms

The Hadamard matrix is a Kronecker matrix, and by using Theorem 3, Sect.5.1, it is not too difficult to obtain:

$$\mathbf{HAD}_{2^n} = 1\,/\,2^{n/2} \prod_{i=0}^{n-1} \left(\mathbf{I}_{2^{n-i-1}} \otimes \mathbf{h}_2 \otimes \mathbf{I}_{2^i} \right) \quad . \tag{5.34}$$

The transformation graph corresponding to (5.34) is shown in Fig.5.3 for $n = 3$. Its most important feature in comparison with, say, the Haar fast transform

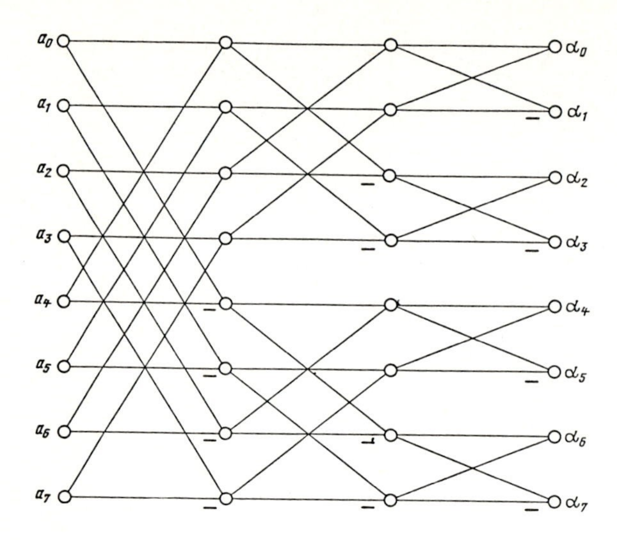

Fig. 5.3. Graph of the fast Hadamard transform algorithm

graph shown in Fig.5.2, is that it gives a so-called transform "in place"; at every stage of the transformation, the calculation is performed on a pair of values and the result is placed at the nodes on the same level as the initial values. Since the graph nodes correspond to the processor memory cells, this means that calculation does not require additional memory.

The Hadamard matrix is symmetrical, i.e., it does not change during transposition. The matrix factors $(\mathbf{I}_{2^{n-1-i}} \otimes \mathbf{h}_2 \otimes \mathbf{I}_{2^i})$ in its factorized representation (5.34) are also symmetrical. Therefore by transposing the matrices in (5.34), one can obtain the inverse version of the factorized representation:

$$\mathbf{HAD}_{2^n} = 1/2^{n/2} \prod_{i=0}^{n-1} \left(\mathbf{I}_{2^i} \otimes \mathbf{h}_2 \otimes \mathbf{I}_{2^{n-1-i}}\right) \quad . \tag{5.35}$$

The Walsh-Paley transform matrix $\mathbf{PAL}_{2^n}$ can be derived by multiplying the $\mathbf{HAD}_{2^n}$ matrices by the bit reversal matrix, see (4.121a), i.e.,

$$\mathbf{PAL}_{2^n} = 1/2^{n/2}\, \mathbf{M}^{rev}_{2^n} \prod_{i=0}^{n-1} \left(\mathbf{I}_{2^{n-i-1}} \otimes \mathbf{h}_2 \otimes \mathbf{I}_{2^i}\right) \quad . \tag{5.36}$$

On the other hand, from the representation of the PAL_{2^n} matrices in the form of the layered Kronecker matrices in Table 4.5,

$$\mathbf{PAL}_{2^n} = 1/2^{n/2} \mathop{\boxminus}_{i=0}^{n} \left[\left(\mathbf{PAL}_{2^{i-1}} \otimes \mathbf{E}_2^1\right)^{\overline{\delta}(i)} \otimes (\mathbf{E}_2^0)^{[n-i]}\right] \quad , \tag{5.37}$$

one can, by a procedure similar to the one used in Sect.5.2 for Haar matrices, obtain the factorized representation of $\mathbf{PAL}_{2^n}$ matrices, which does not require bit reversal in the final stage of transformation, see (5.34):

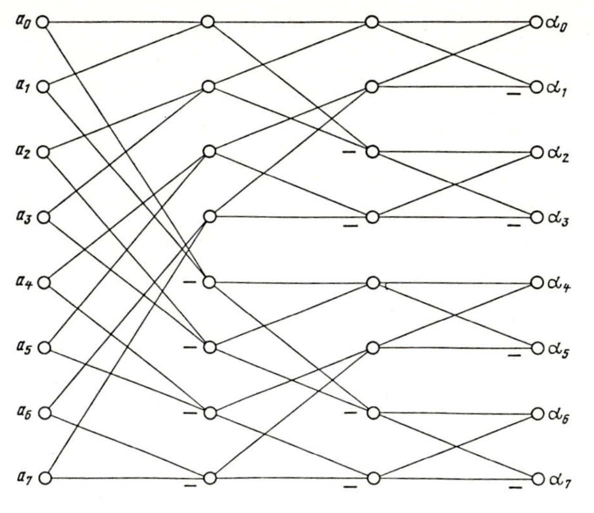

Fig. 5.4. Graph of the fast Paley transform algorithm

$$\mathbf{PAL}_{2^n} = 1/2^{n/2} \prod_{i=0}^{n-1} \left(\mathbf{I}_{2^{n-1-i}} \otimes \begin{bmatrix} \mathbf{I}_{2^i} \otimes \mathbf{E}_2^0 \\ \mathbf{I}_{2^i} \otimes \mathbf{E}_2^1 \end{bmatrix} \right) \quad . \tag{5.38}$$

Figure 5.4 shows the graph corresponding to such a representation for $n = 3$ (for comparison, see Fig.5.2).

Equation (5.37) makes it possible to obtain the factorized representation of bit reversal matrices. Indeed, by denoting

$$\overline{\mathbf{PAL}}_{2^n} = 2^{n/2}\mathbf{PAL}_{2^n} \quad ,$$

we can, as with (5.23 - 35), rewrite

$$\overline{\mathbf{PAL}}_{2^n} = \begin{bmatrix} \overline{\mathbf{PAL}}_{2^{n-1}} \otimes \mathrm{E}_2^0 \\ \overline{\mathbf{PAL}}_{2^{n-1}} \otimes \mathrm{E}_2^1 \end{bmatrix} \quad . \tag{5.39}$$

Let us now apply Theorem 4 of Sect.5.1:

$$\overline{\mathbf{PAL}}_{2^n} = \begin{bmatrix} \mathrm{E}_2^0 \otimes \overline{\mathbf{PAL}}_{2^{n-1}} \\ \mathrm{E}_2^1 \otimes \overline{\mathbf{PAL}}_{2^{n-1}} \end{bmatrix} \begin{bmatrix} \mathbf{I}_{2^{n-1}} \otimes \mathrm{G}_2^0 \\ \mathbf{I}_{2^{n-1}} \otimes \mathrm{G}_2^1 \end{bmatrix} \quad . \tag{5.40}$$

From the definition of the direct matrix product, the matrix $\overline{\mathbf{PAL}}_{2^{n-1}}$ can be factorized out from the first matrix (5.40). Since the resultant matrix $[\mathbf{E}_2^0 / \mathbf{E}_2^1]$ is, according to (4.135), equal to $\mathbf{h}_2$,

$$\overline{\mathbf{PAL}}_{2^n} = \left(\mathbf{h}_2 \otimes \overline{\mathbf{PAL}}_{2^{n-1}} \right) \begin{bmatrix} \mathbf{I}_{2^{n-1}} \otimes \mathrm{G}_2^0 \\ \mathbf{I}_{2^{n-1}} \otimes \mathrm{G}_2^1 \end{bmatrix} \quad , \tag{5.41}$$

or, according to Theorem 3 of Sect.5.1,

$$\overline{\mathbf{PAL}}_{2^n} = \left(\mathbf{h}_2 \otimes \mathbf{I}_{2^{n-1}}\right)\left(\mathbf{I}_2 \otimes \overline{\mathbf{PAL}}_{2^{n-1}}\right)\begin{bmatrix}\mathbf{I}_{2^{n-1}} \otimes \mathbf{G}_2^0 \\ \mathbf{I}_{2^{n-1}} \otimes \mathbf{G}_2^1\end{bmatrix} \quad . \tag{5.42}$$

This formula is the recursive expression for $\overline{\mathbf{PAL}}_{2^n}$ matrices, cf. (5.25). Using it and transforming the matrix product $\mathbf{I}_{2^i} \; \overline{\mathbf{PAL}}_{2^{n-i}}$, which is derived in the ith step of iteration, with the aid of (5.10),

$$\mathbf{I}_{2^i} \otimes \overline{\mathbf{PAL}}_{2^{n-i}} = \mathbf{I}_{2^i}\mathbf{I}_{2^i}\mathbf{I}_{2^i} \otimes (\mathbf{h}_2 \otimes \mathbf{I}_{2^{n-1-i}})(\mathbf{I}_2 \otimes \overline{\mathbf{PAL}}_{2^{n-1-i}}) \begin{bmatrix}\mathbf{I}_{2^{n-1-i}} \otimes \mathbf{G}_2^0 \\ \mathbf{I}_{2^{n-1-i}} \otimes \mathbf{G}_2^1\end{bmatrix}$$

$$= (\mathbf{I}_{2^i} \otimes \mathbf{h}_2 \otimes \mathbf{I}_{2^{n-1-i}})(\mathbf{I}_{2^{i+1}} \otimes \overline{\mathbf{PAL}}_{2^{n-1-i}}) \left(\mathbf{I}_{2^i} \otimes \begin{bmatrix}\mathbf{I}_{2^{n-1-i}} \otimes \mathbf{G}_2^0 \\ \mathbf{I}_{2^{n-1-i}} \otimes \mathbf{G}_2^1\end{bmatrix}\right) , \tag{5.43}$$

we obtain

$$\mathbf{PAL}_{2^n} = 1/2^{n/2} \prod_{i=0}^{n-1} (\mathbf{I}_{2^i} \otimes \mathbf{h}_2 \otimes \mathbf{I}_{2^{n-1-i}}) \prod_{i=0}^{n-1} \left(\mathbf{I}_{2^{n-1-i}} \otimes \begin{bmatrix}\mathbf{I}_{2^i} \otimes \mathbf{G}_2^0 \\ \mathbf{I}_{2^i} \otimes \mathbf{G}_2^1\end{bmatrix}\right) \quad . \tag{5.44}$$

By comparing this expression with (5.35,4.121c), we can conclude that the second product in (5.44) is the factorized representation of the bit reversal matrix:

$$\mathbf{M}_{2^n}^{\mathrm{rev}} = \prod_{i=0}^{n-1} \left(\mathbf{I}_{2^{n-i-1}} \otimes \begin{bmatrix}\mathbf{I}_{2^i} \otimes \mathbf{G}_2^0 \\ \mathbf{I}_{2^i} \otimes \mathbf{G}_2^1\end{bmatrix}\right) \quad . \tag{5.45}$$

The graph of bit reversal corresponding to (5.45) is shown in Fig.5.5 for $n = 3$. Such a factorized representation of the bit reversal matrix can be used when the processor instruction system has no convenient instructions for manipulating individual binary digits which are essential for the direct organization of permutations.

These considerations are also applicable to the matrix $\mathbf{M}_{2^n}^{G/D}$ of permutation from the Gray code to direct binary code. This matrix is essential for the transfer from Hadamard to Walsh matrices:

$$\mathbf{WAL}_{2^n} = \mathbf{M}_{2^n}^{G/D}\mathbf{PAL}_{2^n} = 1/2^{n/2}\mathbf{M}_{2^n}^{G/D}\mathbf{M}_{2^n}^{\mathrm{rev}} \prod_{i=0}^{n-1} (\mathbf{I}_{2^{n-i-1}} \otimes \mathbf{h}_2 \otimes \mathbf{I}_{2^i}) \quad . \tag{5.46}$$

Let us find the factorized form of matrix $\mathbf{M}_{2^n}^{G/D}$ from the representation of $\mathbf{WAL}_{2^n}$ in the form of the layered Kronecker matrices in Table 4.5. To do this, let us introduce the notation

$$\overline{\mathbf{WAL}}_{2^n} = 2^{n/2}\mathbf{WAL}_{2^n} \quad ; \quad \bar{\mathbf{I}}_{2^k} = (\bar{\mathbf{d}}_0)^{[k]} \quad , \tag{5.47}$$

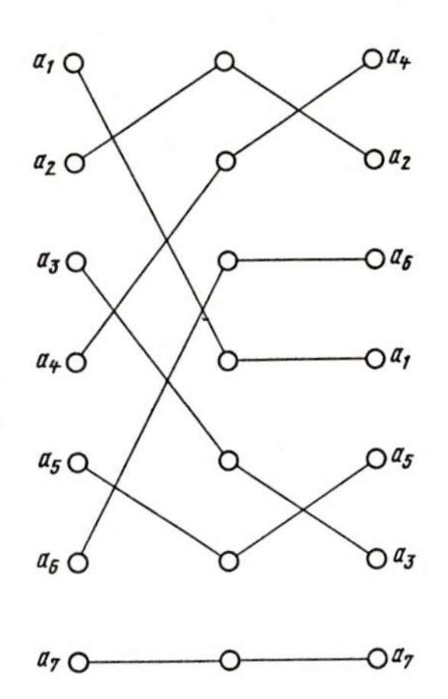

Fig. 5.5. Graph of the bit reversal algorithm

and write

$$\overline{\mathbf{WAL}}_{2^n} = \begin{bmatrix} \overline{\mathbf{WAL}}_{2^{n-1}} \otimes E_2^0 \\ \bar{\mathbf{I}}_{2^{n-1}} \overline{\mathbf{WAL}}_{2^{n-1}} \otimes E_2^1 \end{bmatrix} \quad . \tag{5.48}$$

Further, from Theorems 1 and 2 of Sect.5.1 we have

$$\overline{\mathbf{WAL}}_{2^n} = \begin{bmatrix} \overline{\mathbf{WAL}}_{2^{n-1}}(\mathbf{I}_{2^{n-1}} \otimes E_2^0) \\ \bar{\mathbf{I}}_{2^{n-1}} \overline{\mathbf{WAL}}_{2^{n-1}}(\mathbf{I}_{2^{n-1}} \otimes E_2^1) \end{bmatrix} = \left(\overline{\mathbf{WAL}}_{2^{n-1}} \boxplus \bar{\mathbf{I}}_{2^{n-1}}\overline{\mathbf{WAL}}_{2^{n-1}}\right) \begin{bmatrix} \mathbf{I}_{2^{n-1}} \otimes E_2^0 \\ \mathbf{I}_{2^{n-1}} \otimes E_2^1 \end{bmatrix}$$

$$= \left(\mathbf{I}_{2^{n-1}}\overline{\mathbf{WAL}}_{2^{n-1}} \boxplus \bar{\mathbf{I}}_{2^{n-1}}\mathbf{WAL}_{2^{n-1}}\right) \begin{bmatrix} \mathbf{I}_{2^{n-1}} \otimes E_2^0 \\ \mathbf{I}_{2^{n-1}} \otimes E_2^1 \end{bmatrix} = \left(\mathbf{I}_{2^{n-1}} \boxplus \bar{\mathbf{I}}_{2^{n-1}}\right)$$

$$\times \left(\overline{\mathbf{WAL}}_{2^{n-1}} \boxplus \overline{\mathbf{WAL}}_{2^{n-1}}\right) \begin{bmatrix} \mathbf{I}_{2^{n-1}} \otimes E_2^0 \\ \mathbf{I}_{2^{n-1}} \otimes E_2^1 \end{bmatrix} = \left(\mathbf{I}_{2^{n-1}} \boxplus \bar{\mathbf{I}}_{2^{n-1}}\right)\left(\mathbf{I}_2 \otimes \overline{\mathbf{WAL}}_{2^{n-1}}\right)$$

$$\times \begin{bmatrix} \mathbf{I}_{2^{n-1}} \otimes E_2^0 \\ \mathbf{I}_{2^{n-1}} \otimes E_2^1 \end{bmatrix} \quad . \tag{5.49}$$

Equation (5.49) is the recursive expression for the matrices $\overline{\mathbf{WAL}}_{2^n}$, cf. (5.25). Similarly, by transforming (5.49) as with (5.43,44) we obtain

$$\mathbf{WAL}_{2^n} = 1/2^{n/2} \prod_{i=0}^{n-1} \left(\mathbf{I}_{2^i} \otimes \left[\mathbf{I}_{2^{n-1-i}} \boxplus \bar{\mathbf{I}}_{2^{n-1-i}}\right]\right) \prod_{i=0}^{n-1} \left(\mathbf{I}_{2^{n-1-i}} \otimes \begin{bmatrix} \mathbf{I}_{2^i} \otimes E_2^0 \\ \mathbf{I}_{2^i} \otimes E_2^1 \end{bmatrix}\right) = \prod_{i=0}^{n-1} \left(\mathbf{I}_{2^i} \otimes \left[\mathbf{I}_{2^{n-1-i}} \boxplus \bar{\mathbf{I}}_{2^{n-1-i}}\right]\right) \mathbf{PAL}_{2^n} \quad . \tag{5.50}$$

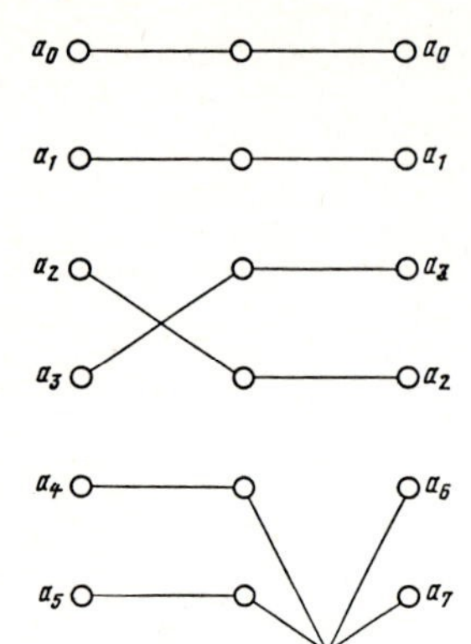

Fig. 5.6. Graph of the permutation algorithm from Gray code to direct binary code

Taking into consideration the connection between WAL_{2^n} and PAL_{2^n} matrices (4.121), we can conclude that

$$\prod_{i=0}^{n-1}\left(\mathbf{I}_{2^i} \otimes \left[\mathbf{I}_{2^{n-1-i}} \boxplus \overline{\mathbf{I}}_{2^{n-1-i}}\right]\right) = \mathbf{M}_{2^n}^{G/D} \quad . \tag{5.51}$$

Thus (5.51) gives the factorized representation of the permutation matrices $\mathbf{M}_{2^n}^{G/D}$. The corresponding permutation graph is shown in Fig.5.6 for $n = 3$.

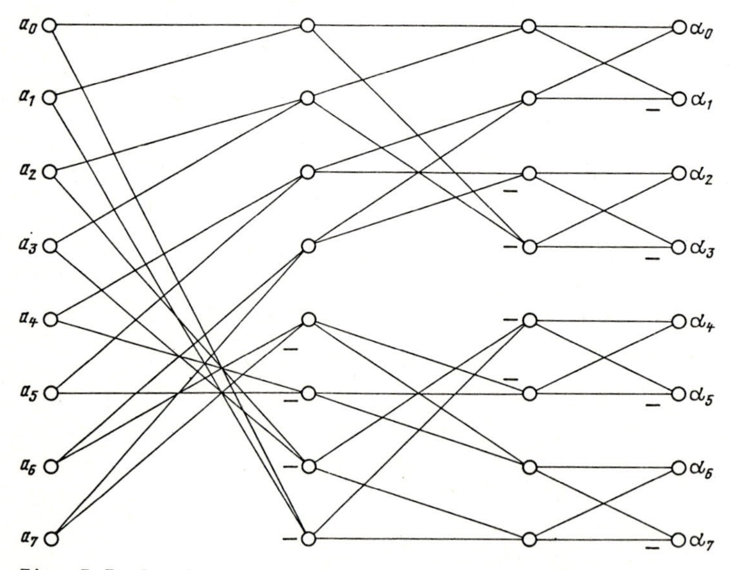

Fig. 5.7. Graph of the Walsh transform algorithm

The matrix $\mathbf{M}_{2^n}^{G/D}$ can be combined with the matrices composing the factorized representation of matrix PAL_{2^n}, and the factorized representation of Walsh transform matrices, which does not require permutation, can be derived:

$$\mathbf{WAL}_{2^n} = 1/2^{n/2} \left\{ \prod_{i=0}^{n-2} \left(\mathbf{I}_{2^{n-i-2}} \otimes \left(\begin{bmatrix} \mathbf{I}_{2^i} \otimes \mathbf{E}_2^0 \\ \bar{\mathbf{I}}_{2^i}(\mathbf{I}_{2^i} \otimes \mathbf{E}_2^1) \end{bmatrix} \oplus \bar{\mathbf{I}}_{2^{i+1}} \begin{bmatrix} \mathbf{I}_{2^i} \otimes \mathbf{E}_2^0 \\ \bar{\mathbf{I}}_{2^i}(\mathbf{I}_{2^i} \otimes \mathbf{E}_2^1) \end{bmatrix} \bar{\mathbf{I}}_{2^{i+1}} \right) \right) \right\} \begin{bmatrix} \mathbf{I}_{2^{n-1}} \otimes \mathbf{E}_2^0 \\ \bar{\mathbf{I}}_{2^{n-1}}(\mathbf{I}_{2^{n-1}} \otimes \mathbf{E}_2^1) \end{bmatrix} . \tag{5.52}$$

The Walsh transform graph corresponding to (5.52) is shown in Fig.5.7 for $n = 3$.

5.4 Fast Discrete Fourier Transform (FFT) Algorithms

The FFT matrix $\mathbf{F}_{2^n} = (1/2^{n/2})\{W_n^{ks}\}$, where $w_n = \exp(2\pi i / 2^n)$, can be easily represented as a layered Kronecker matrix (Table 4.5) if the number of row elements in the matrix is written as a binary number. Indeed,[2] with $s = \sum_{m=0}^{n-1} s_m 2^m$,

$$\mathbf{F}_{2^n} = \left\{ w_n^{k \sum_{m=0}^{n-1} s_m 2^m} \right\} = \left\{ \prod_{m=0}^{n-1} w^{k s_m 2^m} \right\} = \mathop{\boxminus}_{k=0}^{2^{n-1}} \bigotimes_{m=0}^{n-1} \begin{bmatrix} 1 & w_n^{2^m k} \end{bmatrix} . \tag{5.53}$$

To factorize this matrix we write out the Kronecker factor in (5.53) corresponding to $m = 0$ explicitly, and apply Theorem 3, Sect.5.1, to the derived matrix:

$$\mathbf{F}_{2^n} = \mathop{\boxminus}_{k=0}^{2^n-1} \left\{ \left[\bigotimes_{m=1}^{n-1} \begin{pmatrix} 1 & w_n^{2^m k} \end{pmatrix} \right] \otimes \begin{pmatrix} 1 & w_n^k \end{pmatrix} \right\} = \left[\mathop{\boxplus}_{k=0}^{2^n-1} \begin{pmatrix} 1 & w_n^k \end{pmatrix} \right] \left\{ \mathop{\boxminus}_{k=0}^{2^n-1} \left[\left(\bigotimes_{m=1}^{n-1} \begin{bmatrix} 1 & w_n^{2^m k} \end{bmatrix} \right) \otimes \mathbf{I}_2 \right] \right\} . \tag{5.54}$$

By dividing the vertical matrix sum here into two parts in which k runs from 0 to 2^{n-1}-I and from 2^{n-1} to 2^n-1, and noting that with $m > 1$, $w_n^{2^m 2^{n-1}} = I$, we obtain

$$\mathbf{F}_{2^n} = \left[\mathop{\boxplus}_{k=0}^{2^n-1} (1 \quad w_n^k) \right] \left\{ \mathop{\boxminus}_{k=0}^{2^{n-1}-1} \left[\left(\bigotimes_{m=1}^{n-1} [1 \quad w_n^{2^m k}] \right) \otimes \mathbf{I}_2 \right] \boxminus \mathop{\boxminus}_{k=0}^{2^n-1} \left[\left(\bigotimes_{m=1}^{n-1} \right. \right. \right.$$

---

[2] For simplicity of presentation we will omit the factor $2^{-n/2}$ for the time being.

$$\times\left[1\quad w_n^{2^m(2^{n-1}+k)}\right]\right)\otimes \mathbf{I}_2\Big]\Big\} = \left[\mathop{\boxplus}_{k=0}^{2^n-1}(1\quad w_n^k)\right]\left\{(\mathbf{E}_2^0)^T\otimes\left[\mathop{\boxminus}_{k=0}^{2^{n-1}-1}\right.\right.$$

$$\left.\left.\left(\left[\mathop{\otimes}_{m=1}^{n-1}(1\quad w_n^{2^m k})\right]\otimes \mathbf{I}_2\right)\right]\right\} \quad , \tag{5.55}$$

where $(E_2^0)^T$ is the transposed row vector E_2^0 (4.128). Applying Theorem 2, Sect.5.1 to the right-hand factor, we obtain

$$F_{2^n} = \left[\mathop{\boxplus}_{k=0}^{2^n-1}(1\quad w_n^k)\right]\left[(\mathbf{E}_2^0)^T\otimes \mathbf{I}_{2^n}\right]\left\{\mathop{\boxminus}_{k=0}^{2^{n-1}-1}\left[\left(\mathop{\otimes}_{m=1}^{n-1}(1\quad w_n^{2^m k})\right)\otimes \mathbf{I}_2\right]\right\}$$

$$= \left[\mathop{\boxplus}_{k=0}^{2^{n-1}-1}(1\quad w_n^k)\boxplus\mathop{\boxplus}_{k=0}^{2^{n-1}-1}(1\quad w_n^{2^{n-1}+k})\right]\left(\mathbf{I}_{2^n}\boxminus \mathbf{I}_{2^n}\right)$$

$$\times\left\{\mathop{\boxminus}_{k=0}^{2^{n-1}-1}\left[\left(\mathop{\otimes}_{m=1}^{n-1}(1\quad w_n^{2^m k})\right)\otimes \mathbf{I}_2\right]\right\} \quad . \tag{5.56}$$

From Corollary 1, Sect.5.1, it then follows for the first two factors, that

$$F_{2^n} = \left\{\mathop{\boxminus}_{p=0}^{1}\left[\mathop{\boxplus}_{k=0}^{2^{n-1}-1}[1\ (-1)^p w_n^k]\right]\right\}\left\{\mathop{\boxminus}_{k=0}^{2^{n-1}-1}\left[\left(\mathop{\otimes}_{m=1}^{n-1}(1\quad w_n^{2^m k})\right)\otimes \mathbf{I}_2\right]\right\} \quad , \tag{5.57}$$

where the identity $w_n^{2^{n-1}} = -1$ is used.

Continuing in the same manner with respect to the extreme right-hand factor, we obtain

$$F_{2^n} = \prod_{i=0}^{n-1}\left\{\mathop{\boxminus}_{p=0}^{1}\left[\mathop{\boxplus}_{k=0}^{2^{n-1}-1}\left([1\ (-1)^p w_n^{2^i k}]\right)\otimes \mathbf{I}_{2^i}\right]\right\} \quad . \tag{5.58}$$

This formula is the matrix expression for one of the fast Fourier transform algorithms. Indeed, in accordance with it, matrix F_{2^n} is represented in the form of the product of n matrices having only two non-zero elements in each row, i.e., requiring no more than 2×2^n multiplications and 2^n additions during multiplication by the vector.

From the point of view of FFT programming and for purposes of facilitating the comparison of FFT algorithms with fast algorithms of other transforms, it is convenient to transform (5.58) so that the elementary matrices $\mathbf{E}_2^0$, $\mathbf{E}_2^I$, and $\mathbf{d}_{2^i k}$ appear (4.128):

$$F_{2^n} = \prod_{i=0}^{n-1}\left\{\mathop{\boxminus}_{p=0}^{1}\left[\mathop{\boxplus}_{k=0}^{2^{n-1-i}-1}\left([1\ (-1)^p]\begin{bmatrix}1 & 0\\ 0 & w^{2^i k}\end{bmatrix}\right)\otimes \mathbf{I}_{2^i}\right]\right\}$$

$$= \prod_{i=0}^{n-1}\left\{\mathop{\boxminus}_{p=0}^{1}\left[\mathop{\boxplus}_{k=0}^{2^{n-1-i}-1}\left(\mathbf{E}_2^p\,\mathbf{d}_{2^i k}\right)\otimes \mathbf{I}_{2^i}\right]\right\} \quad , \tag{5.59}$$

or, from the properties (5.10,5) of the direct product and the direct sum of matrices:

$$\mathbf{F}_{2^n} = \prod_{i=0}^{n-1} \left\{ \mathop{\boxminus}_{p=0}^{1} \left[\mathop{\boxplus}_{k=0}^{2^{n-1-i}-1} \left(\mathbf{E}_2^p \otimes \mathbf{I}_{2^i}\right)\left(\mathbf{d}_{2^i k} \otimes \mathbf{I}_{2^i}\right)\right]\right\}$$

$$= \prod_{i=0}^{n-1} \left\{ \mathop{\boxminus}_{p=0}^{1} \left[\mathop{\boxplus}_{k=0}^{2^{n-1-i}-1} \left(\mathbf{E}_2^p \otimes \mathbf{I}_{2^i}\right)\right] \left[\mathop{\boxplus}_{k=0}^{2^{n-1-i}-1} \left(\mathbf{d}_{2^i k} \otimes \mathbf{I}_{2^i}\right)\right]\right\} \quad . \tag{5.60}$$

Since the second factor in the vertical sum over p does not depend on p, then according to Corollary 2, Sect.5.1:

$$\mathbf{F}_{2^n} = \prod_{i=0}^{n-1} \left\{ \mathop{\boxminus}_{p=0}^{1} \left[\mathop{\boxplus}_{k=0}^{2^{n-1-i}-1} \left(\mathbf{E}_2^p \otimes \mathbf{I}_{2^i}\right)\right]\right\} \left[\mathop{\boxplus}_{k=0}^{2^{n-1-i}-1} \left(\mathbf{d}_{2^i k} \otimes \mathbf{I}_{2^i}\right)\right] \quad . \tag{5.61}$$

Finally, by replacing the direct sum of equal terms in the first factor with the Kronecker product having an identity matrix, we obtain, reinserting the factor $2^{-n/2}$:

$$\mathbf{F}_{2^n} = 2^{-n/2} \prod_{i=0}^{n-1} \left[\mathop{\boxminus}_{p=0}^{1} \left(\mathbf{I}_{2^{n-1-i}} \otimes \mathbf{E}_2^p \otimes \mathbf{I}_{2^i}\right)\right] \left[\mathop{\boxplus}_{k=0}^{2^{n-1-i}-1} \left(\mathbf{d}_{2^i k} \otimes \mathbf{I}_{2^i}\right)\right] \quad . \tag{5.62}$$

In this expression the first matrix in every ith of n pairs of matrices contains only 0 and 1. This means that in vector multiplication, the operation of multiplication is not required, and the second matrix is a diagonal matrix with $(2^{n-1-i}-1)2^i = 2^{n-1}-2^i$ elements differing from 1. This matrix determines the number of multiplications in the ith step of the transformation. The total number of additions, N_{ad}, and multiplication, N_m, of complex numbers for this FFT algorithm is

$$N_{ad} = n2^n \quad ;$$

$$N_m = \prod_{i=0}^{n-1} (2^{n-1} - 2^i) = 2^{n-1}(n-2) + 1 \quad . \tag{5.63}$$

The FFT algorithm graph corresponding to (5.62) is shown in Fig.5.8a for $n = 3$. The numbers in circles are factors of the argument of the exponential function, $\exp(2\pi i / 8)$.

At present there is a wide diversity of FFT algorithms. The best known of these involve a permutation of the elements of the sequence under transformation or its Fourier spectrum according to the binary inverted numbers of the elements. In contrast with (5.53), these algorithms correspond to the recursive representation of the matrix $\mathbf{F}_{2n}$, which is analogous to the representation of the matrices $\mathbf{MHAD}_{2n}$ and $\mathbf{PAL}_{2n}$ (Table 4.5). It is not difficult to obtain

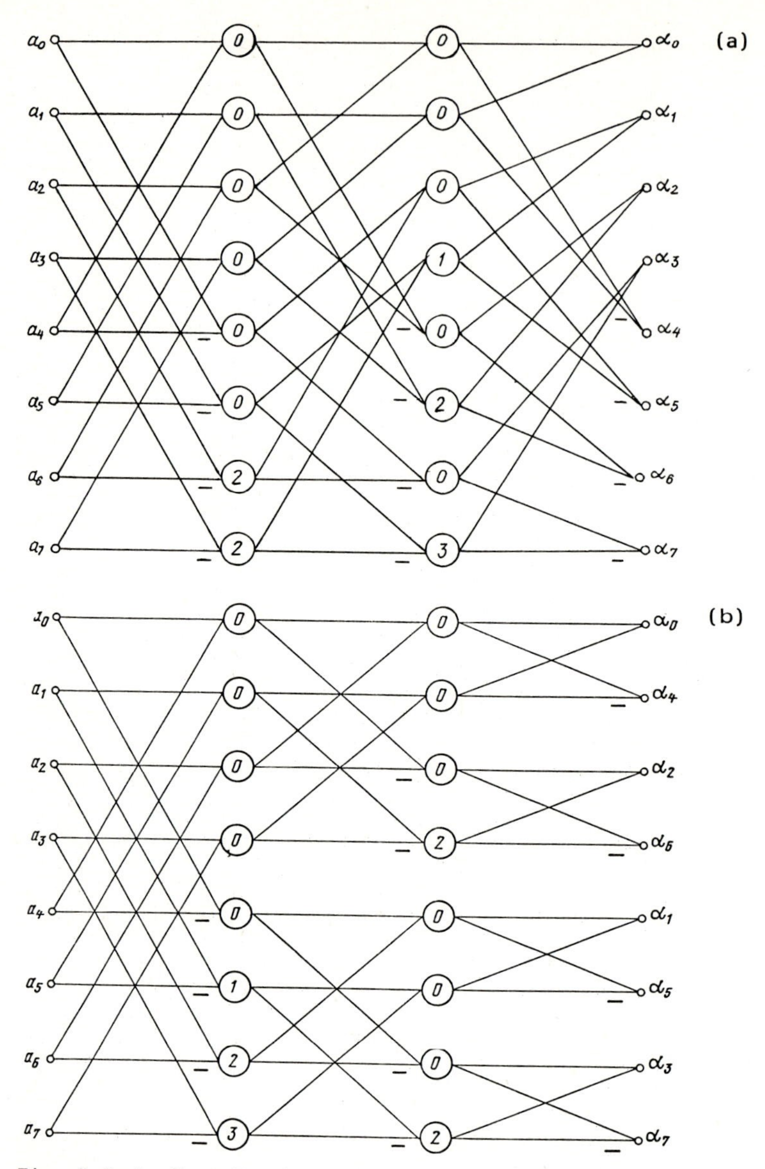

Fig. 5.8a,b. Fast Fourier transform algorithms: (a) algorithm without permutation according to (5.62), (b) "in-place" algorithm

this representation from (5.53) if the vertical sum over k in (5.53) is replaced by a multidimensional sum over the values of the binary digits K_q of the binary code for k, and if it is noted that reversing the order of summation over K_q corresponds to binary inversion of the matrix rows:

$$
\begin{aligned}
\mathbf{F}_{2^n} &= \mathop{\boxminus}_{k=0}^{2^n-1} \Bigg[\bigotimes_{m=0}^{n-1} \Big(1 \quad w_n^{2^m k}\Big) \Bigg] = \mathop{\boxminus}_{k_{n-1}=0}^{1} \mathop{\boxminus}_{k_{n-2}=0}^{1} \dots \mathop{\boxminus}_{k_0=0}^{1} \Bigg[\bigotimes_{m=0}^{n-1} \Big(1 \quad w^{2^m k}\Big) \Bigg] \\
&= \mathbf{M}_{2^n}^{\mathrm{rev}} \Bigg\{ \mathop{\boxminus}_{k_0=0}^{1} \dots \mathop{\boxminus}_{k_{n-1}=0}^{1} \Bigg[\bigotimes_{m=0}^{n-1} \Big(1 \quad w_n^{2^m k}\Big) \Bigg] \Bigg\} = \mathbf{M}_{2^n}^{\mathrm{rev}} \Bigg\{ \mathop{\boxminus}_{k_1=0}^{1} \dots \mathop{\boxminus}_{k_{n-1}=0}^{1} \\
&\times \Bigg[\bigotimes_{m=0}^{n-1} \Big(1 \quad w_n^{2^m [2(k/2)]}\Big) \Bigg] \boxminus \mathop{\boxminus}_{k_1=0}^{1} \dots \mathop{\boxminus}_{k_{n-1}=0}^{1} \Bigg[\bigotimes_{m=0}^{n-1} \Big(1 \quad w_n^{2^m\, 2(k/2)+1}\Big) \Bigg] \Bigg\} \\
&= \mathbf{M}_{2^n}^{\mathrm{rev}} \Bigg\{ \mathop{\boxminus}_{(k/2)=0}^{2^{n-1}-1} \Bigg[\Big(1 \quad w_n^{2^{n-1}[2(k/2)]}\Big) \otimes \bigotimes_{m=0}^{n-2} \Big(1 \quad w_{n-1}^{2^m (k/2)}\Big) \Bigg] \boxminus \mathop{\boxminus}_{(k/2)=0}^{2^{n-1}-1} \\
&\times \Bigg[\Big(1 \quad w_n^{2^{n-1}[2(k/2)+1]}\Big) \otimes \bigotimes_{m=0}^{n-2} \Big(1 \quad w_{n-1}^{2^m (k/2)} w_n^{2^m}\Big) \Bigg] \Bigg\} = \mathbf{M}_{2^n}^{\mathrm{rev}} \Bigg(\mathop{\boxminus}_{(k/2)=0}^{2^{n-1}-1} \\
&\times \Bigg[[1 \;\; 1] \otimes \bigotimes_{m=0}^{n-2} \Big(1 \quad w_{n-1}^{2^m (k/2)}\Big) \Bigg] \boxminus \mathop{\boxminus}_{(k/2)=0}^{2^{n-1}-1} \Bigg\{ [1 \;\; -1] \otimes \bigotimes_{m=0}^{n-2} \Bigg[\Big(1 \quad w_{n-1}^{2^m (k/2)}\Big) \mathbf{d}_{2^m} \Bigg] \Bigg\} \Bigg) \\
&= \mathbf{M}_{2^n}^{\mathrm{rev}} \Bigg\{ \Bigg[\mathbf{E}_2^0 \otimes \Bigg(\mathop{\boxminus}_{(k/2)=0}^{2^{n-1}-1} \bigotimes_{m=0}^{n-2} \Big[1 \quad w_{n-1}^{2^m (k/2)}\Big] \Bigg) \Bigg] \boxminus \Bigg[\mathbf{E}_2^1 \otimes \Bigg(\mathop{\boxminus}_{(k/2)=0}^{2^{n-1}-1} \bigotimes_{m=0}^{n-2} \\
&\times \Big[1 \quad w_{n-1}^{2^m (k/2)}\Big] \Bigg) \Bigg(\bigotimes_{m=0}^{n-2} \mathbf{d}_{2^m} \Bigg) \Bigg] \Bigg\} = \mathbf{M}_{2^n}^{\mathrm{rev}} \Bigg\{ \Bigg[\mathbf{E}_2^0 \otimes \Big(\mathbf{M}_{2^{n-1}}^{\mathrm{rev}} \, \mathbf{F}_{2^{n-1}} \Big) \Bigg] \\
&\boxminus \Bigg[\mathbf{E}_2^1 \otimes \Big(\mathbf{M}_{2^{n-1}}^{\mathrm{rev}} \, \mathbf{F}_{2^{n-1}} \Big) \Bigg(\bigotimes_{m=0}^{n-2} \mathrm{d}_{2^m} \Bigg) \Bigg] \Bigg\} \quad . \qquad (5.64)
\end{aligned}
$$

Here, $(k/2)$ denotes the largest integer less than $k/2$; the identities $w_n^{2^n} = I$ and $w_n^{2^{n-1}} = -1$ are used; and the factor $2^{-n/2}$ is omitted.

By using the same approach in (5.64) as in the derivation of the Haar and Paley fast transform algorithms, we obtain the following version of the FFT algorithm:

$$
\mathbf{F}_{2^n} = 2^{-n/2} \, \mathbf{M}_{2^n}^{\mathrm{rev}} \prod_{i=0}^{n-1} \Bigg[\mathbf{I}_{2^{n-1-i}} \otimes \Bigg(\mathbf{I}_{2^i} \boxplus \bigotimes_{m=0}^{n-2} \mathbf{d}_{2^{m+2}} \Bigg) \Bigg] \Big(\mathbf{I}_{2^{n-1-i}} \otimes \mathbf{h}_2 \otimes \mathbf{I}_{2^i} \Big) \quad . \qquad (5.65)
$$

The graph corresponding to this algorithm is shown in Fig.5.8b. It can be seen that this differs from the algorithm in Fig.5.8a in that transformation by this algorithm is effected, so to speak, "in place"; i.e., at every step of the transform the result can be written in the same place as the input numbers. The algorithm in Fig.5.8a does not have this advantage, but to compensate, it does not require any permutation.

Comparison of (5.63) with (5.34), and of Fig.5.8 with Fig.5.3 reveals the similarities between the factorized representations of Hadamard and FFT matrices with the rows rearranged according to binary inversion: they differ only by the presence in every factor in (5.63) of diagonal matrices. By transposing (5.63) one can obtain the following representation of DFT matrices:

$$\mathbf{F}_{2^n} = 1/2^{n/2}\left\{\prod_{i=0}^{n-1}\left(\mathbf{I}_{2^i}\otimes\mathbf{h}_2\otimes\mathbf{I}_{2^{n-1-i}}\right)\left[\mathbf{I}_{2^i}\otimes\left(\mathbf{I}_{2^{n-1-i}}\boxplus\bigotimes_{j=0}^{n-2-i}\mathbf{d}_{j+2}\right)\right]\right\}\times\mathbf{M}^{rev}_{2^n}\ , \qquad (5.66)$$

which is analogous to (5.35).

This analogy can be found for all of the various representations of DFT and Walsh-Hadamard transforms constructed from (5.63) and (5.34) using the rules of matrix algebra. Some examples of such transforms for DFT matrices have been presented in [5.12]. This analogy also indicates that any W-H T algorithm can be converted into an FFT algorithm and vice versa by adding or removing the diagonal matrices containing the complex exponential factors.

FFT algorithms (5.65,66) require either binary inversion of the order of the spectrum samples, or binary inversion of the elements of the initial transformed sequence. It should be noted that in many picture-processing tasks, especially when the DFT is used for signal filtering in the frequency domain, binary inversion of discrete spectrum samples following FFT is not obligatory, since after direct transformation by (5.65) in these cases the inverse transform is required, and it can be carried out on the basis of (5.66).

It should also be noted that the operations required for the DFT are operations on complex numbers, whereas the elementary operations performed by existing computers and digital processors are operations on real numbers. In order to express the DFT algorithms in terms of these elementary operations, (5.62,64,65) can be transformed so that the matrix elements are real numbers. The method of transformation depends on the form in which complex numbers are represented in the digital processor. We will illustrate this for the common case in which the real and imaginary parts of the signal samples are distributed as adjacent words in the processor's memory. Corresponding to this is a representation of a signal sample vector such that $a = (a_0^{re}, a_0^{im}, a_1^{re}, a_1^{im}, \ldots, a_{n-1}^{re}, a_{n-1}^{im})$. The dimensionality of this vector is twice as large as the number of signal samples. Therefore the size of the matrices composing the factorized representation of the matrices is also doubled so that, for example, (5.65) takes the form:

$$\mathbf{F}_{2^n} = 2^{-n/2}\left(\mathbf{M}^{rev}_{2^n}\otimes\mathbf{I}_2\right)\prod_{i=0}^{n-1}\left[\mathbf{I}_{2^{n-1-i}}\otimes\left(\mathbf{I}_{2^{i+1}}\boxplus\bigotimes_{m=0}^{i-1}\mathbf{four}^{i}_{m}\right)\right]$$

$$\times \left(I_{2^{n-1-i}} \otimes h_2 \otimes I_{2^{i+1}}\right) \quad , \qquad \text{where}$$

$$four_m^i = \begin{bmatrix} \cos(2\pi m / 2^{i+1}) & -\sin(2\pi m / 2^{i+1}) \\ \sin(2\pi m / 2^{i+1}) & \cos(2\pi m / 2^{i+1}) \end{bmatrix} .$$

5.5 Review of Other Fast Algorithms. Features of Two-Dimensional Transforms

With the help of the methods described in the previous section, the factorized representation of matrices and other orthogonal transforms described in Sect.4.11 can be obtained, as well as a number of other useful results. These results are presented in Table 5.1. Of special interest are the so-called truncated Fourier and Walsh transforms and the factorized representation of matrices connecting the matrices of different transforms.

Table 5.1. Factorized representations of unitary transform matrices

Transform	Factorized representation
Haar	$\mathrm{HAR}_{2^n} = D_{2^n}^{\mathrm{HAR}} \prod_{i=0}^{n-1} \left\{ \begin{bmatrix} I_{2^i} \otimes E_2^0 \\ \hline I_{2^i} \otimes E_2^1 \end{bmatrix} \boxplus I_{2^n - 2^{i+1}} \right\}$
Walsh-Hadamard	$\mathrm{HAD}_{2^n} = \frac{1}{2^{n/2}} \prod_{i=0}^{n-1} (I_{2^{n-1-i}} \otimes h_2 \otimes I_{2^i})$
Walsh-Paley	$\mathrm{PAL}_{2^n} = \frac{1}{2^{n/2}} \prod_{i=0}^{n-1} \left\{ I_{2^{n-1-i}} \otimes \begin{bmatrix} I_{2^i} \otimes E_2^0 \\ \hline I_{2^i} \otimes E_2^1 \end{bmatrix} \right\}$
Walsh	$\mathrm{WAL}_{2^n} = \frac{1}{2^{n/2}} \left[\prod_{i=0}^{n-2} \left(I_{2^{n-2-i}} \otimes \left\{ \begin{bmatrix} I_{2^i} \otimes E_2^0 \\ \hline \bar{I}_{2^i}(I_{2^i} \otimes E_2^1) \end{bmatrix} \boxplus \bar{I}_{2^{i+1}} \begin{bmatrix} I_{2^i} \otimes E_2^0 \\ \hline \bar{I}_{2^i}(I_{2^i} \otimes E_2^1) \end{bmatrix} \bar{I}_{2^{i+1}} \right\} \right) \right] \begin{bmatrix} I_{2^{n-1}} \otimes E_2^0 \\ \hline \bar{I}_{2^{n-1}}(I_{2^{n-1}} \otimes E_2^1) \end{bmatrix}$
Truncated Walsh-Hadamard	$P_{2^n}^{(k)} \mathrm{HAD}_{2^n} P_{2^n}^{(l)} = \begin{cases} \frac{1}{2^{n/2}} \left\{ \prod_{i=0}^{n-1-k} (I_{2^i} \otimes S_2^0 \otimes P_{2^{n-1-i}}^{(k)}) \left[\prod_{i=n-k}^{l-1} (I_{2^i} \otimes h_2 \otimes I_{2^{n-1-i}}) \right] (I_{2^l} \otimes S_{2^{n-l}}^T) \right\}; & k+l \geqslant n; \\ \frac{1}{2^{n/2}} \left\{ \left[\prod_{i=0}^{l-1} (I_{2^i} \otimes S_2^0 \otimes P_{2^{n-1-i}}^{(k)}) \right] (I_{2^l} \otimes P_{2^{n-l-k}} \otimes S_{2^k}^T) \right\}; & k+l \leqslant n \end{cases}$

Table 5.1 (continued)

Transform	Factorized representation
Fourier	$\begin{aligned} F_{2^n} &= \frac{1}{2^{n/2}} M_{2^n}^{\mathrm{rev}} \left\{ \prod_{i=0}^{n-1} \left[I_{2^{n-1-i}} \otimes \left(I_{2^i} \boxplus \bigotimes_{j=0}^{i-1} d_{2^{j+2}} \right) \right] \right. \\ &\qquad \left. \times (I_{2^{n-1-i}} \otimes h_2 \otimes I_{2^i}) \right\} \\ &= \frac{1}{2^{n/2}} \left\{ \prod_{i=0}^{n-1} \left[\boxminus_{p=0}^{1} (I_{2^{n-1-i}} \otimes E_2^p \otimes I_{2^i}) \right] \left[\boxplus_{k=0}^{2^{n-1-i}-1} (d_{2^i k} \otimes I_{2^i}) \right] \right\} \end{aligned}$
Truncated Fourier	$\begin{aligned} F_{2^n}^{(k,l)} &= P_{2^n}^{(k)} F_{2^n} P_{2^n}^{(l)} \\ &= \begin{cases} \frac{1}{2^{n/2}} \left\{ \prod_{i=0}^{n-1-k} (I_{2^i} \otimes S_2^0 \otimes P_{2^{n-1-i}}^{(k)}) \left[I_{2^i} \otimes \left(P_{2^{n-1-i}}^{(k)} \boxplus P_{2^{n-1-i}}^{(k)} \bigotimes_{j=0}^{n-1-i} d_{2^{j+2}} \right) \right] \right\} \\ \quad \left\{ \prod_{i=n-k}^{l-1} (I_{2^i} \otimes h_2 \otimes I_{2^{n-1-i}}) \times \left[I_{2^i} \otimes \left(I_{2^{n-1-i}} \boxplus \bigotimes_{j=0}^{n-1-i} d_{2^{j+2}} \right) \right] \right\} \\ \quad (I_{2^l} \otimes S_{2^{n-1}}^T) M_{2^n}^{\mathrm{rev}}, \qquad k+l \geqslant n; \\ \frac{1}{2^{n/2}} \left\{ \prod_{i=0}^{l-1} (I_{2^i} \otimes S_2^0 \otimes P_{2^{n-1-i}}^{(k)}) \times \left[I_{2^i} \otimes \left(P_{2^{n-1-i}}^{(k)} \boxplus P_{2^{n-1-i}}^{(k)} \bigotimes_{j=0}^{n-1-i} d_{2^{j+2}} \right) \right] \right\} \\ \quad (I_{2^l} \otimes P_{2^{n-l-k}} \otimes S_{2^k}^T) M_{2^n}^{\mathrm{rev}}, \qquad k+l \leqslant n \end{cases} \end{aligned}$
Modified Hadamard	$\begin{aligned} \mathrm{MHAD}_{2^n} &= \frac{1}{2^{n/2}} D_{2^n}^{\mathrm{HAR}} \prod_{i=0}^{n-1} \left(\begin{bmatrix} E_2^0 \otimes I_{2^i} \\ \hline E_2^1 \otimes I_{2^i} \end{bmatrix} \boxplus I_{2^n - 2^{i+1}} \right) \\ &= \frac{1}{2^{n/2}} D_{2^n}^{\mathrm{HAR}} \prod_{i=0}^{n-1} [(h_2 \otimes I_{2^i}) \boxplus I_{2^n - 2^{i+1}}] \end{aligned}$
Hadamard-Haar	$\begin{aligned} \mathrm{HDHR}_{2^n}^{(r)} &= (I_{2^r} \otimes D_{2^{n-r}}^{\mathrm{HAR}}) \left[\prod_{i=0}^{r-1} (I_{2^{r-1-i}} \otimes h_2 \otimes I_{2^{n-r+i}}) \right] \\ &\qquad \prod_{i=r}^{n-1} \left[I_{2^r} \otimes \left(\begin{bmatrix} I_{2^{i-r}} \otimes E_2^0 \\ \hline I_{2^{i-r}} \otimes E_2^1 \end{bmatrix} \boxplus I_{2^n - 2^{i-r+1}} \right) \right] \end{aligned}$

Table 5.1 (continued)

Transform	Factorized representation
Bit reversal	$M_{2^n}^{\text{rev}} = \prod_{i=0}^{n-1} \left\{ I_{2^{n-1-i}} \otimes \left[\frac{I_{2^i} \otimes G_2^0}{I_{2^i} \otimes G_2^1} \right] \right\}$
Rearrangement from Gray code to simple binary code	$M_{2^n}^{G/D} = \prod_{i=0}^{n-1} [I_{2^i} \otimes (I_{2^{n-1-i}} \boxplus \bar{I}_{2^{n-1-i}})]$
Transfer from modified Hadamard transform to Walsh-Hadamard transform	$M_{2^n}^{\text{MH-H}} = \prod_{i=1}^{n-1} \prod_{k=0}^{i-1} \{I_{2^{n-1-i}} \otimes [I_{2^i} \boxplus (h_2 \otimes I_{2^k}) \boxplus I_{2^i-2^{k+1}}]\}$
Transfer from Haar transform to Walsh-Paley transform	$M_{2^n}^{\text{HR-P}} = \prod_{i=1}^{n-1} \prod_{k=0}^{i-1} \left[I_{2^{n-1-i}} \otimes \left(I_{2^i} \boxplus \left[\frac{I_{2^k} \otimes E_2^0}{I_{2^k} \otimes E_2^1} \right] \boxplus I_{2^i-2^{k+1}} \right) \right]$
Transfer from modified Hadamard transform to DFT	$M_{2^n}^{\text{MH-F}} = \prod_{i=1}^{n-1} \left\{ \prod_{k=0}^{i-1} \{I_{2^{n-1-i}} \otimes [I_{2^i} \boxplus (h_2 \otimes I_{2^k}) \boxplus I_{2^i-2^{k+1}}]\} \times \left[I_{2^{n-1-i}} \otimes \left(I_{2^i} \boxplus \bigotimes_{j=0}^{i-1} d_{2^{j+2}} \right) \right] \right\}$

5.5.1 Truncated FFT and FWT Algorithms

In Fourier and Walsh discrete transformation it often happens that either the initial sequence contains many zero elements, or that not all of the transform elements must be calculated, or both. This circumstance can be used to reduce further the number of operations required in transformation (as compared with the above-described FFT and FWT algorithms). If FFT and FWT representations are considered as graphs, it becomes clear that the presence of a certain number of zero elements in the initial sequence, or the needlessness of calculating certain transform coefficients, eliminates certain transformation graph edges. This means that some of the multiplication operations (in FFT) and some of the addition operations can be omitted. If the number of zero elements in the transformed sequence or the number of unnecessary transform elements is less than half the length of the sequence, the structure of the transformation graph is not seriously affected, and the saving in the number of operations is not great. But if the number is much greater than half the sequence length, the gain can be considerable.

Let us see how the FFT and FWT transform structures change in this case, and evaluate the possible gains. In view of the above-noted kinship between FWT and FFT we shall only look at the latter.

The presence of zero elements in the transformed sequence and of unnecessary transform coefficients can be conveniently described by multiplying the transform matrix from right and left by diagonal truncating matrices (let us call them $\mathbf{P}_{2^n}^{(k)}$ and $\mathbf{P}_{2^n}^{(\ell)}$ respectively), which have zeros at those elements of the diagonal whose numbers correspond to the position of zero elements or unnecessary elements.

From the standpoint of possible modifications of transform algorithms, it is sufficient to consider cases in which the quantity of non-zero or required elements is an integer power of 2; the truncating matrices can then be presented in the form of Kronecker matrix products

$$\mathbf{P}_2 = \begin{bmatrix} 1 & 0 \\ 0 & 0 \end{bmatrix} \tag{5.67}$$

and identity matrices

$$\mathbf{I}_2 = \begin{bmatrix} 1 & 0 \\ 0 & 1 \end{bmatrix} \quad . \tag{5.68}$$

Let us consider the simplest case, in which the last elements of the initial sequence and of its transformation are cut off. Then the truncating matrix of size 2^n, which contains 2^ℓ units at the beginning of the diagonal, is written in the following form:

$$\mathbf{P}_{2^n}^{(\ell)} = \mathbf{P}_2^{[n-\ell]} \otimes \mathbf{I}_2^{[\ell]} = \mathbf{P}_{2^{n-\ell}} \otimes \mathbf{I}_2 \quad . \tag{5.69}$$

The factorized form of matrices truncated from both sides,

$$\mathbf{F}_{2^n}^{(k,\ell)} = \mathbf{P}_{2^n}^{(k)} \mathbf{F}_{2^n} \mathbf{P}_{2^n}^{(\ell)} \quad , \tag{5.70}$$

can be considered in two steps: first by multiplying $\mathbf{F}_{2^n}$ from the left by $\mathbf{P}_{2^n}^{(k)}$, and then by multiplying the resulting matrix from the right by $\mathbf{P}_{2^n}^{\ell}$. Then by using methods of matrix combination similar to those used in Sect.5.3, we can derive the following expression for the factorized FFT matrix representation (5.66) [5.12-14], for $k+\ell \geqslant n$:

$$F^{(k,\ell)}_{2^n} = 1/2^{n/2} \left\{ \prod_{i=0}^{n-1-k} \left(I_{2^i} \otimes S_2^0 \otimes P^{(k)}_{2^{n-1-k}} \right) \left[I_{2^i} \otimes \left(P^{(k)}_{2^{n-1-i}} \boxplus P^{(k)}_{2^{n-1-i}} \right.\right.\right.$$

$$\left.\left.\left. \times \bigotimes_{j=0}^{n-1-i} d_{2^{j+2}} \right) \right] \right\} \left\{ \prod_{i=n-k}^{\ell-1} \left(I_{2^i} \otimes h_2 \otimes I_{2^{n-1-i}} \right) \left[I_{2^i} \otimes \left(I_{2^{n-1-i}} \right.\right.\right.$$

$$\left.\left.\left. \boxplus \bigotimes_{j=0}^{n-1-i} d_{2^{j+2}} \right) \right] \right\} \left(I_{2^\ell} \otimes S^T_{2^{n-\ell}} \right) M^{rev}_{2^n} \quad , \text{ where} \tag{5.71}$$

$$S_2^0 = \begin{bmatrix} 1 & 1 \\ 0 & 0 \end{bmatrix} \quad ,$$

$$\left(S_{2^k} \right)^T = \left[\left(S_2^0 \right)^T \right]^{(k)} = \left(\begin{bmatrix} 1 & 0 \\ 1 & 0 \end{bmatrix} \right)^{(k)} \quad . \tag{5.72}$$

Thus the truncated transform consists of a cascade of $(n-k)+(\ell+k-n)+2$ members: one member is a binary inversion; one is the periodic repetition of non-zero input samples [multiplication matrix $(I_{2^\ell} \otimes S^T_{n-\ell})$]; and there are $(\ell+k-n)$ members of non-truncated transforms (matrix product from $i=n-k$ to $i=\ell-1$) and $(n-k)$ transform members "with gaps" (matrix product from $i=0$ to $i=n-1-k$).

When $k+\ell \leqslant n$, the transform degenerates: the matrices of the non-truncated transform, constructed on h_2, disappear:

$$F^{(k,\ell)}_{2^n} = 1/2^{n/2} \left\{ \prod_{i=0}^{\ell-1} \left(I_{2^i} \otimes S_2^0 \otimes P^{(k)}_{2^{n-1-i}} \right) \left[I_{2^i} \otimes \left(P^{(k)}_{2^{n-1-i}} \boxplus P^{(k)}_{2^{n-1-i}} \right.\right.\right.$$

$$\left.\left.\left. \times \bigotimes_{j=0}^{n-1-i} d_{2^{j+2}} \right) \right] \left(I_{2^\ell} \otimes P_{2^{n- \, -k}} \otimes S^T_{2^k} \right) M^{rev}_{2^n} \quad . \tag{5.73}$$

The graph of the truncated transform (5.71) is shown in Fig.5.9 for $n=5$, $k=3$, $\ell=3$. The numbers at the nodes denote the factors of the argument of complex exponents $\exp(2\pi i/32)$. It is not difficult to calculate the number of operations effected by the truncated transform. The number of additions $N^{(k,\ell)}_{ad}$, and of multiplications $N^{(k,\ell)}_{mul}$, is, when $k+1 \geqslant n$:

$$N^{(k,\ell)}_{ad} = 2^n(\ell + k - n + 1) - 2^k \quad ,$$

$$N^{(k,\ell)}_{mul} = 2^n + 2^{n-1}(k + \ell - n) - (2^k + 2^\ell) + 1 \quad ; \tag{5.74}$$

and when $k+\ell \leqslant n$:

$$N^{(k,\ell)}_{ad} = 2^k(2^\ell - 1) \quad , \quad N^{(k,\ell)}_{mul} = (2^k - 1)(2^\ell - 1) \quad . \tag{5.75}$$

The saving in the number of operations resulting from transform truncation, as compared with calculating non-truncated transforms, is, when $k+\ell > n$:

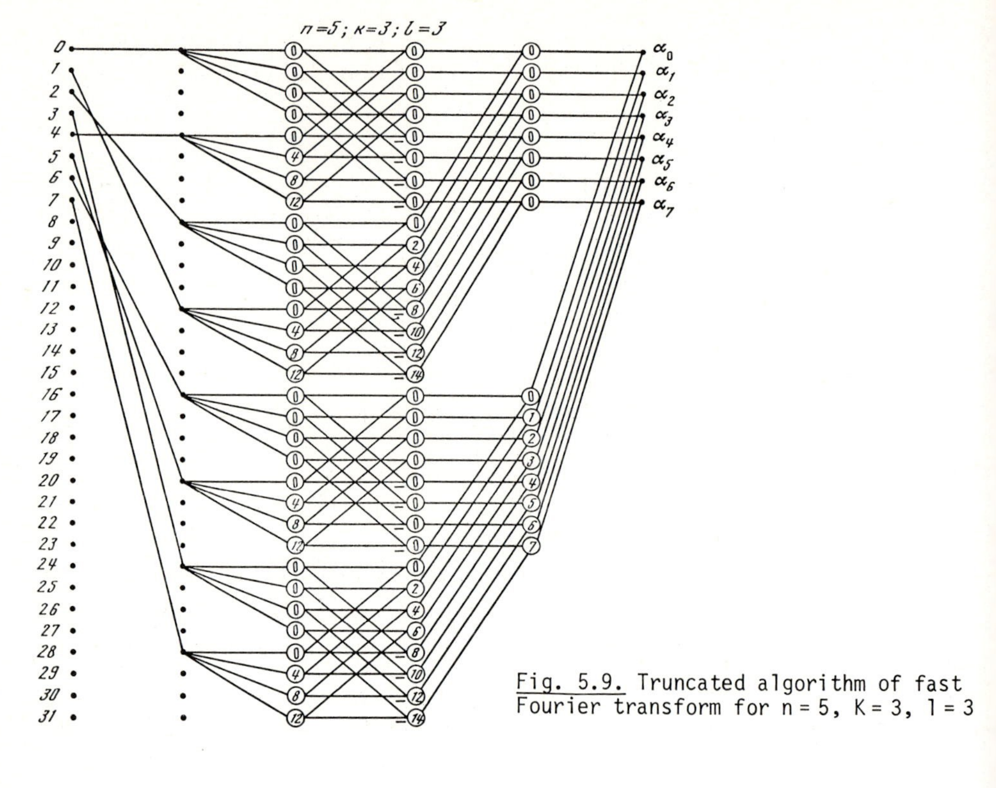

Fig. 5.9. Truncated algorithm of fast Fourier transform for n = 5, K = 3, l = 3

$$g_{\mathrm{ad},n}^{(k,\ell)} = N_{\mathrm{ad}}^{(n,n)} / N_{\mathrm{ad}}^{(k,\ell)} = n / (k + \ell - n + 1 - 2^{k-n}) \quad ,$$

$$g_{\mathrm{mul},n}^{(k,\ell)} = N_{\mathrm{mul}}^{(n,n)} / N_{\mathrm{mul}}^{(k,\ell)} = (n - 2 + 2^{1-n}) / (k + \ell - n + 2 - 2^{k-n+1} - 2^{\ell-n+1} + 2^{-n}) \quad ; \tag{5.76}$$

and when $k + \ell \leqslant n$:

$$g_{\mathrm{ad},n}^{(k,\ell)} = n / (2^{\ell+k-n} - 2^{k-n}) \quad ,$$

$$g_{\mathrm{mul},n}^{(k,\ell)} = (n - 2 + 2^{1-n}) / (2^{\ell+k-n+1} - 2^{\ell-n+1} + 2^{k-n+1} + 2^{1-n}) \quad . \tag{5.77}$$

With large n the saving in the number of additions grows with increasing n approximately as $n / (\ell + k - n + 1)$, and the saving in the number of multiplications as $n / (\ell + k - n + 2)$. With large n and large redundancy the saving can be considerable.

5.5.2 Transition Matrices Between Various Transforms

In digital signal processing it is sometimes necessary to find the signal representation coefficients on one basis when the representation coefficients on the other basis are known. To do this it is sufficient to multiply the coefficient column matrix by the corresponding transition matrix:

$$\boldsymbol{\alpha}^{(2)} = \mathbf{M}^{1-2}\,\boldsymbol{\alpha}^{(1)} \quad . \tag{5.78}$$

By representing the orthogonal transform matrix in the form of layered Kronecker matrices it is fairly easy to find the transition matrix between these transforms. Let us illustrate this, using as an example the links between the DFT matrices F_{2^n} and the matrices of the modified Hadamard transform MHAD_{2^n} (Sect.4.11). According to (5.64),

$$\bar{\mathbf{F}}^{rev}_{2^k} = \left(\mathbf{E}_2^0 \otimes \bar{\mathbf{F}}^{rev}_{2^{n-1}}\right) \boxminus \left(\mathrm{E}_2^1 \otimes \bar{\mathrm{F}}^{rev}_{2^{n-1}} \bigotimes_{j=0}^{n-2} \mathrm{d}_{2^{j+2}}\right) \quad , \tag{5.79}$$

where $\bar{\mathbf{F}}^{rev}_{2k} = 2^{k/2}\mathrm{F}_{2^k}\,\mathbf{M}^{rev}_{2^k}$, and according to Theorem 3 of Sect.5.1,

$$\begin{aligned}\bar{\mathbf{F}}^{rev}_{2^n} &= \left(\bar{\mathbf{F}}^{rev}_{2^{n-1}} \boxplus \bar{\mathrm{F}}^{rev}_{2^{n-1}} \bigotimes_{j=0}^{n-2} \mathbf{d}_{2^{j+2}}\right)\left[\left(\mathrm{E}_2^0 \otimes \mathrm{I}_{2^{n-1}}\right) \boxminus \left(\mathrm{E}_2^1 \otimes \mathbf{I}_{2^{n-1}}\right)\right] \\ &= \left(\bar{\mathrm{F}}^{rev}_{2^{n-1}} \boxplus \bar{\mathbf{F}}^{rev}_{2^{n-1}} \bigotimes_{j=0}^{n-2} \mathbf{d}_{2^{j+2}}\right)\left(\mathrm{h}_2 \otimes \mathbf{I}_{2^{n-1}}\right) \quad .\end{aligned} \tag{5.80}$$

Using (5.80) as the recursive formula for $\mathbf{F}^{rev}_{2^n}$, we obtain

$$\bar{\mathbf{F}}^{rev}_{2^n} = \left(1 \boxplus \boxplus_{i=0}^{n-1} \bar{\mathbf{F}}^{rev}_{2^i} \bigotimes_{j=0}^{i-1} \mathbf{d}_{2^{j+2}}\right)\left\{\prod_{i=0}^{n-1}\left[\left(\mathrm{h}_2 \otimes \mathbf{I}_{2^i}\right) \boxplus \dot{\mathbf{I}}_{2^n-2^{i+1}}\right]\right\} \quad . \tag{5.81}$$

If we compare the expression in braces on the right-hand side of (5.81) with the factorized Haar matrix representation (5.29), and also compare the Haar matrix representation with the matrix of the modified Hadamard transform in the form of the layered Kronecker matrices (Table 4.5), we see that this expression is the factorized representation of the modified Hadamard matrix, with an accuracy determined by the diagonal matrix $\mathbf{D}^{HAR}_{2^n}$:[3]

$$\mathbf{MHAD}_{2^n} = 1/2^{n/2}\,\mathrm{D}^{HAR}_{2^n} \prod_{i=0}^{n-1}\left[\left(\mathbf{h}_2 \otimes \mathbf{I}_{2^i}\right) \boxplus \mathrm{I}_{2^n-1^{i+1}}\right] \quad . \tag{5.82}$$

Then, the first factor in (5.81) is the matrix of the transition from the modified Hadamard transform to DFT with binary inversion:

$$\mathbf{M}^{MH-F}_{2^n} = \mathbf{1} \boxplus \boxplus_{i=0}^{n-1} \bar{\mathrm{F}}^{rev}_{2^i} \bigotimes_{j=0}^{i-1} \mathbf{d}_{2^{j+2}} \quad . \tag{5.83}$$

---

[3] Of course this representation can also be derived directly from the expression for $\mathbf{MHAD}_{2^n}$ in Table 4.5.

Note that to factorize this matrix, it follows from (5.83) that

$$\mathbf{M}^{\text{MH-F}}_{2^n} = \mathbf{M}^{\text{MH-F}}_{2^{n-1}} \boxplus \bar{\mathbf{F}}^{\text{rev}}_{2^{n-1}} \bigotimes_{j=0}^{n-2} \mathbf{d}_{2^{j+2}} = \mathbf{M}^{\text{MH-F}}_{2^{n-1}} \boxplus \mathbf{M}^{\text{MH-F}}_{2^{n-1}} \mathbf{MHAD}_{2^{n-1}} \bigotimes_{j=0}^{n-2} \mathbf{d}_{2^{j+2}}$$

$$= \left(\mathbf{I}_2 \otimes \mathbf{M}^{\text{MH-F}}_{2^{n-1}}\right)\left(\mathbf{I}_{2^{n-1}} \boxplus \mathbf{MHAD}_{2^{n-1}}\right)\left(\mathbf{I}_{2^{n-1}} \boxplus \bigotimes_{j=0}^{n-2} \mathbf{d}_{2^{j+2}}\right) \quad . \qquad (5.84)$$

Using this formula recurrently, we obtain

$$\mathbf{M}^{\text{MH-F}}_{2^n} = \prod_{i=1}^{n-1}\left[\mathbf{I}_{2^{n-1-i}} \otimes \left(\mathbf{I}_{2^i} \boxplus \mathbf{MHAD}_{2^i}\right)\right]\left[\mathbf{I}_{2^{n-1-i}} \otimes \left(\mathbf{I}_{2^i} \boxplus \bigotimes_{j=0}^{i-1} \mathbf{d}_{2^{j+2}}\right)\right] . \qquad (5.85$$

Finally, by applying the expression for $\mathbf{MHAD}_{2^i}$ after the obvious transforms, we obtain the expression

$$\mathbf{M}^{\text{MH-F}}_{2^n} = \prod_{i=1}^{n-1}\left(\prod_{k=0}^{i-1} \left\{\mathbf{I}_{2^{n-1-i}} \otimes \left[\mathbf{I}_{2^i} \boxplus \left(\mathbf{h}_2 \otimes \mathbf{I}_{2^k}\right) \boxplus \mathbf{I}_{2^i-2^{k+1}}\right]\right\}\right)$$

$$\times \left[\mathbf{I}_{2^{n-1-i}} \otimes \left(\mathbf{I}_{2^i} \boxplus \bigotimes_{j=0}^{i-1} \mathbf{d}_{2^{j+2}}\right)\right] \quad . \qquad (5.86)$$

The inverse transition matrix is equal to the product of these matrices, taken in reverse order and with $\mathbf{d}_{2^j}$ replaced by $\mathbf{d}^{-1}_{2^j}$:

$$\mathbf{d}^{-1}_{2^j} = \begin{bmatrix} 1 & 0 \\ 0 & \exp(-2\pi i)/2^j) \end{bmatrix} \quad , \qquad \text{i.e.,} \qquad (5.87)$$

$$\mathbf{M}^{\text{F-MH}}_{2^n} = \prod_{i=1}^{n} \left[\mathbf{I}_{2^{i-1}} \otimes \left(\mathbf{I}_{2^{n-i}} \boxplus \bigotimes_{j=0}^{n-i-1} \mathbf{d}^{-1}_{2^{j+2}}\right)\right]\left(\prod_{k=0}^{n-i-1} \left\{\mathbf{I}_{2^{i-1}} \otimes \left[\mathbf{I}_{2^{n-i}}\right.\right.\right.$$

$$\left.\left.\left.\boxplus \left(\mathbf{h}_2 \otimes \mathbf{I}_{2^{n-i-1-k}}\right) \boxplus \mathbf{I}_{2^{n-i}-2^{n-1-k}}\right]\right\}\right) \quad . \qquad (5.88)$$

The transition from the modified Hadamard transform to DFT with binary inversion is shown in Fig.5.10. For convenience the graph of the modified Hadamard transform (Table 5.1) is shown on the left-hand side. By comparing this combination of graphs with the FFT graph in Fig.5.8 we can see that the MHAD_{2^n} graph is the truncated FFT graph, and that graph $\text{M}^{\text{MH-F}}_{2^n}$ supplements it to yield the FFT graph. Hence it follows that the algorithms for calculating the DFT by means of an FFT or by the modified Hadamard transform with subsequent transition to DFT are equivalent in terms of the number of operations.

In view of the similarity between the factorized representations of the matrices $\mathbf{F}_{2^n}$ and $\mathbf{HAD}_{2^n}$, the matrix $\mathbf{M}^{\text{MH-F}}_{2^n}$ can be converted into the matrix $\mathbf{M}^{\text{MH-H}}_{2^n}$ of the transition from the modified Hadamard transform to the Walsh-Hadamard transform by extracting the diagonal matrices from $\mathbf{M}^{\text{MH-F}}_{2^n}$. Finally,

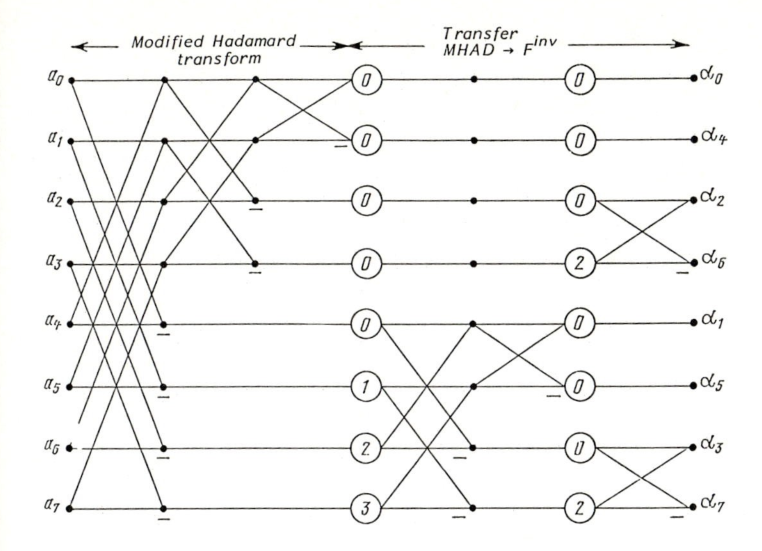

Fig. 5.10. Representation of fast Fourier transform algorithm through a modified Hadamard transform algorithm (MHT), and algorithm of transition from MHT to DFT

by following the same procedure as that used for the matrices F_{2^n}, it is not difficult to obtain the matrix for the transition from the Haar transform to the Paley transform. These matrices are shown in Table 5.1.

5.5.3 Calculation of Two-Dimensional Transforms

As was already mentioned, the 2-D transforms described in Sect.4.11 are defined as twofold applications of the corresponding 1-D transforms. The amount of directly addressable (core) memory available to the arithmetic unit is of great importance in organizing 2-D transforms. If the core memory is sufficient to store the entire 2-D array, the 2-D transformation can be carried out with the fast 1-D algorithm and the corresponding memory addressing during transformation on the rows and columns. It is interesting to note that from the point of view of fast algorithm theory, organizing the calculations is equivalent to stretching the 2-D array in the 1-D approach by joining the rows of the array together and representing the 2-D transformation matrices as Kronecker matrix products of the 1-D transformation matrices.

If, however, as often happens, the core memory is insufficient to store the 2-D array, but sufficient to store one row, the transformation takes place in three stages: transformation row by row, with the result stored in the external memory device (EMD); transposition; and transformation of the transposed array row by row. The computer processing time is determined by the time of access to the EMD, and is also increased by the need to transpose. To accelerate the transposition of the 2-D array in the EMD, the following approach is used; the array is divided into the largest possible sub-arrays that will fit into core and each sub-array is then transposed. Following this, the order of sub-array distribution within the 2-D array is itself transposed. For convenience in transposing the sub-array arrangement, it is convenient to have a supplementary buffer store with a capacity equal to the volume of a sub-array.

Transposition is not obligatory in transforming 2-D signals that do not fit into core. If the same principle of addressing is used as in 2-D transformation in the core memory, algorithms for 2-D transformation can be constructed without transposition [5.15].

5.6 Combined DFT Algorithms

In picture processing using the DFT, the initial signal is usually represented as a sequence of real numbers. Moreover, as stated in Sect.4.8, in calculating convolutions with the DFT the signal must sometimes be evenly supplemented. As a result, the transformed sequence has a twofold or even fourfold redundancy (twofold because the numbers are real, and fourfold because the sequence is also even). This redundancy can be used to achieve a corresponding acceleration of DFT calculation.

5.6.1 Combined DFT Algorithms of Real Sequences

There are two methods of using the twofold redundancy of DFT sequences of real numbers. The first is to combine the transformation of two sequences. The second is to separate one sequence with an even number of members into two sub-sequences, transform these sub-sequences jointly and recalculate the result for the entire sequence.

Let us begin by describing the first method, since the second is in essence similar to it.

Let $\{a_k\}$ and $\{b_k\}$ be two sequences of real numbers of length $N(k = 0, 1, \ldots, N-1)$. Let us form the sequence

$$\{c_k = a_k + ib_k\} \tag{5.89}$$

and find its DFT

$$\begin{aligned} \gamma_s &= 1/\sqrt{N} \sum_{k=0}^{N-1} c_k \exp(i2\pi\, ks/N) = 1/\sqrt{N} \sum_{k=0}^{N-1} (a_k + ib_k) \exp(i2\pi\, ks/N) \\ &= \alpha_s + i\beta_s = \alpha_s^{re} - \beta_s^{im} + i(\alpha_s^{im} + \beta_s^{re}) \quad , \end{aligned} \tag{5.90}$$

where

$$\alpha_s = 1/\sqrt{N} \sum_{k=0}^{N-1} a_k \exp(i2\pi\ ks/N) \quad ;$$

$$\beta_s = 1/\sqrt{N} \sum_{k=0}^{N-1} b_k \exp(i2\pi\ ks/N) \quad . \tag{5.91}$$

The superscripts re and im denote real and imaginary parts of the corresponding numbers.

Since

$$\alpha_{N-s} = \alpha_s^* \quad , \qquad \beta_{N-s} = \beta_s^* \quad , \tag{5.92}$$

(Table 4.1, line 5), then

$$\gamma_{N-s} = \alpha_{N-s} + i\beta_{N-s} = \alpha_s^* + i\beta_s^* \quad . \tag{5.93}$$

Therefore

$$\gamma_s + \gamma_{N-s} = (\alpha_s + \alpha_s^*) + i(\beta_s + \beta_s^*) = 2\alpha_s^{re} + i2\beta_s^{re} \quad ;$$

$$\gamma_s - \gamma_{N-s} = (\alpha_s - \alpha_s^*) + i(\beta_s - \beta_s^*) = 2\beta_s^{im} + i2\alpha_s^{im} \quad , \tag{5.94}$$

i.e.,

$$\alpha_s = 1/2(\gamma_s + \gamma_{N-s}^*) \quad , \qquad \beta_s = -1/2i(\gamma_s - \gamma_{N-s}^*) \quad . \tag{5.95}$$

Equation (5.95) shows how by performing $(N+2)$ multiplication operations on real numbers, one can find the Fourier coefficients of sequences $\{a_k\}$ and $\{b_k\}$ from the transform of the derived sequence $\{a_k + ib_k\}$. [Only the coefficients for $s = 0, 1, \ldots, N/2$ need be calculated by (5.95); the other α_s and β_s can be found by taking complex conjugates.]

This calculation procedure is shown in Fig.5.11a. A DFT combination method such as this is useful in transforming 2-D arrays, when it is convenient to choose $\{a_k\}$ and $\{b_k\}$ to be neighbouring rows of the array.

In the transformation of 1-D real arrays, the second method of decreasing the number of operations is more convenient. Let $\{a_k\}$ be a sequence of real numbers of length $2N (k = 0, 1, \ldots, 2N-1)$. Its discrete Fourier transform must be found:

$$\alpha_s = (2N)^{-1/2} \sum_{k=0}^{2N-1} a_k \exp(i2\pi\ ks/2N) \quad . \tag{5.96}$$

Let us isolate the odd and even terms from the sum (5.96):

$$\alpha_s = (2N)^{-1/2} \left\{ \sum_{k=0}^{N-1} a_{2k} \exp(i2\pi\ 2ks/2N) + \sum_{k=0}^{N-1} a_{2k+1} \exp[i2\pi\ (2k+1)s/2N] \right\}$$

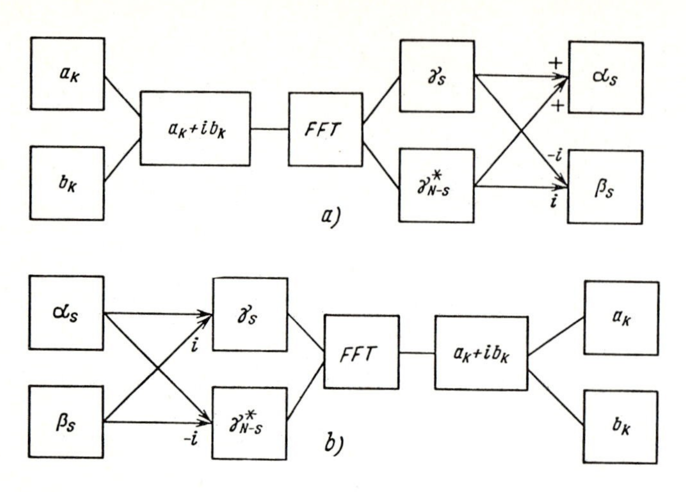

Fig. 5.11a,b. Layout of combined Fourier transform algorithms for sequences of real numbers: (a) direct transform, (b) inverse transform

$$= (2N)^{-1/2} \left\{ \sum_{k=0}^{N-1} a_{2k} \exp(i2\pi\ ks / N) + \left[\sum_{k=0}^{N-1} a_{2k+1} \exp(i2\pi\ ks / N) \right] \times \exp(i\pi s / N) \right\} \quad . \tag{5.97}$$

Thus, the DFT of the entire sequence $\{a_k\}$ can be found by calculating the DFT of two of its sub-sequences which are composed respectively of even and odd members of the initial sequence, and by summing up the results according to the formula:

$$\alpha_s = 1 / \sqrt{2}\ [\alpha_s^e + \exp(i\pi s / N)\alpha_s^0] \quad , \tag{5.98}$$

where α_s^e and α_s^0 are the DFTs of $\{a_{2k}\}$ and $\{a_{2k+1}\}$, respectively. We can find α_s^e and α_s^0 by using the first combined DFT method (5.95). As a result we obtain

$$\alpha_s = 2 / 2\sqrt{2}\ [(\gamma_s + \gamma_{N-s}^*) - i \exp(i\pi\ s / N)(\gamma_s - \gamma_{N-s}^*)] \quad , \tag{5.99}$$

where

$$\gamma_s = 1 / \sqrt{N} \sum_{k=0}^{N-1} (a_{2k} + ia_{2k+1}) \exp(i2\pi\ ks / N) \quad , \tag{5.100}$$

$s = 0, 1, \ldots, N$. The values α_s for the remaining s can be found from (5.92).

Comparison of (5.95) with (5.99) shows that the second method requires more complicated supplementary calculation.

The algorithms described for combined Fourier transforms can easily be reversed to derive the algorithms for calculating the DFT of complex-conjugate sequences. Indeed, it can be seen from Fig.5.11a that given two sequences $\{\alpha_s\}$ and $\{\alpha_s\}$ such that $\alpha_s = \alpha^*_{N-s}$, $\beta_s = \beta^*_{N-s}$, the series γ_s can be constructed from them by the following rule:

$$\gamma_s = \alpha_s + i\beta_s \quad ; \quad \gamma_{N-s} = \alpha^*_s + i\beta^*_s \quad , \quad s = 0, 1, \ldots, N/2 \quad . \tag{5.101}$$

Then the real part of the Fourier transform of this sequence is the Fourier transform of $\{\alpha_s\}$ and the imaginary part is the Fourier transform of $\{\beta_s\}$. The layout of this algorithm is shown in Fig.5.11b.

Similary, given one sequence of length $2N\{\alpha_s = \alpha^*_{2N-s}\}$, two sequences can be formed from it:

$$\alpha^e_s = 1/\sqrt{2}\,(\alpha_s + \alpha^*_{N-s}) \quad ; \quad \alpha^e_{N-s} = \alpha^{e*}_s \quad ; \quad \alpha^0_s = 1/\sqrt{2}\exp(-i\pi s/N)$$
$$(\alpha_s - \alpha^*_{N-s}) \quad ; \quad \alpha^0_{N-s} = \alpha^{0*}_s \quad , \tag{5.102}$$

and then the approach described above for two separate sequences can be used.

These algorithms of combined transforms can also be used, with the corresponding saving in computing time, to calculate the 2-D DFT of real or complex-conjugate arrays. This is carried out as two 1-D transforms. In fact, in transforming 2-D sequences of real numbers, the first Fourier transformation can be achieved by combining the DFTs of a pair of rows of the array, while the second Fourier transformation of the resulting array of complex numbers is carried out only halfway up the columns; the second half is found to be the complex conjugate of the first half. A complex-conjugate array is transformed in reverse order: the first Fourier transformation is carried out only halfway up the array, and then supplemented with numbers that are the complex conjugates of the results obtained; the second Fourier transformation is then effected using the algorithm described for the combined transformation of two sequences with paired complex-conjugate members.

5.6.2 Combined SDFT (1/2, 0) Algorithms of Even and Real Even Sequences

For the SDFT(1/2, 0) of even sequences with an even number of terms, the following relations are valid (Table 4.3, lines 7, 13):

$$\{a_k = a_{2N-1-k}\} \xleftarrow{\text{SDFT}(1/2,\,0)}\!\!\rightarrow \{\alpha^{1/2,0}_s = -\alpha^{1/2,0}_{2N-s}\} \quad , \tag{5.103a}$$

$$\{a_k \exp[i2\pi N(k+1/2)/(2N)] = i(-1)^k a_k\} \xleftarrow{\text{SDFT}(1/2,0)}\!\!\rightarrow \{\alpha^{1/2,0}_{s+N}\}, \tag{5.103b}$$

$$\alpha^{1/2,0}_{s+2N} = -\alpha^{1/2,0}_s \quad . \tag{5.103c}$$

They can be used to construct combined SDFT algorithms (1/2, 0) of even sequences. Indeed, let $\{a_k\}$, $\{b_k\}$ be two even sequences with an even number of terms:

$$a_k = a_{2N-k-1} \quad , \quad b_k = b_{2N-k-1} \quad , \quad k = 0, 1, \ldots, 2N-1 \quad . \tag{5.104}$$

Let us form the sequence

$$c_k = a_k + i(-1)^k b_k \quad . \tag{5.105}$$

Its SDFT(1/2, 0) is

$$\gamma_s^{1/2,0} = \alpha_s^{1/2,0} + \beta_{s+N}^{1/2,0} \quad , \tag{5.106}$$

where $\{\alpha_s^{1/2,0}\}$ and $\{\beta_s^{1/2,0}\}$ are the SDFTs (1/2, 0) corresponding to $\{a_k\}$ and $\{b_k\}$.

According to (5.103a,c),

$$\gamma_{2N-s}^{1/2,0} = \alpha_{2N-s}^{1/2,0} + \beta_{2N-s+N}^{1/2,0} = -\alpha_s^{1/2,0} - \beta_{s-N}^{1/2,0} = -\alpha_s^{1/2,0} + \beta_{s+N}^{1/2,0} \quad . \tag{5.107}$$

Therefore

$$\alpha_s^{1/2,0} = 1/2(\gamma_s^{1/2,0} - \gamma_{2N-s}^{1/2,0}) \quad , \quad \beta_{s+N}^{1/2,0} = 1/2(\gamma_s^{1/2,0} + \gamma_{2N-s}^{1/2,0}) \quad ,$$

$$= 0, 1, \ldots, N-1 \quad . \tag{5.108}$$

The values $\alpha_s^{1/2,0}$ and $\beta_s^{1/2,0}$ for other s can be found from (5.103a).

As a result of the combination step, the time necessary to transform even sequences slightly exceeds the time necessary to transform half-length sequences.

The procedure described above is shown in Fig.5.12a. It can easily be reversed to give the procedure for inverse transformation (Fig.5.12b).

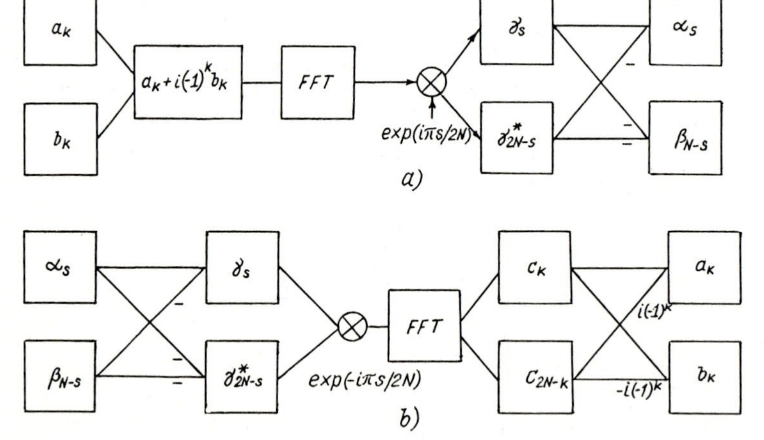

Fig. 5.12a,b. Layout of combined cosine transform algorithms: (a) direct transform, (b) inverse transform

If the sequences being transformed are real, then the coefficients of their SDFT(1/2, 0) are also purely real (Table 4.3, line 8). Therefore the real and imaginary parts $\alpha_s^{1/2,0}$ and $\beta_s^{1/2,0}$ in (5.107) are the SDFT(1/2,0) of the corresponding real and imaginary parts $\{a_k\}$ and $\{b_k\}$.

Thus, by forming two complex even sequences from two pairs of real even sequences using the approach based on the combined SDFT(1/2, 0), real even sequences can be transformed nearly four times as fast as one complex sequence of the same length.

5.7 Recursive DFT Algorithms

When processing pictures and other 2-D signals it is sometimes necessary to calculate the spectra of successive, strongly overlapping signal fragments, or so-called local spectra. The spectrum of each fragment is best calculated recurrently, using the spectrum of the previous fragment to process the next. Let the fragments consist of $N_1 \times N_2$ elements and be overlaid on the picture at intervals p and q in the two coordinates. Let us link the spectra of two adjacent overlapping fragments, and suppose that spectrum $\alpha_{r,s}^{(0,0)}$ is the 2-D spectrum of the first of these fragments:

$$\alpha_{r,s}^{(0,0)} = 1 / \sqrt{N_1 N_2} \sum_{k=0}^{N_1-1} \sum_{k=0}^{N_2-1} \alpha_{k,\ell} \times \exp[i2\pi(kr / N_1 + \ell s / N_2)] \quad . \quad (5.109)$$

Then the spectrum $\alpha_{r,s}^{(p,q)}$ of the second fragment, shifted relative to the first along the coordinate axes by p and q, can be written as (Fig.5.13)

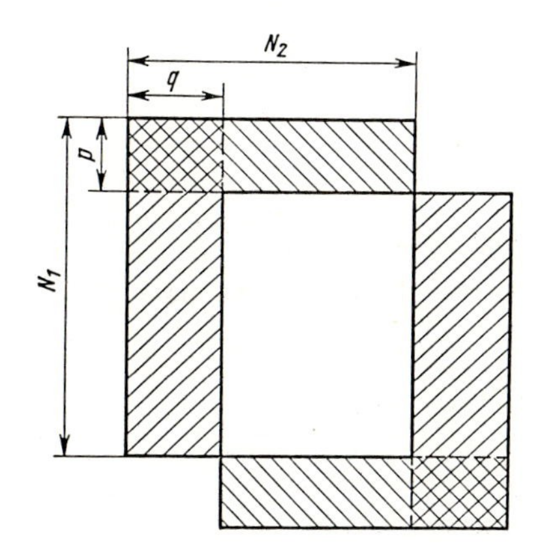

Fig. 5.13. Two fragments of a picture, shifted along the coordinates of recursion by p and q picture elements

$$\alpha_{r,s}^{(p,q)} = 1/\sqrt{N_1N_2} \sum_{k=0}^{N_1-1} \sum_{\ell=0}^{N_2-1} a_{k+p,\ell+q} \exp[i2\pi(kr/N_1 + \ell s/N_2)]$$

$$= 1/\sqrt{N_1N_2} \left[\sum_{k=p}^{N_1+p-1} \sum_{\ell=q}^{N_2+q-1} a_{k,\ell} \exp i2\pi (kr/N_1 + \ell s/N_2) \right]$$

$$\times \exp[-i2\pi(pr/N_1 + qs/N_2)] = 1/\sqrt{N_1N_2} \left\{ \sum_{k=0}^{N_1-1} \sum_{\ell=0}^{N_2-1} a_{k,\ell} \right.$$

$$\times \exp[i2\pi(kr/N_1 + \ell s/N_2)] - \sum_{k=0}^{p-1} \sum_{\ell=0}^{q-1} a_{k,\ell} \exp[i2\pi(kr/N_1 + \ell s/N_2)]$$

$$- \sum_{k=p}^{N_1-1} \sum_{\ell=0}^{q-1} a_{k,\ell} \exp[i2\pi(kr/N_1 + \ell s/N_2)] - \sum_{k=0}^{p-1} \sum_{\ell=q}^{N_2-1} a_{k,\ell} \exp[i2\pi$$

$$(kr/N_1 + \ell s/N_2)] + \sum_{k=N_1}^{N_1+p-1} \sum_{\ell=q}^{N_2-1} a_{k,\ell} \exp[i2\pi(kr/N_1 + \ell s/N_2)]$$

$$+ \sum_{a=N_1}^{N_1+p-1} \sum_{\ell=N_2}^{N_2+q-1} a_{k,\ell} \exp[i2\pi(kr/N_1 + \ell s/N_2)] + \sum_{k=p}^{N_1-1} \sum_{\ell=N_2}^{N_2+q-1}$$

$$\left. \times a_{k,\ell} \exp[i2\pi(kr/N_1 + \ell s/N_2)] \right\} \times \exp[-i2\pi(pr/N_1 + qs/N_2)], \quad (5.110)$$

or, after grouping and substituting the variables,

$$\alpha_{r,s}^{(p,q)} = \alpha_{r,s}^{(0,0)} \exp[-i2\pi(pr/N_1 + qs/N_2)] + 1/\sqrt{N_1N_2} \left\{ \sum_{k=0}^{p-1} \sum_{=0}^{q-1} \right.$$

$$\times (a_{k+N_1,\ell+N_2} - a_{k,\ell}) \times \exp[i2\pi(kr/N_1 + \ell s/N_2)] + \sum_{k=p}^{N_1-1} \sum_{\ell=0}^{q-1}$$

$$\times (a_{k,\ell+N_2} - a_{k,\ell}) \times \exp[i2\pi(kr/N_1 + \ell s/N_2)] + \sum_{k=0}^{p-1} \sum_{=q}^{N_2-1} (a_{k+N_2,\ell} - a_{k,\ell})$$

$$\left. \times \exp[i2\pi(kr/N_1 + \ell s/N_2)] \right\} \exp[-i2\pi(pr/N_1 + qs/N_2)] \quad . \quad (5.111)$$

Thus, $\alpha_{r,s}^{(p,q)}$ is calculated from $\alpha_{r,s}^{(0,0)}$ and three truncated DFTs..

In particular cases of shifting in only one direction (say along ℓ),

$$\alpha_{r,s}^{(0,q)} = \left(\alpha_{r,s}^{(0,0)} + \left\{ 1/\sqrt{N_1N_2} \sum_{k=0}^{N_1-1} \sum_{=0}^{q-1} (a_{k,\ell+N_2} - a_{k,\ell}) \times \exp[i2\pi \right.\right.$$

$$\left.\left. \times (kr/N_1 + \ell s/N_2)] \right\} \right) \exp(-i2\pi q\, s/N_2) \quad . \quad (5.112)$$

If the shift is equal to one element, as with so-called slide processing,

$$\alpha_{r,s}^{(0,\ )} = \left[\alpha_{r,s}^{(0,0)} + 1/\sqrt{N_1N_2}\sum_{k=0}^{N_1-1}(a_{k,N_2} - a_{k,0}) \times \exp(i2\pi\ kr/N_1)\right] \times \exp(-i2\pi\ s/N_2) \quad . \tag{5.113}$$

In the 1-D case, clearly

$$\alpha_r^{(p)} = \left[\alpha_r^{(0)} + 1/\sqrt{N}\sum_{k=0}^{p-1}(\alpha_{k+N} - \alpha_k) \times \exp(i2\pi\ kr/N)\right]\exp(-i2\pi p\ r/N) \quad , \tag{5.114}$$

and with slide processing,

$$\alpha_r^{(1)} = [\alpha_r^{(0)} + 1/\sqrt{N}\,(\alpha_N - \alpha_0)]\exp(-i2\pi\ r/N) \quad . \tag{5.115}$$

If the above relations are used, there is a definite saving in the number of operations required for the calculation of local spectra. The extent of this saving depends on the size of the fragments, the degree to which they overlap, and the available capacity of the memory device used to store intermediate results.

5.8 Fast Algorithms for Calculating the DFT and Signal Convolution with Decreased Multiplication

We noted above that in existing digital processors, the operations of adding and multiplying two numbers are not equally fast. Usually multiplications take several times as long as additions. The recently developed algorithms for spectral analysis and signal convolution are therefore of great interest, as they require fewer multiplications than algorithms based on the DFT.

The simplest of these is the Winograd algorithm of fast signal convolution. Its purpose is to decrease the number of multiplications at the expense of a certain increase in the number of additions.

Let us write the formula for discrete convolution (4.16) in the following way:

$$b_k = \sum_{n=0}^{N-1} h_n a_{k-n} \quad . \tag{5.116}$$

We will consider all of the required $N+k-1$ samples of $\{a_n\}$ to be given. Also, let N be an even number; if N were odd, we would have to consider the sum over the N-1 largest terms.

Let us first calculate the paired products in the sequences $\{h_n\}$ and $\{a_{k-n}\}$:

$$\eta = \sum_{n=0}^{(N/2)-1} h_{2n}h_{2n+1} \quad ; \tag{5.117}$$

$$\zeta = \sum_{n=0}^{(N/2)-1} a_{k-2n} a_{k-(2n+1)} \quad . \qquad (5.118)$$

Then, as is easily checked,

$$b_k = \sum_{n=0}^{(N/2)-1} (h_{2n} + a_{k-(2n+1)})(h_{2n+1} + a_{k-2n}) - \eta - \zeta_k \quad . \qquad (5.119)$$

Thus, calculating the sequences $\{h_n\}$ and $\{a_n\}$ reduces to calculating the convolutions of two shorter sequences composed of the paired sums of the elements in the initial sequence. As for the supplementary members η in (5.117) and ζ_k in (5.118), η does not depend on k and must only be calculated once for all the convolution elements, and ζ_k can be calculated recurrently for odd and even k. Indeed, with even $k(k = 2k_1)$,

$$\zeta_k = \zeta_{2k_1} = \sum_{n=0}^{(N/2)-1} a_{2(k_1-n)} a_{2(k_1-n)+1} = a_{2k_1} a_{2k_1+1} + \sum_{n=0}^{(N/2)-1} a_{2(k_1-1-n)}$$
$$\times a_{2(k_1-1-n)+1} - a_{2(k_1-1-n)+1} - a_{2(k_1-N/2)} a_{2(k_1-N/2)+1} = a_k a_{k+1}$$
$$- a_{k-N} a_{k-N+1} + \zeta_{k-2} \quad . \qquad (5.120)$$

Precisely the same formula is applicable for odd k.

Let us evaluate the number of operations required to calculate the k values for convolution (5.116).[4] To calculate the k sums (5.119), $k(3N/2) - 1$ additions and $kN/2$ multiplications must be performed. Calculating η requires $(N/2) - 1$ additions and $(N/2)$ multiplications. To calculate $k/2$ values of even ζ_k requires $(N/2) - 1$ additions for ζ_0 plus two additions and two multiplications for each of the remaining $(k/2) - 1$ values of ζ_k, i.e., $2(k + (N/2) - 3$ additions and $2(k + (N/2) - 2$ multiplications. The total number of additions is

$$N_{ad} = 3NK\,[1 + (2k + 3N - 7)/3NK]/2 \quad , \qquad (5.121)$$

and the total number of multiplications is

$$N_{mul} = NK\,[1 + (4k + 3N - 8)/kN]/2 \quad . \qquad (5.122)$$

From this it is clear that in comparison with direct calculation (5.116), which requires $k(N - 1)$ additions and kN multiplications, the algorithm described requires approximately half as many multiplications and one and a half times as many additions. According to the evaluation in [5.16], this algorithm

---

[4] We will consider the times required for addition and subtraction to be equal.

is more efficient than the DFT when

$$k < 10(1 + \log_2 N) \quad . \tag{5.123}$$

This is, of course, a rough evaluation, and the limits for the applicability of the algorithms described are determined by the relation between addition time and multiplication time in the given processor.

The described method can in principle also be used to calculate 2-D convolutions:

$$b_{k,\ell} = \sum_{n=0}^{N_1-1} \sum_{m=0}^{N_2-1} h_{n,m} a_{k-n,\ell-m} \quad , \tag{5.124}$$

but an additional gain due to two-dimensionality is difficult to achieve here, since convolution is usually calculated one-dimensionally: first all possible values of k are run through, and then ℓ is changed by unity.

To calculate 2-D convolutions, NUSSBAUMER and QUANDALLE have recently proposed the use of so-called polynomial transforms [5.17,18], in which a substantial part of the multiplications is replaced by cyclical shift. According to the data in [5.19], the use of these transforms in calculating convolutions for an array of 256×256 elements gives almost double the saving in the number of multiplications and a 30% decrease in the number of additions in comparison with algorithms using the DFT. True, the actual gain is not so large: a FORTRAN program runs only 20% faster [5.19].

An algorithm for calculating the DFT with a decreased number of multiplications was also developed by WINOGRAD for cases in which the number of signal samples is a simple number or a power of a simple number [5.20,21]. It is based upon representation of a DFT as convolutions by means of special permutation, and upon fast calculation of the convolutions with a decreased number of multiplications. At present such algorithms are known for numbers $N \in \{2, 3, 4, 5, 7, 8, 9, 16\}$ [5.21]. If N is a composite number it may break down into simple factors from the stated row. If so, the factors of the DFT matrix factorize as in Sect.5.4 into several DFTs with a simple base; these are executed with the accelerated Winograd algorithm [5.22,23].

A class of algorithms used to calculate signal convolution and generally not requiring multiplication operations is based on so-called number theoretic transforms [5.20]. In these algorithms calculation does not take place as in ordinary mathematics, but in a finite field. As a consequence, multiplication operations are replaced by shifts, and moreover, there are no rounding errors. These advantages, however, are achieved at the expense of a strong link between the size of the field and the length of the sequences, a circumstance considerably limiting the applicability of such algorithms.

6 Digital Statistical Methods

This chapter describes digital methods of statistical measurement and simulation in picture processing. Section 6.1 discusses basic approaches to the statistical description of pictures from the standpoint of quality criteria. Digital methods of measuring the main statistical characteristics of pictures — the grey-level histogram, correlation functions, and spectra — are described in Sects.6.2,3. Section 6.4 presents efficient methods of generating pseudorandom numbers with prescribed statistical properties. Some approaches to measuring noise in noise-distorted pictures are given in Sect.6.5, together with the appropriate practical algorithms.

6.1 Principles of the Statistical Description of Pictures

According to the fundamentals of information theory and the theory of optimal signal reception, in processing problems pictures should be considered elements of statistical ensemble [6.1-6]. The content of the random variables forming this ensemble is determined by the type of information that the picture contains for the user.

From this standpoint it is convenient to distinguish between two types of pictures: those containing local information and those containing structural information, or texture.

Pictures of the first type carry information about many objects; about their presence or absence, their coordinates, their mutual arrangement, orientation, number, form, etc. Examples include pictures of the earth's surface used for geological surveys, X-ray and radiographic pictures used in medical diagnostics to discover and localize pathological changes in internal organs, etc. It is characteristic of pictures conveying local information that they can be divided into objects and background constituents.

Pictures containing structural information, or texture carry information about one object, and this information is contained within certain parameters characterizing the picture as a whole. It is not possible to separate textural pictures into objects and background constituents. This group includes pictures obtained in the analysis of microstructures in biology and materials research, as well as pictures of forest covering, agricultural crops, and sea surfaces, obtained by remote sensing of natural resources.

Classical methods and models from the theory of random processes [6.5,6] can be used in the statistical description of textural pictures.

From the point of view of this theory, a picture is fully described statistically by its multidimensional probability density—the probability that the picture will fall into a given element of the volume of signal space. As the dimensionality of picture signal space is usually very large (of the order of several hundred thousands to millions), it is necessary to use various simplified statistical models for the analytical definition of such multidimensional distributions.

The simplest model is the spatially homogenous random process with independent samples (the 2-D analogue of the stationary 1-D process). The multidimensional probability density of this process is the product of the 1-D probability densities of the values of the picture elements.

To take into account the mutual statistical dependence of the picture elements, Gaussian and Markovian models for correlated random fields are used. Detailed information on such models can be found in [6.7]. There have also been recent attempts to develop other statistical models that would take the peculiarities of pictures into account [6.8,9].

The statistical description of local information pictures is more complicated. It must be based on the statistical description of the objects imaged, their informational parameters, and of the process of transforming objects into pictures. The background constituents of such pictures must be considered to be fixed in many processing problems. This results in algorithms which adapt themselves to the background. Examples of such algorithms are described in Chaps.7-9.

The sections which follow in this chapter are devoted to digital methods of measuring the most important statistical characteristics used in classical picture models, as well as to the statistical characteristics of the random distortions which are encountered when correcting image-forming systems.

6.2 Measuring the Grey-Level Distribution

The simplest and at the same time one of the most important statistical characteristics of signals is the probability distribution of their values. With digital signals one can speak of the frequency with which the various values are encountered. This frequency, seen as a function of the values of the signal, is called the histogram of signal values [6.4].

The histogram describing the frequency with which the values of individual signal samples appear is called the *one-dimensional histogram*. The histogram characterizing the frequency with which the values of several signal samples simultaneously appear is called the *multidimensional histogram*.

The N-dimensional histogram, or the histogram of an N-component vector signal, can be constructed using a simple algorithm: as each signal sample is measured, a constant is added to the processor memory cell with the address

$$A = \sum_{i=1}^{n-1} m_i \prod_{k=0}^{i-1} M_k + m_0 + A_0 \quad ; \tag{6.1}$$

this constant may, for example, be inversely proportional to the number of samples being analyzed. Here A_0 is the initial address array of the histogram array; m_i is the number of the quantized value of the signal in the ith dimension; $m_i = 0, 1, \ldots, M_i - 1$; M_i is the maximum number of signal quantization levels in the ith dimension; and n is the dimensionality of the histogram and of the vector value of the signal being measured.

The histogram is formed as an N-dimensional array of data numbered in a suitable system with complex bases $\{M_0, M_1, M_2, \ldots, M_{n-1}\}$.

This procedure can be mathematically described as the averaging of the Kronecker δ-functions:

$$h(\overline{m}_0, \overline{m}_1, \ldots, \overline{m}_{n-1}) = 1/N \sum_{k=0}^{N-1} \delta\left[\prod_{i=0}^{n-1} (m_{i,k} - \overline{m}_i)\right] \quad , \tag{6.2}$$

where $h(\overline{m}_0, \overline{m}_1, \ldots, \overline{m}_{n-1})$ is an n-dimensional histogram, and $\{m_{ik}\}$ are the quantized values of the kth signal sample, or, for the 2-D signal,

$$h(\overline{m}_0, \overline{m}_1. \ldots, \overline{m}_{n-1}) = 1/(N_1 N_2) \sum_{k=0}^{N_1-1} \sum_{\ell=0}^{N_2-1} \delta\left[\prod_{i=0}^{n-1} (m_{i,k,\ell} - \overline{m}_i)\right] \quad . \tag{6.3}$$

This expression can also be regarded as a digital convolution, see (4.16,24), and in the 2-D case, a separable 2-D convolution. Clearly, to calculate the histograms, N addition operations must be carried out (in the 2-D case, $N_1 N_2$ operations).

The histogram can be used as a characteristic not only of the entire observed picture, but also of its separate parts or fragments. In the latter case it is said to be *local*.

In certain picture-processing problems, local histograms of overlapping picture fragments must be measured (see Sect.8.3 on sliding equalization).

In these cases it is prudent to take advantage of the possibility of writing (6.3) as a recurrence relation between histograms of neighbouring fragments. Indeed, if (6.3) is considered to be an expression for the local histogram of the (r,s)th fragment

$$h^{(r,s)}(\bar{m}_0, \bar{m}_1, \ldots, \bar{m}_{n-1}) = 1 / (N_1 N_2) \sum_{k=rk_0}^{N_1+rk_0-1} \sum_{\ell=s\ell_0}^{N +s\ell_0-1} \delta\left[\prod_{i=0}^{n-1} (m_{i,k,\ell} - \bar{m}_i)\right], \tag{6.4}$$

where k_0, ℓ_0 are the intervals between fragments along the two coordinates, then it can be expressed in term of another histogram, for example of the (r - 1, s)th fragment:

$$h^{(r,s)} (\bar{m}_0, \bar{m}_1, \ldots, \bar{m}_{n-1}) = h^{(r-1,s)}(\bar{m}_0, \bar{m}_1, \ldots, \bar{m}_{n-1}) + 1 / (N_1 N_2)$$

$$\times \sum_{k=(r-1)k_0}^{rk_0-1} \sum_{\ell=s\ell_0}^{N_0+s\ell_0-1} \left\{\delta\left[\prod_{i=0}^{n-1} (m_{i,k+N_1,\ell} - \bar{m}_i)\right] - \delta\left[\prod_{i=0}^{n-1} (m_{i,k,\ell} - \bar{m}_i)\right]\right\} . \tag{6.5}$$

The meaning of this formula is clear: the histogram of the given fragment can be derived from the histogram of the neighbouring fragment if one adds to it the difference of histograms calculated from those parts of the given fragment and the neighbouring fragment that do not simultaneously belong to both fragments.

As with recursive calculation of the DFT, the specific method of calculating local histograms recursively depends on the capacity of the memory available. The minimum buffer memory volume required is clearly $\Pi_{i=0}^{n-1} M_i$ memory cells for one histogram. In that case the recursion calculation of the histogram of one fragment with recursion along k, as in (6.5), requires $2k_0N_2$ plus $(N_1N_2) / N_t$ addition-subtraction operations, where N_t is the total number of fragments; i.e., with large N_t, approximately in factor of $N_1 / 2k_0$ fewer such operations than with direct calculation. Note that to decrease the number of initial fragments to one in this case, it is advantageous to move from one fragment to another in a zigzag manner: in the direction of increasing k for even s, and in the reverse direction for odd s.

Histograms measured on small picture fragment are usually fairly jagged functions. With an increase in the volume of measurement the histogram usually becomes smoother. However, it is sometimes necessary to derive a smooth histogram from small volume of measurements. There are three commonly used smoothing methods.

6.2.1 Step Smoothing

The range of values of the histogram argument is divided into a small number of intervals, and the histogram value within each small interval is replaced by average values over the larger interval. A smoothed histogram of this kind can be constructed immediately if, before measuring the histogram, one quantizes the signal value on a small number of levels (the same number as the number of divisions in smoothing).

6.2.2 Smoothing by Sliding Summation

The values of the smoothed histograms $\bar{h}(m_0, \bar{m}_1, \dots, \bar{m}_{n-1})$ are derived from the initial histogram by digital convolution with a certain smoothing function $w(r_0, r_1, \dots, r_{n-1})$:

$$\bar{h}(\bar{m}_0, \bar{m}_1, \dots, \bar{m}_{n-1}) = \sum_{r_0=-R_0}^{R_0} \sum_{r_1=-R_1}^{R_1} \dots \sum_{r_{n-1}=-R_{n-1}}^{R_n-1} \times w(r_0, r_1, \dots, r_{n-1})\, h(\bar{m}_0 - r_0, \bar{m}_1 - r_1, \dots, \bar{m}_{n-1} - r_{n-1}) \quad . \quad (6.6)$$

The simplest and most commonly used smoothing function is the rectangular "window":

$$w(r_0, r_1, \dots, r_{n-1}) = \begin{cases} 1, & |r_0| \leqslant R_0 \quad , |r_1| \leqslant R_1 \quad , \dots, |r_{n-1}| \leqslant R_{n-1} \ ; \\ 0 & \text{in the opposite case} \end{cases} \quad (6.7)$$

Step smoothing and smoothing by sliding summation are basically equivalent. This is easy to understand if we remember that step smoothing gives a function with a correspondingly lower number of samples, in view of which it can be treated as filtering (as with sliding summation) and discretization.

6.2.3 Smoothing with Orthogonal Transforms

The coefficients of the histogram representation $h(\bar{m}_0, m_1, \dots, \bar{m}_{n-1})$ are calculated on a certain orthogonal basis $\{\varphi_{s_0,s_1,\dots,s_{n-1}}(\bar{m}_0, \bar{m}_1, \dots, \bar{m}_{n-1})\}$:

$$\eta_{s_0,s_1,\dots,s_{n-1}} = \sum_{\bar{m}_0=0}^{M_0-1} \sum_{\bar{m}_1=0}^{M_1-1} \sum_{\bar{m}_{n-1}=0}^{M_{n-1}-1} h(\bar{m}_0, \bar{m}_1, \dots, \bar{m}_{n-1}) \times \varphi_{s_0,s_1,\dots,s_{n-1}}(\bar{m}_0, \bar{m}_1, \dots, \bar{m}_{n-1}) \quad . \quad (6.8)$$

Some of the coefficients $\{\eta_{s_0,s_1,\dots,s_{n-1}}\}$ (usually those with low values) are replaced by zeros. The smoothed histogram is obtained by inverse transformation after rejecting the coefficients $\{\eta_{s_0,s_1,\dots,s_{n-1}}\}$.

The histogram smoothed in this way can also be derived immediately during the measuring process if the approach described in [6.10] is used. It is based on the following considerations. Let us substitute (6.2) into (6.8):

$$\eta_{s_0,s_1,\dots,s_{n-1}} = 1/N \sum_{\overline{m}_0=0}^{M_0-1} \sum_{\overline{m}_1=0}^{M_1-1} \dots \sum_{\overline{m}_{n-1}=0}^{M_{n-1}-1} \sum_{k=0}^{N-1} \delta\left(\prod_{i=0}^{n-1} (m_{i,k} - \overline{m}_i)\right)$$

$$\times \varphi_{s_0,s_1,\dots,s_{n-1}}(\overline{m}_0, \overline{m}_1, \dots, \overline{m}_{n-1}) = 1/N \sum_{k=0}^{N-1} \varphi_{s_0,s_1,\dots,s_{n-1}}(m_{0,k}, m_{1,k}, \dots, m_{n-1,k}) \quad . \tag{6.9}$$

This means that the coefficients $\{\eta_{s_0,s_1,\dots,s_{n-1}}\}$ of the expansion $h(\overline{m}_0, \overline{m}_1, \dots, \overline{m}_{n-1})$ on the basis $\{\varphi_{s_0,s_1,\dots,s_{n-1}}(m_0, m_1, \dots, m_{n-1})\}$ can be found by averaging over all the signal samples projected onto the basis functions, calculated each time from the values of the observed samples. If a smoothed histogram is wanted, the values $\eta_{s_0,s_1,\dots,s_{n-1}}$ simply need not be calculated and are replaced by zeros for those $\{s_i\}$ which have corresponding coefficients. This saves computer time and memory.

Smoothing by orthogonal transformation is of especial interest when dealing with multidimensional distributions, if step smoothing is undesirable (for example because it gives an interrupted, step-like function). Then the procedure described for obtaining smoothed histograms using averaged basis functions can result in significant savings, since instead of $\Pi_{i=0}^{n-1} M_i$ memory cells, only $\Pi_{i=0}^{n-1} s_i$ cells are required to accumulate a histogram (s_i is the number of non-zero representation coefficients on the ith coordinate).

6.3 The Estimation of Correlation Functions and Spectra

Correlation functions and power spectra are commonly used to describe so-called stationary ergodic random processes (in the 2-D case, spatially homogeneous ergodic random fields). For such processes these concepts are defined in the following way[1] [6.11].

The cross-correlation functions of processes a(t) and b(t) are

$$R_{a,b}(\tau) = \lim_{T\to\infty} 1/T \int_{t_1}^{t_1+T} a(t)b(t+\tau)dt \quad . \tag{6.10}$$

[1] For multidimensional signals, the arguments t and s are understood to be vectorial.

The auto-correlation function of process a(t) is

$$R_a(\tau) = \lim_{T\to\infty} 1/T \int_{t_1}^{t_1+T} a(t)a^*(t+\tau)dt \quad . \tag{6.11}$$

The power spectrum of process a(t) is

$$A(f) = \int_{-\infty}^{\infty} R_a(\tau)\exp(i2\pi f\tau)d\tau \quad . \tag{6.12}$$

By applying sampling theory (Sect.3.3), we can obtain the corresponding definitions for discrete processes, i.e., for the results of continuous process discretization:

$$R_{a,b}(n) = \lim_{N\to\infty} 1/N \sum_{k=0}^{N-1} a_k b_{k+n}^* \quad ; \tag{6.13}$$

$$R_a(n) = \lim_{N\to\infty} 1/N \sum_{k=0}^{N-1} a_k a_{k+n}^* \quad ; \tag{6.14}$$

$$A(s) = \sum_{n=0}^{N_n-1} R_a(n)\exp(i2\pi\, ns/N_n) \quad , \tag{6.15}$$

where $\{a_k\}$, $\{b_k\}$, $\{R_a(n)\}$, $\{A(s)\}$ are the samples of the processes, the auto-correlation function, and the power spectrum, respectively, and N_n is the number of samples of the correlation function, considered as a signal.

These formulae assume that the signals being analyzed are of infinite length. For digital signals the number of samples, N, can only be considered more or less large, and for larger values than N the sample values are not defined. For this reason the formulae with finite N for correlation functions

$$R_{a,b}(n) = 1/N \sum_{k=0}^{N-1} a_k b_{k+n}^* \quad , \tag{6.16}$$

$$R_a(n) = 1/N \sum_{k=0}^{N-1} a_k a_{k+n}^* \tag{6.17}$$

are considered a means of statistically estimating continuous correlation functions from their digital representations, and the power spectrum, which is derived with the aid of discrete Fourier transforms from the estimated correlation function, is considered an estimate of the continuous power spectrum.

In calculating with (6.16,17), boundary effects are encountered, as in general with digital filtering. Therefore the same methods of predetermining signal samples as were considered in Sect.4.5 are also applied here. If the missing samples are replaced by zeros, the resulting estimate of the correlation functions by (6.16,17) is biased, since some of the terms drop out (the larger n is, the more drop out). In this case it is advisable to use a modified formula of the form

$$R_{a,b} = 1/(N-n) \sum_{k=0}^{N-n-1} a_k b^*_{k+n} \quad ; \qquad (6.18)$$

it gives the best estimate of the correlation functions of the continuous signal for large n with fixed N.

The equations (6.16,17) for correlation functions are related to (4.16) for digital convolutions. Therefore, the correlation functions are calculated by the same algorithms and by methods of signal predetermination similar to those used in calculating convolutions: direct calculation by (6.16,17), fast algorithms with a decreased number of multiplications (as in Sect.5.8; see also [6.12]), and calculation using DFT and SDFT (Sect.4.8). With SDFT, an estimate of the values of the correlation functions of signals can be obtained at any point between their samples. In [6.13,14] methods using Walsh transforms for calculating correlation functions have been developed.

In evaluating the power spectrum of a picture, the squared modulus of its DFT or (if the value of the spectrum at arbitrary points is required) SDFT is usually used, owing to the fact that for the calculation of DFT or SDFT it is possible to take advantage of FFT algorithms. According to sampling theory the frequency resolution of such a method is equal to the width of the spectrum (for 2-D signals – to the area of the spatial spectrum), divided by the number of samples in the sequence obtained by discretization.

If the signals being analyzed are considered to be the realization of a certain ensemble of random signals, then to obtain the spectra characterizing the signal ensemble as a whole, it is necessary to smooth the spectrum estimates found for the separate realizations. To do this, methods are used which are analogous to those applied for the smoothing of distribution estimates. The following are the most important of these [6.11].

6.3.1 Averaging Local Spectra

The signal being analyzed is divided into fragments whose size corresponds to the given resolution of the frequency analysis. The smoothed estimate of the spectrum, A(s), is found to be the average

$$A(s) = 1/k \sum_{i=0}^{k-1} A_i(s) \qquad (6.19)$$

of the estimates of the spectrum, $A_i(s)$, for each fragment (local spectra).

6.3.2 Masking (Windowing) the Process by Smooth Functions

If the realization of the process being analyzed is not long enough for the above method to be used, it is multiplied by a smooth function, the so-called window function, which decreases more or less rapidly towards the edges. Then the power spectrum of the masked process is found.

6.3.3 Direct Smoothing of Spectra

Direct smoothing of spectra is achieved by convolving the derived estimate of the power spectrum with a normalized smooth function extending over several samples:

$$A(s) = \sum_{k=-k_0}^{k_0} A(s+k)w(k) \quad ; \tag{6.20}$$

$$\sum_{k=-k_0}^{k_0} w(k) = 1 \quad . \tag{6.21}$$

This method is less effective for computing, but it can be useful in cases when both smoothed and unsmoothed estimates of the spectrum are required.

6.4 Generating Pseudorandom Numbers

In digital picture processing and digital simulation of image-forming systems it is frequently necessary to create sequences of pseudorandom numbers with given statistical properties (Sect.7.2). The usual means of obtaining such sequences is first to generate, with the aid of sufficiently simple algorithms, independent pseudorandom numbers with uniform distributions, and then to subject them to linear and nonlinear transformations so that they exhibit the desired statistical properties [6.15-17]. The most common requirements are that the Gaussian distribution law be valid and that a given correlation function or power spectrum hold for the numbers.

Constructing an algorithm that generates pseudorandom numbers with uniform distribution is a fairly complicated and extremely interesting problem. By means of simple algorithms that are realized with a relatively small number of instructions, number sequences must be found which from the standpoint of the problem to be solved can be considered random sequences not described by simple laws. Usually, the main problem is to ensure the statistical independence of the numbers in the sequence.

To obtain sequences of pseudorandom numbers with a uniform distribution, the so-called congruent method [6.17] is commonly used. In it every successive number ξ_k in a sequence is obtained from every previous ξ_{k-1} using the simple relation

$$\xi_k = (c_1\xi_{k-1} + c_2) \bmod c_3 \quad , \tag{6.22}$$

where c_1, c_2, and c_3 are certain constants.

The initial number ξ_0 usually has little influence on the quality of the resulting sequence. The constant c_3 is defined by the word-length of the processor used. The periodicity of the sequences depends on it, and therefore it is desirable to select as large a value for it as possible. The constant c_2 has little influence on the properties of the sequence and can even be chosen to be zero [6.17]. The choice of the constant c_1 is the most critical Recommendations regarding this choice can be found in [6.17].

Common statistical criteria are applied to verify the extent to which the statistical properties of the pseudorandom numbers satisfy the given requirements. For example, we may consider whether their distribution law can be considered uniform, or whether the numbers are independent or non-correlated.

In picture-processing problems and in the simulation of picture transformation systems, high demands are made on the spatial correlation of the pseudorandom 2-D sequence used. To check the independence of the derived numbers, it is most convenient to apply the properties of the visual system to find a regular picture structure. To do this, the values of the sequence elements are represented as the brightness of the picture elements, and thus the field of pseudorandom numbers is converted into a picture with the aid of the corresponding devices — filmwriters or displays. If observations of such a picture does not reveal a regular structure, the pseudorandom numbers are considered independent.

To obtain a Gaussian distribution from independent pseudorandom numbers with uniform distribution, it is simplest to use the central limit theorem of probability theory, according to which the sum of a sufficiently large number of independent random values has a distribution which approaches a Gaussian distribution as the number increases. The corresponding linear transformation of the uniform distribution is conveniently performed by the discrete Fourier transform, which is implemented by one of the FFT algorithms [6.18]. Let us take a closer look at this method.

Let $\{\xi_k^{re}\}$, $\{\xi_k^{im}\}$, $k = 0, 1, \ldots, N-1$ be two sequences of equally distributed non-correlated numbers. Let us form a sequence of complex numbers $\{\xi_k^{re} + i\xi_k^{im}\}$ and multiply every member of this sequence by the samples of an even sequence $\{h_k = h_{N-k}\}$. Then, by discrete Fourier transformation of this modified sequence we obtain the complex numbers

$$\eta_\ell = \eta_\ell^{re} + i\eta_\ell^{im} = 1/\sqrt{N} \sum_{k=0}^{N-1} h_k(\xi_k^{re} + i\xi_k^{im}) \exp(i2\pi\, k\ell/N) \quad , \tag{6.23}$$

where the real and imaginary parts have a distribution that is close to Gaussian distribution.

It can be shown [6.18] that the multidimensional characteristic function calculated for the numbers $\{\eta_\ell^{re}\}$ and $\{\eta_\ell^{im}\}$ approaches the characteristic Gaussian distribution function as $O(1/\sqrt{N})$. Thus for large N, the quality of the derived Gaussian distribution can be very high.

To illustrate this, the solid line in Fig.6.1 shows on the Gaussian scale a plot of the empirical distribution of pseudorandom numbers of which there are 262,144, obtained by the method described for sequences with 2048 numbers ($N = 2048$). The dashed line is the exact Gaussian distribution. Figure 6.1 shows that departures from the Gaussian distribution are appreciable only for probabilities below 10^{-4}.

In [6.18] it is also shown that the auto-correlation functions of the real and imaginary parts of the transformed sequence are proportional to the DFT of the set of coefficients $\{h_k^2\}$:

$$R(\ell_1 - \ell_2) = \sigma_\xi^2/N \sum_{k=0}^{N-1} h_k^2 \exp[i2\pi\, k(\ell_1 - \ell_2)/N] \quad , \tag{6.24}$$

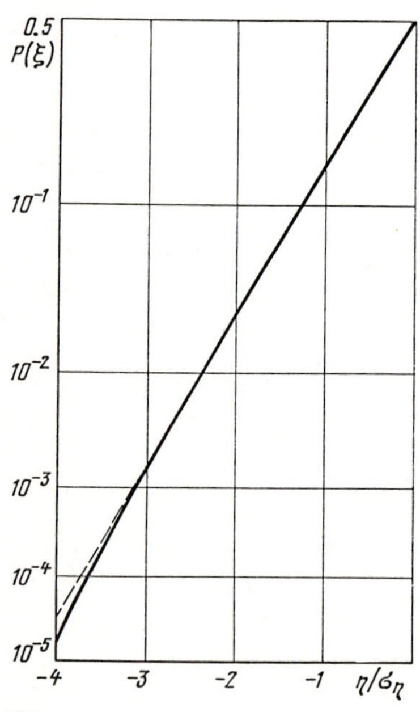

Fig.6.1. The empirical distribution function (solid line) and theoretical distribution function (dashed line) of Gaussian pseudorandom numbers

where σ_ξ^2 is the second moment of the pseudorandom numbers of the initial sequence, and the real and imaginary parts are mutually uncorrelated. Hence it follows that the coefficients h_k^2 must be chosen as samples of the required power spectrum of the sequence. With the method described, the two mutually uncorrelated Gaussian number sequences with the given power spectrum or correlation functions can be obtained from two sequences of uncorrelated, uniformly distributed numbers.

In applying this method, one should bear in mind the following normalizing relations:

$$\sigma^2_{\eta^{re}} = \sigma^2_{\eta^{im}} = \sigma_\xi^2 / N \sum_{k=0}^{N-1} h_k^2 \quad , \tag{6.25}$$

$$\overline{\eta} = 1 / N \sum_{\ell=0}^{N-1} \eta_\ell = h_0 / \sqrt{N} \, (\overline{\xi}_0^{re} + i\overline{\xi}_0^{im}) \quad . \tag{6.26}$$

The first of these relations defines the connection between the variances σ^2 of the numbers of the initial and transformed Gaussian sequences, and the second defines the mean value $\overline{\eta}$ of the transformed Gaussian sequence. For the Gaussian sequences to have zero means, it is sufficient to set

$$\xi_0^{re} = \xi_0^{im} = 0 \quad . \tag{6.27}$$

To obtain the given average value it is generally necessary to assign the appropriate determined (and not random) value to the zero element of the initial sequence.

The method described has a number of advantages over other methods of generating Gaussian sequences [6.16,17]. The approximation to the normal distribution law is substantially improved while the amount of computer time expended is the same, and the initial sequences of independent numbers are used much more economically: to generate one point in the Gaussian distribution, exactly one number of the initial sequence of numbers with uniform distribution is used. This last point is especially important in picture processing and simulation, when an array must be formed out of millions of samples without repetition.

6.5 Measuring Picture Noise

Being formed by imaging systems, pictures are usually distorted by random interference, or noise. If these distortions are to be taken into account, the statistical characteristics of the noise must be known. Sometimes these characteristics can be determined from the structure and specifications of the imaging system. For example, the noise from the grainy texture of film in photographic systems is determined by the type of film and the development process, while noise in television systems is determined by the power of the signal in the communication channel. If these data are not available, one has to evaluate the noise characteristics from previously obtained picture or set of pictures distorted by the same kind of noise. In such cases the statistical parameters of the noise must be deduced from measurements of the statistical characteristics of the observed video signal.

At first glance there may appear to be an inherent contradiction in this task: in order to evaluate the noise parameters, noise must be extracted from the observed signal, and this can only be achieved if the statistical parameters of the noise and the signal are known. This vicious circle is avoided not by separating the signal and the noise in order to determine the statistical characteristics of the noise, but by separating their characteristics on the basis of measurements of the characteristics of the observed noisy signal.

This task of separation can be solved as a deterministic problem if the characteristics of the non-distorted signal are precisely known, or as a statistical problem of evaluating parameters. In the latter case the signal characteristic being analyzed must be considered a random value if it is a number, or a random process if it is a sequence, and the characteristics found for the observed signal must be considered their realizations.

As in statistical decision theory, the optimal procedure for evaluating the noise parameters in this approach must, in principle, be constructed on the basis of statistical models of the characteristics being analyzed. These must be constructed and firmly established for each characteristic chosen.

Fortunately, in the majority of practical cases the interference that must be suppressed in the signal consists of very simple objects, statistically speaking; they are described by a small number of parameters, and the characteristics of the undistorted signal depend mainly on the background picture and are weakly dependent on any particular features. Therefore the problem of evaluating the noise parameters can be solved by relatively simple means even if the statistical properties of the video signal characteristics being measured are, a priori, given in a rather crude form. One need merely select from all of the available statistical characteristics of the signal those in which distortion of the signal by noise should be most easily discoverable from the anomalous behaviour of these characteristics.

It is not our aim here to construct an exhaustive theory of anomaly detection and estimation, but we shall describe two quite universal methods of detection which are based on general, a priori preconditions of smoothness in the characteristics of the undistorted signal: the prediction method and the voting method [6.19]. On the theoretical level these methods are related to recently developed robust methods of parameter evaluation [6.20].

6.5.1 The Prediction Method

For every given element of the sequence under analysis a value is found which is predicted from the preceding elements that have already been studied, and the observed and predicted values are then compared. If the differences exceed a certain threshold, it is concluded that anomalous bursts of noise are present.

The number of elements used for forecasting and the method for determining the predicted value and threshold must be assigned a priori for the given class of background pictures.

6.5.2 The Voting Method

The voting method is a generalization of the method of median smoothing [6.7, 21], and is based on the construction of rank statistics. Every element of the sequence being analyzed is considered simultaneously with a certain number 2n of its neighbouring elements (n on each side). These (2n + 1) values are ranked in increasing or decreasing order, and a check is made of whether the value of the given element falls within some number k of values from the end (i.e., the largest or the smallest values) of the ordered sequence. The values n and k are assigned a priori on assumptions concerning the characteristics of the undistorted signal. If this check yields a positive result, it is concluded that the value of the given element is anomalously large (or, small). The). voting method is based on the assumption that the "normal" characteristic being analyzed is, as a rule, locally monotonic, and that deviations from local monotonicity, if any, are not large.

6.5.3 Measuring the Variance and the Auto-Correlation Function of Additive Wideband Noise

Fluctuating and additive noise which is independent of the video signal is described by its variance and auto-correlation function. If, as often happens, the noise is uncorrelated or only weakly correlated, a simple algorithm can be constructed to define its variance and its correlation function. This algorithm is based on the measurement of the correlation function of the observed picture.

Owing to the additivity and independence of the noise, the auto-correlation function $R_0(k,\ell)$ measured from Q observed pictures of size $N_1 \times N_2$ elements is the sum of the auto-correlation function $R_p(k,\ell)$ of the noise-free picture, the noise auto-correlation function $R_\xi(k,\ell)$, and the realization of a certain random process $\varepsilon(k,\ell)$ which characterizes the measurement error in a realization of finite size:

$$R_0(k,\ell) = R_p(k,\ell) + R_\xi(k,\ell) + \varepsilon(k,\ell) \quad . \tag{6.28}$$

The variance of the random process $\varepsilon(k,\ell)$ is inversely proportional to the square root of the size of the sample QN_1N_2 on which the measurement $R_0(k,\ell)$ is made [6.11]. This value is usually sufficiently large for the measurement on single picture to be adequate, since in the pictures used in practice the number of samples usually exceeds hundreds of thousands. Therefore the random error $\varepsilon(k,\ell)$ in (6.28) is sufficiently small for $R_\xi(k,\ell)$ to be evaluated as

$$R_\xi(k,\ell) = R_0(k,\ell) - R_p(k,\ell) \quad . \tag{6.29}$$

Let us first look at the case of uncorrelated noise, for which

$$R_\xi(k,\ell) = \sigma_\xi^2 \, \delta_{k,\ell} \quad , \tag{6.30}$$

where σ_ξ^2 is the noise variance and $\delta_{k,\ell}$ is the Kronecker δ-function. Thus, the auto-correlation function of the observed picture differs from the auto-correlation function of the noise-free picture only at the origin of coordinates, and the difference is equal to the noise variance

$$\sigma_\xi^2 = R_0(0,0) - R_p(0,0) \quad . \tag{6.31}$$

For all other values of (k,ℓ) it can be used as an estimate of $R_p(k,\ell)$:

$$R_0(k,\ell) = R_p(k,\ell) \quad . \tag{6.32}$$

As measurements of the auto-correlation functions show [6.22,23] close to the origin ($k=0$, $\ell=0$) they are the extremely slowly changing functions of k and ℓ. Therefore the value $R_p(0,0)$, which is needed to calculate the variance of uncorrelated noise by (6.31), can be evaluated with great accuracy by extrapolating from values $R_p(k,\ell) = R_0(k,\ell)$ at the points k,ℓ close to zero. Thus, to determine the variance of additive uncorrelated noise in pictures, it is sufficient to measure the auto-correlation function $R_0(k,\ell)$ of the observed picture in a small area around the point 0, and find by extrapolation the estimate $\hat{R}_p(0,0)$ of $R_p(0,0)$. The quantity

$$\sigma_\xi^2 = R_0(0,0) - \hat{R}_p(0,0) \quad . \tag{6.33}$$

is then a good estimate of the variance. Experiments show that a good estimate is obtained even with extrapolation on a 1-D cross-section of the auto-correlation function [6.24].

This approach can also be used to estimate the variance and the auto-correlation function of weakly correlated noise, i.e., noise with an auto-correlation function $R_\xi(k,\ell)$, that differs from zero only in a small area near the origin, where the values of the auto-correlation function of the noise-free picture can be satisfactorily extrapolated from the values $R_0(k,\ell)$ at those points where $R_\xi(k,\ell)$ vanishes.

In both cases, the noise is measured with a minimum of a priori information about the correlational properties of the noise and the picture: the approximate size of the domain within which the non-zero values $R_\xi(k,\ell)$ are concentrated, and smoothness of $R_p(k,\ell)$ in the vicinity of this domain.

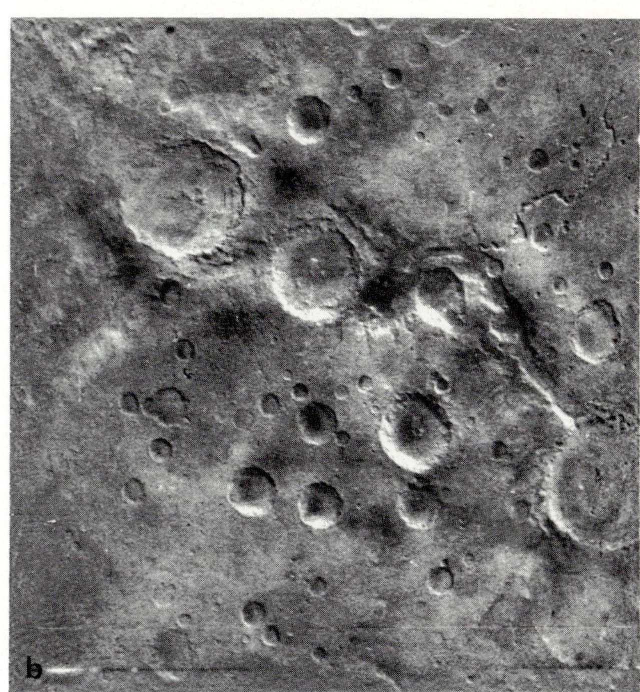

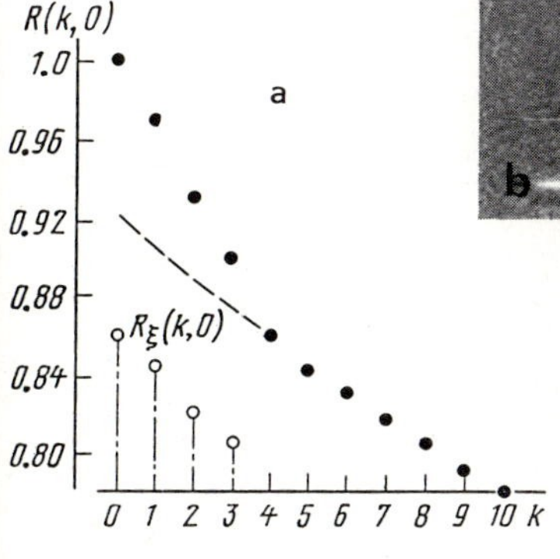

Fig.6.2a,b. Measurement of additive wide-band noise in pictures. The dots in (a) denote the one-dimensional cross-section of the empirical correlation function of the picture shown in (b); the dashed line shows extrapolation of the correlation function to the points 0,1,2,3; and the open circles show the estimation of the noise correlation function

Figure 6.2a illustrates this method. The points show the value of the 1-D cross-section $R_0(k,0)$ of the auto-correlation function of the pictures in Fig.6.2b for $k = 0, 1, \ldots, 10$ [6.24]. These points clearly reveal a discontinuity in the auto-correlation function at the point $k = 4$. The dashed line shows the extrapolated values $R_0(k,0)$ of the picture auto-correlation function close to zero, constructed from the values in the domain $k > 3$; the dot-dashed lines show the difference $R_0(k,0) - \hat{R}_0(k,0)$, which serves as an estimate of the auto-correlation function.

6.5.4 Evaluation of the Intensity and Frequency of the Harmonic Components of Periodic Interference and Other Types of Interference with Narrow Spectra

Periodic (moiré) interference arises primarily in television and photographic transmission systems in which the video signal is transmitted by a radio channel. Sometimes it appears as a result of the discretization of pictures containing details with high-frequency periodic structures (moiré effect), or as an interference effect in pictures received in coherent optical imaging systems.

The characteristic feature of this kind of interference is that its spectra in the Fourier basis contain only a small number of components that differ significantly from zero. This type of interference also includes interference with narrow spectra on other bases. At the same time, the spatial frequency spectra of noise-free pictures in the Fourier basis and some other well-known bases are usually more or less smooth and monotonic functions. Therefore the presence of narrow-band noise appears in the form of anomalous large and localized deviations, or bursts, in the distorted picture spectrum. In contrast to the above-considered case of fluctuating noise, which gives bursts at the origin of the auto-correlation function, the location of the bursts in this case is not usually known. These anomalous bursts can be localized by applying the above-described methods of prediction or voting.

To do this the mean squared modulus of the spectral components of the noisy signal $\langle|\overline{\beta_s}|^2\rangle$ in the chosen basis is found by averaging over all the observed pictures having the same type of periodic interference. If 1-D filtering is carried out, say, along the picture rows, then averaging can also take place over all the picture rows being filtered. Then, the localized components of noise are discovered by means of the prediction method or the voting method, that is, those spectral components $\langle|\overleftarrow{\beta_{r,s}}|^2\rangle$ of the observed signal are noted that are distorted by noise.

Owing to the additivity of noise, the values $\langle|\overline{\beta_{r,s}}|^2\rangle$ are equal to the sum of the spectral components of the noise-free signal, $\langle|\alpha_{r,s}|^2\rangle$, and of the noise $|\overline{\varkappa_{r,s}}|^2$ (see Sect.7.5). Consequently,

$$|\overline{\varkappa_{r,s}}|^2 = <|\overline{\beta_{r,s}}|^2> - <|\alpha_{r,s}|^2> \quad . \tag{6.34}$$

Using the a priori smoothness of the spectrum of the noise-free signal, the values $<|\alpha_{r,s}|^2>$, which are necessary for calculation by (6.34), can be found by interpolation from the values closest to the values $<|\beta_{r,s}|^2>$ not distorted by noise.

6.5.5 Evaluation of the Parameters of Pulse Noise, Quantization Noise and Strip-Like Noise

The term "pulse noise" is used to describe anomalously large distortions of individual video signal samples. These distortions can be recognized by applying prediction or voting methods directly to the video signal (for more details see Sect.7.4).

Quantization noise is typical of digital systems. It is defined in terms of the number of quantization levels of the video signal (Sect.3.10). This number can be determined by constructing a histogram of the signal values and calculating the number of values for which the histogram differs from zero. Thus, by using the prediction or voting methods to find "gaps" in the histogram, which is measured on a sufficiently large set of pictures, and by measuring the depth of the "gaps", one can estimate the probability that these or other quantization levels will be missing.

Part 2

Picture Processing

7 Correcting Imaging Systems

In principle, picture processing starts with the correction of distortions introduced by the imaging system. This chapter treats basic problems of correction and the appropriate methods and algorithms.

In Sect.7.1 the principal types of distortions are characterized and the problem of correction is formulated. Methods of suppressing additive noise and pulsed noise are described in Sects.7.2,3, while Sect.7.4 deals with the correction of so-called linear distortions such as defocusing and blur. Some digital methods of correcting nonlinear brightness distortions are described in Sect.7.5.

7.1 Problem Formulation

The initial assumption in correcting imaging systems is that there exists a certain function, say $a(t_1, t_2)$, which describes the picture at the output of an ideal imaging system, and that the action of a real imaging system can be described by a certain transformation F of the ideal picture into the one actually observed:

$$b(t_1, t_2) = F[a(t_1, t_2)] \quad . \tag{7.1}$$

Then, if certain parameters of the transformation F are known, the problem of correction becomes a matter of finding a correcting transformation Φ of the observed picture such that its result

$$\hat{a}(t_1, t_2) = \Phi[b(t_1,t_2)] \tag{7.2}$$

is as close as possible to the ideal picture as measured by certain similarity criteria or metrics. Thus defined, this problem is called picture restoration [7.1-5].

The method by which the problem is solved depends on the transformation F.

In practice, in describing transformations of signals, and especially such complex signals as pictures, a hierarchic description of transformations is used as a combination of certain elementary transformations (Sect.2.5). In addition to the linear transforms and pointwise, nonlinear transforms (discussed in Sect.2.6) which, individually or in various combinations, usually though not necessarily describe so-called deterministic distortions in imaging systems, models of additive, pulse, and multiplicative noise are used to describe random perturbations of the signal.

The model of *additive* noise is used when the signal, at the output of the system or in some intermediary stage of transformation, can be considered the sum of a useful signal and a certain random signal (noise). The graininess of a photographic film and the fluctuating noise in television systems are described in this manner.

If the noise effect does not appear throughout the length of the signal, but only at randomly distributed points, where the signal values are replaced by random values, the noise is called *pulse* noise or pepper-and-salt noise. Pulse noise is characteristic of picture-transmitting systems in radio channels using nonlinear types of signal modulation (frequency modulation, pulse-position modulation, etc.) and of digital transmitting systems and picture storage [7.6,7].

The *multiplicative* model is used when the useful signal can be considered to be multiplied by a random signal. Examples are, to a first approximation, the effect of the noise of photomultipliers and of speckle noise in coherent optical and other holographic imaging systems.

Since the logarithm of a product is equal to the sum of the logarithms of the factors, and since the exponent of a sum is equal to the product of the exponents of the terms, the additive and multiplicative models can, in certain cases, be reduced to each other by introducing logarithmic or exponential transforms [7.8].

Linear and nonlinear distortions can also have a random character. They are then described as transformations whose parameters (e.g., frequency response of the distorting linear systems) are random variables or functions.

There is now an extensive literature on the various aspects of picture-restoring problems, particularly those concerning the elimination of defocusing (blur), and ways of solving them, both by analogue means (mainly using coherent optics) and by digital means (see the reviews [7.2,5,9-12].

We cannot hope to give an exhaustive presentation of all that has been achieved solving these problems in this relatively short chapter. Therefore, we shall present only a few characteristic examples illustrating the possibilities of applying digital methods to solve certain specific practical problems. Special attention is paid to the little-discussed adaptive approach to the synthesis of correction algorithms. This approach is based on concepts relating to the statistical description pictures conveying local information, which were considered in Sect.6.1.

7.2 Suppression of Additive Noise by Linear Filtering

In many cases, the effect of noise in an imaging system can be described as the addition to a noise-free video signal $a(t_1, t_2)$ of a random signal $\xi(t_1, t_2)$ which is independent of it — the additive noise. The problem of restoring the picture then consists in obtaining from the total observed signal

$$b(t_1, t_2) = a(t_1, t_2) + \xi(t_1, t_2) \tag{7.3}$$

an estimate $\hat{a}(t_1, t_2)$ of the signal $a(t_1, t_2)$ that is as close as possible to $a(t_1, t_2)$ as measured by some metric which is defined by the criteria for accuracy of picture reproduction. As the noise is random, this evaluation is the best only in the statistical sense.

The simplest way of suppressing additive noise is by linear filtering of the noisy signal. The parameters of the necessary filter are usually found by applying the principles of optimal (Wiener) filtering, in which the mean squared error averaged over the ensemble of pictures and interferences is minimized; these are considered to be statistically independent, spatially homogeneous random fields [7.6,14-16]. The choice of parameters for the Wiener filter in discrete picture representation on an arbitrary basis is discussed in [7.3,16].

As already noted in Sect.6.1, the model of pictures as homogeneous random fields corresponds more or less satisfactorily to pictures containing structural information with a homogeneous texture. Pictures conveying local information require separate statistical descriptions of the objects being interpreted and those in the background. With such pictures it is in many cases more natural not to average the correction error over the ensemble of backgrounds, but to find the minimum error for the given background. Let us find the laws for determining the optimum parameters of the linear filter, taking these considerations into account. Let $\varphi_s(k)$ be the basis functions of a certain orthonormal basis; and let $\{\alpha_s\}$, $\{\varkappa_s\}$, $\{\beta_s\}$ be the sequence representation coefficients of the samples of a noise-free video signal $\{a_k\}$, the noise $\{\xi_k\}$ and their sum $\{b_k = a_k + \xi_k\}$ on this basis. Further, let N be the number of samples in the picture being processed, or its sub-pictures[1], and $\lambda_{s,n}$ be the coefficients of the discrete representation of the linear filter, which is used to derive the signal estimate (Sect.4.1);

$$\hat{\alpha}_s = \sum_{n=0}^{N-1} \lambda_{s,n}\beta_n \quad . \tag{7.4}$$

We now seek coefficients $\lambda_{s,n}$ such that the ensemble average of the squared difference between the restoration and the original

$$\langle\overline{|\varepsilon|^2}\rangle = \left\langle \overline{\sum_{k=0}^{N-1} \left|a_k - \sum_{s=0}^{N-1} \alpha_s\varphi_s(k)\right|^2} \right\rangle = \left\langle \overline{\sum_{k=0}^{N-1} \left|\sum_{s=0}^{N-1} (\alpha_s - \hat{\alpha}_s)\,\varphi_s(k)\right|^2} \right\rangle , \tag{7.5}$$

averaged over the noise ensemble and the restored signal is minimum. The bar over this expression indicates averaging over the noise ensemble; summation over k gives the mean square of the error along the realization of the signal; and the angle brackets denote averaging over the signal ensembles, including averaging over any random parameters of the objects being interpreted and (but not necessarily) over the ensemble of background pictures. Since the basis $\{\varphi_s(k)\}$ is orthonormal, by Parseval's relation (2.34)

$$\langle\overline{|\varepsilon|^2}\rangle = \left\langle \overline{\sum_{s=0}^{N-1} |\alpha_s - \hat{\alpha}_s|^2} \right\rangle \quad . \tag{7.6}$$

Substituting α_s from (7.4) we obtain

$$\langle\overline{|\varepsilon|^2}\rangle = \left\langle \sum_{s=0}^{N-1} |\alpha_s|^2 \right\rangle + \left\langle \sum_{s=0}^{N-1}\sum_{n=0}^{N-1}\sum_{m=0}^{N-1} \lambda_{sn}\lambda_{sm}^{*}\overline{\beta_n\beta_m^{*}} \right\rangle - \left\langle \sum_{s=0}^{N-1} \alpha_s \sum_{m=0}^{N-1} \lambda_{sm}^{*}\overline{\beta_m^{*}} \right\rangle$$

$$- \left\langle \sum_{s=0}^{N-1} \alpha_s^{*} \sum_{n=0}^{N-1} \lambda_{sn}\overline{\beta_n} \right\rangle \quad . \tag{7.7}$$

The optimum value λ_{sn} is the solution of the system of equations

$$\left\{\frac{\partial\langle\overline{|\varepsilon|^2}\rangle}{\partial\lambda_{sn}} = 0\right\} \quad , \quad \left\{\frac{\partial\langle\overline{|\varepsilon|^2}\rangle}{\partial\lambda_{sm}^{*}} = 0\right\} \quad , \qquad \text{or} \tag{7.8}$$

$$\left\{\sum_{m=0}^{N-1} \lambda_{sm}^{*}\langle\overline{\beta_n\beta_m^{*}}\rangle = \langle\alpha_s^{*}\overline{\beta_n}\rangle\right\} \quad , \quad \left\{\sum_{n=0}^{N-1} \lambda_{sn}\langle\overline{\beta_n\beta_m^{*}}\rangle = \langle\alpha_s\overline{\beta_m^{*}}\rangle\right\} \quad . \tag{7.9}$$

If, as usually happens, the noise is zero-mean, then

$$\overline{\beta}_n = \overline{\alpha_n + \varkappa_n} = \alpha_n \quad , \tag{7.10}$$

[1] For the sake of brevity we shall use 1-D notation. To switch to the 2-D case, the indices k, s, and n need merely be replaced by pairs of indices.

As the equations (7.9) for λ_{sn} and λ_{sm} are equivalent, only one needs to be solved. By substituting (7.10) into (7.9) we obtain the following system of equations for λ_{sn}:

$$\left\{\sum_{n=0}^{N-1} \lambda_{s,n} \langle \overline{\beta_n \beta_m^*} \rangle = \langle \alpha_s \alpha_m^* \rangle \right\} \quad . \tag{7.11}$$

For the so-called eigenbasis, for which

$$\langle \alpha_s \alpha_m^* \rangle = \langle |\alpha_s|^2 \rangle \delta_{s,m} \quad , \quad \overline{\varkappa_n \varkappa_m^*} = \overline{|\varkappa_m|}^2 \delta_{n,m} \quad , \tag{7.12}$$

(7.11) is also degenerate and it has solutions

$$\lambda_{s,n} = \langle |\alpha_s|^2 \rangle / \langle |\overline{\beta_s}|^2 \rangle \ \delta_{s,n} = \lambda_s \delta_{s,n} \tag{7.13a}$$

or since $\langle |\overline{\beta_s}|^2 \rangle = \langle |\alpha_s|^2 \rangle + |\overline{\varkappa_s}|^2$, for signal-independent noise,

$$\lambda_{s,n} = \frac{\langle |\overline{\beta_s}|^2 \rangle - |\overline{\varkappa_s}|^2}{\langle |\overline{\beta_s}|^2 \rangle} \quad . \tag{7.13b}$$

This filter is called a scalar or masking filter, since it is described by a diagonal matrix, and the representation coefficients of the filtered signal $\hat{\alpha}_s$ are found by simple multiplication of the coefficients β_s of the noisy signal by λ_s, or by "masking" by the sequence $\{\lambda_s\}$:

$$\hat{\alpha}_s = \lambda_s \beta_s \quad . \tag{7.14}$$

Since the matrix of the representation coefficients (7.13) is diagonal, only one multiplication operation is necessary for each $\hat{\alpha}_s$, and not N operations, as in the general case for the filter of (7.11). In this respect (7.14) is preferable to (7.11). Operating in the same way as with (7.11), we obtain the result that for an arbitrary basis, (7.13) determines the best filter among the masking filters described by diagonal matrices. The point of this equation is simple: the greater the intensity of the noise present in the spectral components of the signal being filtered in comparison with the intensity of the signal, the more the filter weakens these components.

Equations (7.13) are the analogue of the classical Wiener filter formula [7.6,14-16]:

$$H(f) = P_s(f) / [P_s(f) + P_n(f)] \quad , \tag{7.15}$$

where H(f) is the transfer function of the optimal filter, and $P_s(f)$ and $P_n(f)$ are the power spectral densities (energy spectra) of the signal and the noise. This filter is optimal as regards the minimum mean-square error when separating a stationary Gaussian process from its sum with stationary Gaussian noise. The eigenbasis of the ensemble of discrete signals $\{a_k\}$, defined by (7.12), is, correspondingly, the discrete analogue of the kernel of the continuous Fourier transform, which corresponds to stationary random processes (in the 2-D case, spatially homogeneous ones).

However, (7.13a,b) differ in two major respects from the formula (7.15) for Wiener filtering. First, according to (7.13) the optimal filter is not defined by the statistical power spectral density of the signal and the interference, understood as spatially homogeneous random fields, but by the square of the moduli of the spectra of the observed noisy picture, averaged over the ensemble of noise realizations, over the random parameters of the objects of interpretation, and possibly, though not obligatorily, over the ensemble of the background of the picture. If, as is commonly the case, we are interested in the filtering quality for a given background, and the area occupied by the objects of interest in the picture is small in comparison with the area of the picture being processed, then the spectrum $\langle|\beta_s|^2\rangle$, averaged over the random parameters of these objects and the noise, practically does not differ from the spectrum $|\beta_s|^2$, averaged only over noise. The averaging of the spectrum over noise and the evaluation of the power spectral density of the noise $|\varkappa_s|^2$ can in certain cases be carried out over the observed noisy picture (see Sect.6.5). This means that the optimal filter (7.13) can be constructed from the results of measurement of the parameters of the observed noisy picture, and it turns out to be adaptable to this picture.

Second, if the criteria of filtering quality make it necessary to average the filtering error over the sub-picture samples of the picture being processed, the filter is then locally adaptive. Such local criteria are natural for spatially non-homogeneous pictures.

Experience shows that filtering of additive noise by means of the optimal Wiener linear filters usually does not result in very high quality. For example, using a filter of type (7.15) to suppress noise with a uniform power spectrum ($|\varkappa_s|^2$ = const) on a basis of complex exponential functions results in weakening of the high spectral components of the picture, i.e., in smoothing so that the quality of a picture after Wiener filtering seems to be even worse than that of the noisy picture before filtering [7.2,17]. This is probably explained by the fact that classical Wiener filtering does not take into account the spatial inhomogeneity of real pictures and ignores the individuality of the background pictures.

A characteristic example of successful use of linear filters to suppress additive noise is the filtering of strongly correlated additive noise, whose power spectrum has only a small number of components differing markedly from zero, i.e., noise whose realization is a combination of a small number of basis functions with random intensity. An example of this is periodic sinusoidal interference, which is typical of certain picture-transmission systems. An example of a picture with such interference is shown in Fig.7.1.

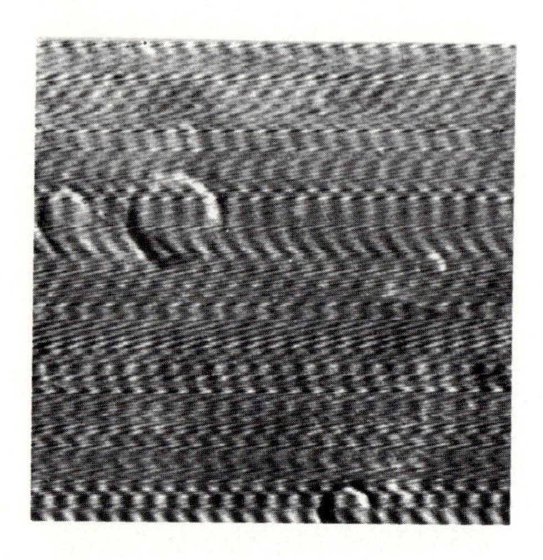

Fig.7.1. Picture with periodic noise

The filter (7.13), designed to suppress such interference, allows the spectral components of the video signal to pass without being weakened where the noise intensity is zero, and significantly weakens the components where the noise intensity is high. When the intensity of the noise components is much higher than the corresponding signal components $\langle|\alpha_s|^2\rangle$, the filter (7.13) is approximated well by the "rejection" filter, which fully suppresses the spectral components of the signal that are strongly affected by noise:

$$\lambda_s = \begin{cases} 1, & \overline{|\varkappa_s|^2} = 0; \\ 0, & \langle|\alpha_s|^2\rangle \leq \overline{|\varkappa_s|^2} \end{cases} \quad . \tag{7.16}$$

In terms of calculation the rejector filter is even simpler than filter (7.13).

The best basis for sinusoidal periodic interference is, in principle, the basis of complex exponential functions, because in this case the spectral components of the interference are the most concentrated. For periodic interference in the form of rectangular waves, the Walsh functions can be used.

If the interference is essentially 1-D, as for example in Fig.7.1, then both the interference analysis and the filtering can be 1-D.

The spectral composition and intensity of periodic interference are not usually known beforehand. Therefore the first stage in processing is to evaluate the statistical parameters of the signal and the noise (see Sect.6.5). To construct a rejection filter (7.16), one need only find the location of noise bursts. To construct filter masks (7.13), it is also necessary to evaluate the intensity of the spectral components of the noise and signal.

Direct moiré filtering consists in calculating orthogonal transforms on the chosen basis of the spectrum of the observed signal $\{\beta_s\}$, masking it by coefficients $\{\lambda_s\}$ (7.14), and restoring the signal by inverse transformation. With rejection filtering, the spectral components of the signal that qualified as noisy at the detection stage are simply set to zero. In another version of the rejection filter, individual components of the signal being filtered that were identified as noisy during the detection stage are calculated and substracted from the signal. If the number of such components is not large, this method can be more economical than methods using direct and inverse signal transformation, even when fast algorithms are used.

Figure 7.2a illustrates these methods of filtering periodic interference. It shows on a logarithmic scale a graph of the intensity of the horizontal spectral components of the Walsh transform for the picture shown in Fig.7.1. Figure 7.2b shows the plot of the coefficients $\{\lambda_s\}$ of the masking filter. The result of using this filter in the Walsh transform domain for the picture shown in Fig.7.1 is presented in Fig.7.3.

Experiments with filtering periodic interference in pictures show that rejection filtering gives results that do not differ visually from those obtained with a masking filter. However, with rejection filtering more care must be taken to extend the signal appropriately, in order to combat boundary effects. Moreover, the presence in the picture of highly contrasting details with sharp edges (such as the edge between a planet and space in space pictures; deep shadows; etc.) can lead, in rejection filtering, to the appearance of periodic distortions in the region of these edges, because the assumptions concerning the smoothness of the spectrum are not valid for these details. In such cases, special methods are employed to remove these details from the picture before filtering and to restore them afterwards. One such method is described in Sect.9.5.

Another case in which additive interference with a localized spectrum effectively undergoes linear filtering is strip-type interference with an uneven background, which arises in certain phototransmission systems [7.18,19].

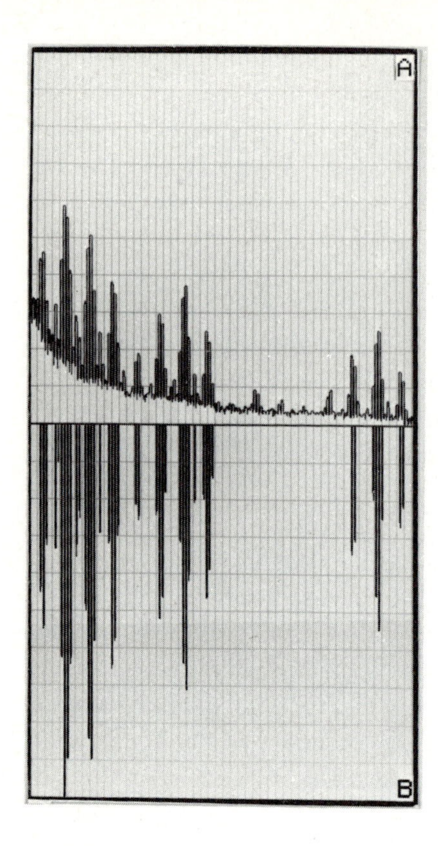

◄Fig.7.2a,b. (a) Graph of the intensity of the horizontal spectral components of the Walsh transform for the picture shown in Fig.7.1. (b) Plot of the coefficients $\{\lambda_s\}$ of the optimal masking filter

Fig.7.3. Result of filtering in the Walsh transform domain of the picture shown in Fig.7.1

An example of a picture with such interference is shown in Fig.7.4. The discrete Fourier spectrum of the interference is concentrated in the very low spatial frequency domain: the spectrum of the strips is concentrated at the frequencies $\alpha_{0,s}$ and $\alpha_{r,0}$, depending on their direction (r and s are the numbers along the axes in the DFT frequency plane), and the spectrum of the background is concentrated in the coefficients near the origin ($r=0$, $s=0$).

It is convenient to filter such interference by processing not in the spectral, but in the spatial domain. Thus, in [7.18,19] filtering with twofold signal processing by means of 1-D recursion filters of type (4.29)

$$\hat{a}^{(1)}_{k,\ell} = \left(a_{k,\ell} - 1/(2N_1+1) \sum_{m=-N_1}^{N_1} a_{k+m,\ell}\right) + \bar{a} \quad ,$$

$$\hat{a}_{k,\ell} = \left(\hat{a}^{(1)}_{k,\ell} - 1/(2N_2+1) \sum_{n=-N_2}^{N_2} a^{(1)}_{k,\ell+n}\right) + \bar{a} \quad , \tag{7.17}$$

is described where $\bar{a}$ is a constant which is equal to half the maximum value

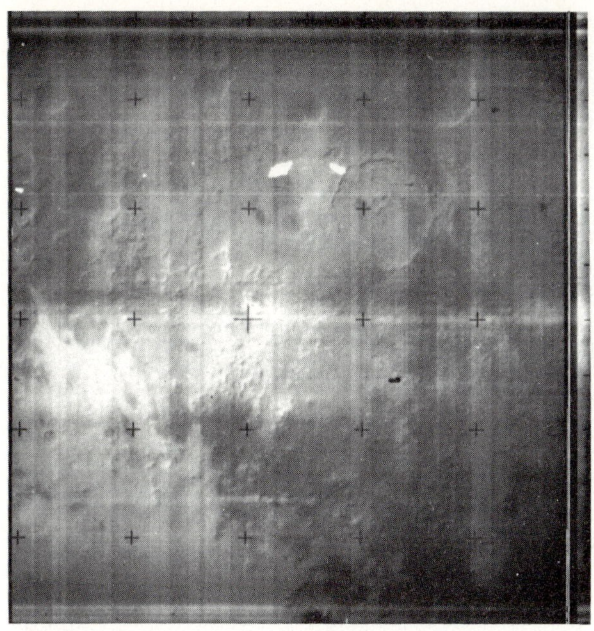

Fig.7.4. Example of a picture distorted by strip-type interference and unevenness of background

of the video signal and is used to evaluate the unknown mean for the entire frame, i.e., $\alpha_{0,0}$. Each 1-D filter equalized the local average signal value and suppressed the strips in the corresponding direction. In view of the size of the background spots the parameters N_1 and N_2 were chosen to be 256 (in certain cases 128) elements with a general frame size of 1024×1024 elements. In the spectral domain, these two 1-D filters fully suppressed the components $\beta_{0,0}$, replacing it by a quantity proportional to $\bar{a}$, and weakened the spectrum components $\alpha_{1,r}$ and $\alpha_{s,1}$ by nearly three times (with $N_1 = 256$). The cross-section of the filter frequency response (7.17) is shown in Fig.7.5. The result of processing the picture in Fig.7.4 in this way is presented in Fig.7.6. On the basis of these photographs it must be emphasized once again that in rejection filtering of interference with concentrated spectra, it is necessary

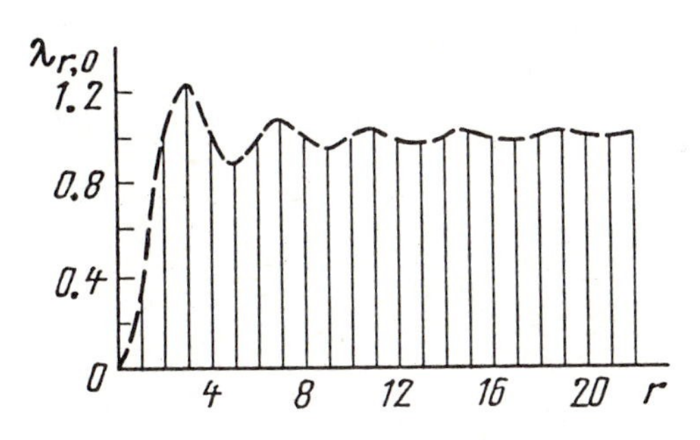

Fig.7.5. Cross-section of the frequency transfer function of filter (7.17)

Fig.7.6. Result of filtering the picture shown in Fig.7.4

to remove details with sharp boundaries from the picture. In the given case these were the benchmarks (Figs.7.4,6). Before the photograph shown in Fig.7.4 was filtered, the crosses were removed using the non-linear procedure of detection and smoothing described in Sect.9.5, and after filtering they were restored [7.18-20].[2]

7.3 Filtering of Pulse Interference

As was described in Sect.7.1, pulse (pepper-and-salt) interference causes, with a certain probability (error probability), the value of the video signal to be replaced by a random magnitude. In pictures this interference has the appearance of isolated contrasty points. Figure 7.7 shows a picture with an error probability of 0.3 per picture element.

The statistical properties of pulse interference differ sharply from those of the picture. Slowly changing videosignals that differ little in value from element to element are characteristic of pictures, while rapid, abrupt changes

[2] Note that the photographs of the surface of Mars transmitted by "Mars 5" and processed in this way were later used to synthesize a colour picture [7.21]. The average value of the signal influences the average colour tone of the colour picture in this case. It was therefore made more precise by means of data from direct astronomical observation of Mars from Earth.

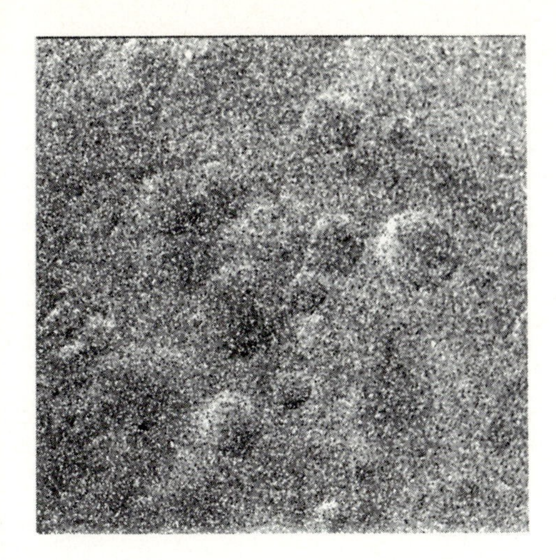

Fig.7.7. Picture distorted by pulse noise

are rare and form extended contours. Pulse interference, in contrast, appears as large, single, isolated bursts. This is why it is visually very easy to distinguish bursts of noise from the picture, even though such noise has a strong interference effect.

Owing to this sharp difference between noise and picture, the filtering algorithm for pulse noise is extremely simple [7.22]. It consists of two operations: finding the distorted picture elements and correcting the value of the video signal in these elements. As the video signal value in each picture element is likely to be close to that of neighbouring elements, noise bursts can be detected by the prediction method, i.e., by comparing the value of the video signal at every element with the value predicted for it from neighbouring elements. If the modulus of the difference exceeds a certain threshold δ_1, noise is considered to be present. The rule for correcting erroneous values can also be based on the smoothness of the picture. By applying one of these rules, one can replace the distorted value of the video signal by the predicted one with the addition of a certain constant δ_2, which is chosen to prevent blurring arising from averaging the video signal. The point is that as a result of correction, the contrast (difference) of the given element relative to that of the surrounding elements decreases to the value of δ_2. Thus the constant δ_2 must be below the threshold for visual detection of single-element details and above the visual threshold for the detection of extended edges. Then, the elimination of pulse noise does not lead to the visual impression that extended edges present in the picture have been lost.

Since the video signal value used for prediction can in turn be distorted, the detection algorithm must be iterative and start with a high value of the detection threshold, δ_1. Experiments have shown that to obtain high-quality filtering, 3 - 4 iterations are sufficient. Processing can be speeded up with the help of the recursive filter, and by using for prediction not all the elements in the region, but only the corrected elements preceding the one in question.

These considerations are confirmed by the formal solution of a problem of filtering pulse noise [7.22].

The digital realization of the procedure described is simple. The value of the video signal $a_{k,\ell}$ at the element (k,ℓ) of the digital picture is transformed as follows:

$$b_{k,\ell} = \begin{cases} a_{k,\ell} & , \text{ if } |a_{k,\ell} - \bar{a}_{k,\ell}| < \delta_1 \quad ; \\ \bar{a}_{k,\ell} + \delta_2 \operatorname{sign}(a_{k,\ell} - \bar{a}_{k,\ell}) & , \text{ if } |\bar{a}_{k,\ell} - a_{k,\ell}| \geq \delta_1 \end{cases} \quad , \tag{7.18}$$

where $\bar{a}_{k,\ell}$ is the value of the video signal at the picture element under analysis, predicted from the neighbouring elements that have already been corrected.

Simple linear prediction can be used to calculate this predicted value; the value is found to be the weighted sum of the values of the neighbouring elements. Good results are obtained from the following formula for prediction in terms of the four closest picture elements to the left and above on the raster (the idea is that processing takes place successively from left to right along the rows, and from top to bottom beginning with the zero line):

$$\bar{a}_{k,\ell} = 0.3(b_{k,\ell-1} + b_{k-1,\ell}) + 0.2(b_{k-1,\ell-1} + b_{k+1,\ell-1}) \quad . \tag{7.19}$$

In [7.22] it is shown that the detection threshold δ_1 is determined by the variance of the prediction error $(a_{k,\ell} - \bar{a}_{k,\ell})$ of the video signal over the region and by the probability of the appearance of pulse noise per picture element. These parameters are usually not known a priori. However, both they and the threshold value δ_1 can usually be evaluated from the observed distorted picture by measuring the histogram of the modulus of the difference signal $|a_{k,\ell} - a_{k,\ell}|$ over the unprocessed picture. This histogram usually has two parts: one which falls rapidly and one in which the values of the histogram change much more slowly (Fig.7.8). The narrow, rapidly falling part is determined by natural variations in the video signal. The slow part of the histogram, which falls nearly linearly, originates in pulse interference. The coordinate of the boundary between these parts (the break in the histogram) is a good estimate of the threshold δ_1. As for the constant δ_2, it is usually of the order of 3 - 4 percent of the range of the video signal values.

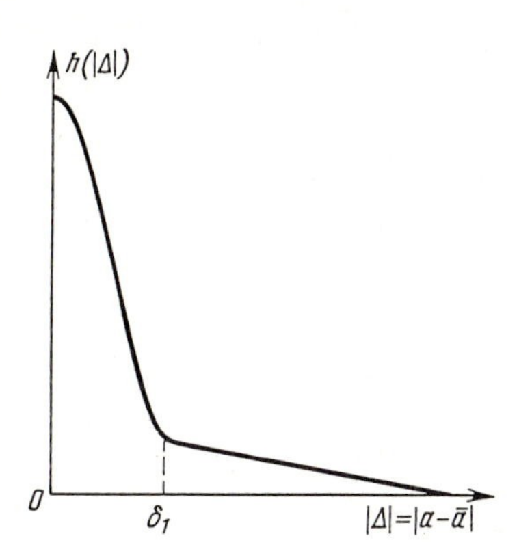

Fig.7.8. Plot of the histogram of difference signal module for the picture shown in Fig.7.7

Fig.7.9. Result of filtering the picture shown in Fig.7.7

Figure 7.9 shows the result of filtering the interference in the picture in Fig.7.7 by applying the algorithm described. Despite its simplicity, this algorithm is a very effective means of improving the quality of pictures distorted by pulse interference. With it, the demands placed on systems of picture transmission along channels with interference can be reduced and the quality of reception markedly increased [7.7]. It has found practical application in processing photographs from unmanned spacecraft [7.18,19,23].

7.4 Correction of Linear Distortion

The majority of imaging systems can be considered, at least to a first approximation, to be linear and shift-invariant. One of the basic characteristics of such a system is its impulse response or the Fourier transformation of the latter—the frequency response, which shows how the system transmits the spatial frequencies present in the picture. The ideal image-generating system is one that does not alter the spatial frequency spectrum of the signal, and hence has uniform frequency response within the spatial frequency domain occupied by the picture spectrum. The characteristics of real imaging systems, be they optical, photographic or television systems, differ from this ideal. As a result, pictures produced by such systems undergo distortion, called *linear distortion*. This distortion usually arises because the imaging system attenuates the high spatial frequencies of the picture. Visually this leads to a reduction of picture sharpness.

Since the Fourier spectrum of the signal at the output of a shift-invariant filter is equal to the product of the spectrum of the input signal and the frequency response of the filter, (2.114), in the absence of noise the problem of correcting linear distortions reduces to finding a linear filter, the frequency response of which is the inverse of that of the system in the frequency domain occupied by the picture. If the system weakens certain spatial frequencies, then the more the frequencies are weakened, the more the restoring filter must strengthen them.

Usually, however, the signal being corrected contains noise. The action of the inverse filter can lead to enhancement of this noise, which will be worse the greater the original attenuation. Therefore inverse filtering, though it improves one indicator of picture quality, namely sharpness, can worsen another, namely noisiness. Clearly, there is an optimum degree of correction at which the sharpness is sufficiently restored and yet the noise is not too great. To find this level, criteria must be selected for the accuracy of picture reproduction.

If the difference between the restored and the ideal picture is evaluated on the basis of the least-squares criterion applied to the modulus of the error, averaged over ensembles of pictures and interferences treated as spatially homogeneous random fields, then the continuous frequency response $H_r(f_1, f_2)$ of the optimum restoring filter is defined in terms of the frequency response $H(f_1, f_2)$ of the original system by [7.2]:

$$H_r(f_1, f_2) = \frac{H^*(f_1, f_2)}{|H(f_1, f_2)|^2 + \varepsilon^2} , \qquad (7.20)$$

where * denotes complex conjugation and ε^2 is a correction term equal to the ratio of the Fourier power spectrum of the noise $P_n(f_1, f_2)$ and that of the picture ensemble $P_p(f_1, f_2)$. This is known as the Wiener filter.

As the pictures conveying local information require separate averaging of the correction error over random parameters of the objects being interpreted and the background, the classical Wiener restoring filter is not the best, as in simple filtering of additive noise. We will therefore consider how a linear correcting filter should be constructed if it is to be optimal in terms of the criteria of type (7.5), given the need to average the error separately over the objects being interpreted and the background. We now analyze this task for the case of discrete representation of signals and operators.

Let the signal undergoing correction, $\mathbf{b} = \{\beta_s\}$, be the result of applying a linear operator Λ to a certain undistorted signal $\mathbf{a} = \{\alpha_s\}$, together with additive independent noise $\mathbf{x} = \{\varkappa_s\}$:

$$\mathbf{b} = \Lambda\mathbf{a} + \mathbf{x} \quad . \tag{7.21}$$

We will consider the linear distortion to be shift-invariant, so that with proper attention to the boundary effects, the operator Λ is described on the DFT or cosine transform basis (Sects.4.6,7) by the diagonal matrix $\Lambda = \{\lambda_s\}$. We therefore assume that the observed signal is also represented on this basis. We will limit ourselves to finding the optimal scalar correction filter $H = \{\eta_s\}$. For such a filter, the average value of the squared modulus of the correction error for each of the N or sub-picture samples is, if the value is averaged over the ensemble of noise and over the variations in the parameters of the objects and (if necessary) of the background:

$$\langle \overline{|\varepsilon|^2} \rangle = \left\langle \sum_{s=0}^{N-1} \overline{|\alpha_s - \eta_s(\lambda_s\alpha_s + \varkappa_s)|^2} \right\rangle \quad . \tag{7.22}$$

This will be minimal if, as for (7.13b),

$$\eta_s = \begin{cases} (1/\lambda_s)(\langle\overline{|\beta_s|^2}\rangle - \overline{|\varkappa_s|^2})/\langle\overline{|\beta_s|^2}\rangle & , \lambda_s \neq 0 \quad , \\ 0, & \lambda_s = 0 \quad , \end{cases} \tag{7.23}$$

where $\langle\overline{|\beta_s|^2}\rangle$ and $\overline{|\varkappa_s|^2}$ have the same meanings and definition as in (7.13b). Equation (7.23) is analogous to (7.20) in the same way that (7.13) was to (7.15). The differences between (7.23) and (7.20) are also the same.

Another useful correction criterion is that the signal spectrum must be restored faithfully [7.3]. This requires the power spectrum of the signal after correction to coincide with the spectrum of the undistorted signal. In our treatment this means that after correction, the signal must have a spectrum coinciding with the spectrum of the undistorted signal, averaged over the variations in the object's features and (if necessary) over the background variations. In this case the optimum correction filter is defined by

$$\eta_s = \begin{cases} (1/\lambda_s)((\langle\overline{|\beta_s|^2}\rangle - \overline{|\varkappa_s|^2})/|\beta_s|^2)^{1/2} & , \lambda_s \neq 0 \quad , \\ (\langle|\alpha_s|^2\rangle / |\beta_s|^2)^{1/2} \quad , & \lambda_s = 0 \quad , \end{cases} \tag{7.24}$$

where $\langle|\alpha_s|^2\rangle$ are the values of the spectrum of the undistorted picture for those numbers s for which $\lambda_s = 0$; the values are averaged over the variations in the object's features and (if necessary) the background. They must be known a priori or be found by interpolation of the values $|\lambda_s|^{-2}(\langle\overline{|\beta_s|^2}\rangle - \overline{|\varkappa_s|^2})$ from the neighbouring points on the assumption that the spectrum is smooth, as was done in the diagnosis of noise with a narrow spectrum (Sect.6.5).

Experimental data indicate that good results are obtained in correcting picture distortion if one uses, as the a priori given undistorted picture spectrum $\langle|\alpha_s|^2\rangle$, a certain typical spectrum for the given class of background pictures which corresponds to the picture being corrected. For example, if a page of printed text is being corrected, an averaged spectrum of texts with the same type-face as the text being corrected can be used. If an aerial photograph is being corrected, the averaged spectrum of aerial photographs showing similar landscapes and having the same scale are used.

In our treatment, such a typical spectrum can be considered to be the evaluation of the picture spectrum, averaged over the variations in the object's features. Denoting it by $|\tilde{\alpha}_s|^2$, we obtain for the criterion of signal-spectrum restoration:

$$\eta_s = ((|\lambda_s|^2)^{1/2} / \lambda_s)(|\tilde{\alpha}_s|^2 / |\beta_s|^2)^{1/2} \quad . \tag{7.25}$$

Hence it follows that to correct linear distortions one need only know the phase characteristics of the distortion systems. Specifically, if the system does not distort the phase of the spectral components of the picture, then

$$\eta_s = (|\tilde{\alpha}_s|^2 / |\beta_s|^2)^{1/2} \quad , \tag{7.26}$$

i.e., the frequency response of the correction filter does not depend on the distorting system. This means that if only the intensity values of the spectral components of the picture are distorted, correction is possible even with an unknown distortion law.

Image-forming systems that do not distort the signal phase form a fairly important and widespread class. It includes systems of observing through a turbulent atmosphere [7.3], systems with Gaussian apertures (that is, practically all systems in which the picture is formed by electron beams). An example of correcting blurred images by such a method is shown in Fig.7.10.

In correcting linear distortions it should be taken into consideration that the correction usually takes place before picture synthesis. The photographic or other recording device used to reproduce the corrected picture also has a frequency response that differs from the ideal one, and thus must be taken into account during correction. Let the continuous frequency response of the system proceeding the part where correction occurs be denoted by $H_1(f_1, f_2)$, and the frequency response of the rest of the system by $H_2(f_1, f_2)$. Then the continuous frequency response of the correcting filter, which is optimal in terms of the mean-square, is defined by

$$H_k(f_1, f_2) = \frac{H_1^*(f_1, f_2)\, H_2^*(f_1, f_2)}{|H_1(f_1, f_2)|^2 |H_2(f_1, f_2)|^2 + |H_2(f_1, f_2)|^2 \varepsilon^2} \quad . \tag{7.27}$$

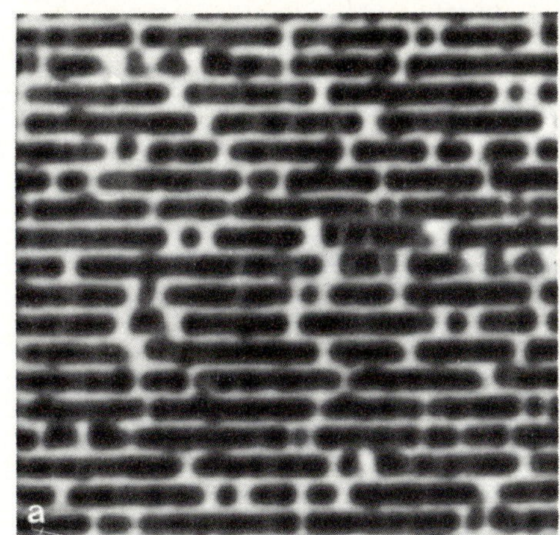

Fig.7.10a,b. Correction of defocusing by an unknown distortion law: (a) picture defocused in a system with a Gaussian aperture, (b) result of correction

It should also be noted that noise in digital processing includes not only noise present in the video signal being corrected but also the noise of quantization and rounding errors which is associated with the limited accuracy of calculation in the processor.

There are two possible ways of realizing the correction filter digitally: by processing the discrete Fourier spectrum using the FFT algorithm (Sect.4.8), or by using digital filters in space domain (Sects.4.2,3). In the first case, samples of the frequency response (7.6) are multiplied by samples of the discrete spectrum of the video signal being corrected, after which the inverse DFT is used. To weaken the boundary effects of filtering it is useful to extend the signal evenly (Sect.4.5)[3] and use combined DFT and SDFT algorithms (Sect.5.6) to reduce processing time. In the second case, the frequency response (7.6) is used to find the samples of the impulse response of the digital filter (Sects.4.3,4, where their relationship is discussed). It is also convenient here to apply even signal extension to weaken boundary effects.

The choice between these two methods of realization is determined by the size of the calculation and the capacity of the memory. In practice, if the digital correction filter cannot be satisfactorily approximated by a separable and recursive one (Sect.4.3), then processing in the spectral domain using

---

[3] In restoring pictures, a cyclicity characteristic of DFT can sometimes also be taken into account in the restoration algorithm, and the missing signal samples can be determined in accordance with the restoration criteria [7.24].

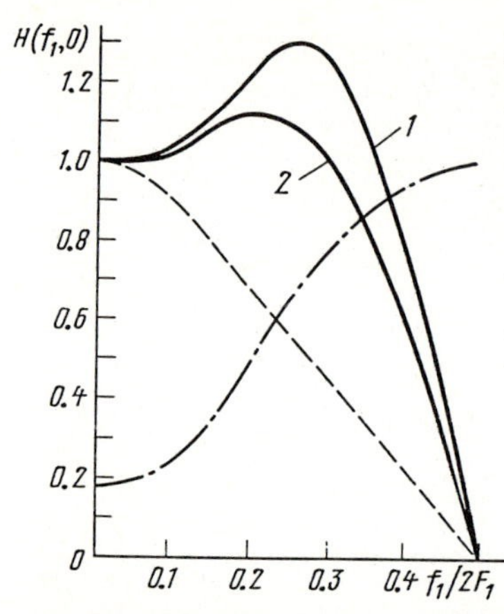

Fig.7.11. Correction of a phototelevision system [7.18]. The dashed line shows the cross-section of the frequency response of the system undergoing correction, and the dot-dashed line the correction frequency response (7.29) for $g = 4$, $N_1 = N_2 = 1$. Curve 1 is the frequency response of a system with correction filter, and Curve 2 the frequency response of the corrected system with account taken of the photo-recording device used to reproduce the corrected picture

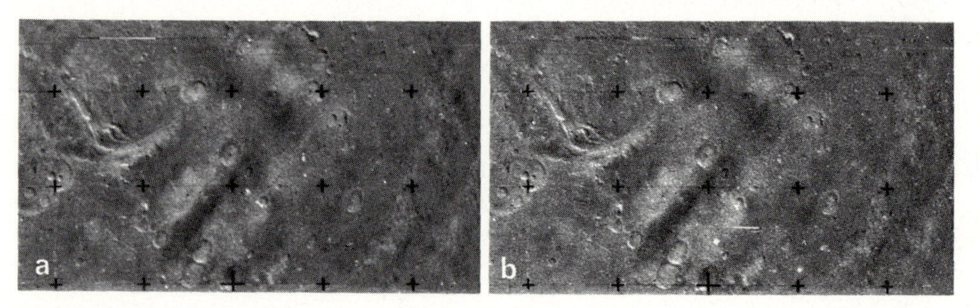

Fig.7.12a,b. Correction of linear distortions in a picture transmission system: (a) uncorrected picture, (b) corrected picture

FFT algorithms usually requires less computer time. Methods similar to those described in Sect.5.8 can somewhat extend the applicability of digital filtering in the space domain.

The picture sharpness correction shown in Fig.7.10 was obtained using spectral processing.

Figures 7.11,12 present the result of sharpening pictures transmitted from the "Mars 4" and "Mars 5" space probes [7.18,19]. A simple separable recursive digital filter of type (4.29) was used. The filter transformed the samples of the video signal under correction, $a_{k,\ell}$, by the formula

$$b_{k,\ell} = a_{k,\ell} + g\left(a_{k,\ell} - \frac{1}{(2N_1 + 1)(2N_2 + 1)} \sum_{m=-N_1}^{N_1} \sum_{n=-N_2}^{N_2} a_{k+m,\ell+n}\right) \quad . \tag{7.28}$$

The gain g of the difference signal, and the size of the region $(2N_1 + 1)$ $(2N_2 + 1)$ over which averaging took place, were chosen so that the frequency response of the correction filter gave a good approximation to the continuous frequency response of the filter (7.28), equal to (4.40)

$$\hat{H}_r(f_1, f_2) = 1 + g\left(1 - \frac{\mathrm{sinc}[\pi(2N_1 + 1)\, f_1 / 2F_1]}{\mathrm{sinc}(\pi f_2 / 2F_2)} \times \frac{\mathrm{sinc}[\pi(2N_2 + 1)\, f_2 / 2F_2]}{\mathrm{sinc}(\pi f_2 / 2F_2)}\right). \tag{7.29}$$

Here $2F_1$, $2F_2$ are the dimensions of the rectangle limiting the spatial spectrum of the picture in accordance with which signal discretization takes place, and $H_0(f_1, f_2)$ is the frequency response of the device employed to reproduce the picture in the processing system.

Thus, as a result of digital correction, the pass-band of the spatial frequencies at the 0.7 level (the level of half-power) increased by more than twice. The effect of this on the visual evaluation of picture sharpness can be judged from Fig.7.12, which shows the picture before and after correction.

It was possible to use a separable recursive filter for this task because the distorted characteristic of the system has a fairly simple form. This filter does not give complete correction. Thus, at "medium" frequencies it leads to a certain overcorrection. But the picture processing time using such a filter is shorter by one or two orders of magnitude than that for processing in the spectral domain with FFT algorithms.

We have considered correction of the response of shift-invariant (spatially homogeneous) systems. The correction of spatially inhomogeneous systems is more complicated. As already noted in Sect.4.3, in certain cases such systems can be digitally corrected by using non-uniform discretization to make them spatially homogeneous (for more details see [7.2,10]).

If the spatially inhomogeneous system does not distort the phase spectrum of the signal, it can be corrected by using a filter of type (7.26). Moreover, spatially inhomogeneous distortions can be corrected by using region-by-region picture processing and choosing the region in such a way that within it the aperture of the distorting system can be considered unchanged.

This section has only described selected methods of correcting linear picture distortion. This problem is being intensively studied at present. More detailed information about the result achieved and the approaches used can be found in [7.1,3,5,26,27].

7.5 Correction of Amplitude Characteristics

One of the simplest forms of video signal distortion in imaging systems is non-linear transformation of the signal, as a result of which the observed brightness of the picture at the output of the system differs from the brightness that would be obtained at the output of an ideal system, provided the difference at each point depends on the value of the video signal only at that point. This kind of distortion is described by the amplitude characteristics of the system, i.e., by functions showing the dependence of the value of the output signal on the value of the input signal:

$$b = \mathbf{w}(a) \quad . \tag{7.30}$$

To describe photographic and television systems, so-calles gamma characteristics are used which link the density of the photographic negative at the output of the system with the illumination or light intensity at the input. To describe television systems the characteristics "light – signal", "signal – light", etc., are also used.

The ideal amplitude characteristic of a system, $w_{id}(a)$, is considered known. It is usually a linear function. Thus the problem of correcting the amplitude characteristics consists in finding a correction transform $w_c(b)$ which converts the characteristic of a real system into $w_{id}(a)$.

Since correction is usually carried out before picture synthesis, to find $w_c(b)$ it is necessary to divide the general characteristic being corrected into two: the characteristic of the system up to the part where correction can take place [denote by $w_1(a)$], and the characteristic of the remaining part of the system, $w_2(a)$. For example, $w_2(a)$ may characterize the distortion introduced by the picture synthesizer, during the transformation of an electric signal into grey levels at the film in a photographic recording device. The correction of these distortions usually requires digital processing as the last stage before picture reproduction.

In the absence of noise in the signal being corrected, the correction function $w_c(b)$ is defined by the following simple relation:

$$\mathbf{w}_2\{\mathbf{w}_c[\mathbf{w}_1(a)]\} = \mathbf{w}_{id}(a) \quad . \tag{7.31}$$

In reality it must be remembered when finding $w_c(b)$ that the signal being corrected is usually distorted not only by certain non-linear transformations, but also by random distortions, or noise. Correction can compensate for non-linear distortion, but at the same time the noise is increased. The correction function $w_c(b)$ must therefore be chosen so as to minimize the difference between the corrected signal $\mathbf{w}_2\{w_c[w_1(a) + \xi]\}$, where ξ is random interference, and the required signal $w_{id}(a)$ relative to some definite criterion (metric).

In digital correction the main source of noise is quantization noise. The distribution of the levels quantization in the processor is fixed by wordlength and the type of number representation. For this reason, as a result of

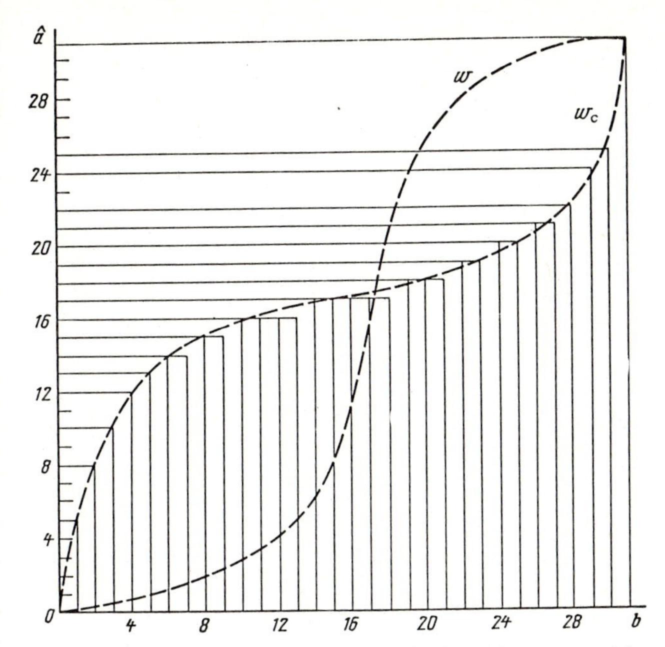

Fig.7.13. Quantization effects during the correction of amplitude characteristics in imaging systems. W shows the characteristics of non-linear amplitude distortion, and W_c the correction function

the non-linear transformation of the quantized signal, values can be obtained during correction that do not coincide with the quantization scale, with the consequence that the quantization noise increases. This phenomenon is illustrated in Fig.7.13, in which the signal being corrected, b, and the result of correction, $\hat{a}$, are quantized uniformly on 32 levels. Curve w represents the function undergoing correction, and curve $\mathbf{w}_c$ the correction function. The quantized corrected values are shown by horizontal lines.

As Fig.7.13 shows, in the parts where the derivative of the function $w_c(\cdot)$ is small, the quantization levels of the initial signal coalesce, i.e., a single value of the transformed signal corresponds to different values of the initial signal. Where the derivative is large, the number of quantization intervals of the corrected signal is greater than that of the initial one. In consequence, the actual number of quantization levels of the initial and transformed signal decreases, and the maximum error caused by representing a continuous signal by a quantized signal increases.

The problem of finding an optimal correction of non-linear distortions that allows for the quantization effect is related to the problem of optimal quantization, which was discussed in Sects.3.8,9. According to the approach

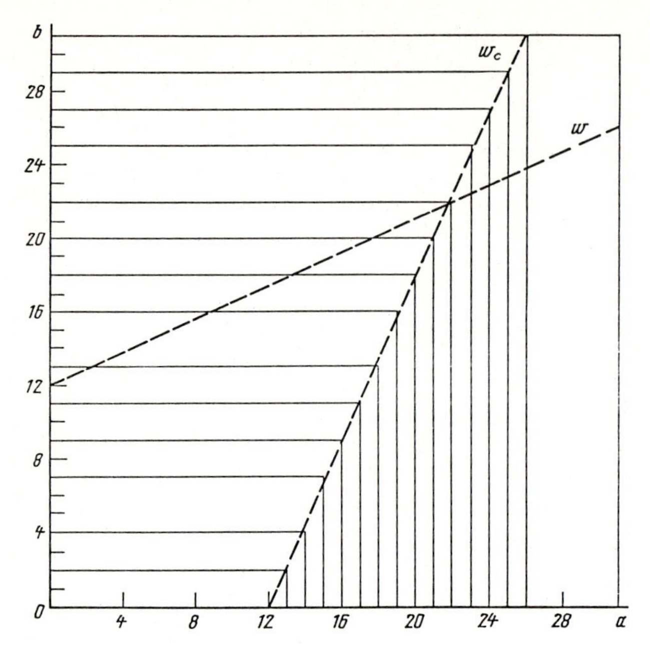

Fig.7.14. Quantization effects during contrast enhancement

proposed there, the optimal correction sequence must be constructed as follows:

1) For a given scale of quantization of the quantity $\mathbf{w}_1(a)$, the boundaries a^r of the quantization intervals of the quantity a are determined (the scale can be given for example by the characteristic of the quantizing device which transforms the initial continuous signal into a digital one).

2) For the quantization intervals that are obtained, the values of the levels $\hat{a}^r$ are found that minimize the difference between these and the values of a for each interval.

3) From the function $\mathbf{w}_2^{-1}(a)$, which is inverse to $\mathbf{w}_2(a)$, the numbers P of the quantized signal values set by the picture synthesizer that are closest to $\mathbf{w}_2^{-1}(\hat{a}^r)$ are found. The resultant P(r) look-up table is the required optimum correcting table.

In addition to the selection of the optimal correction functions, there is yet another possibility of decreasing quantization noise during picture correction, namely, the addition of pseudorandom noise to the signal under correction (Sect.3.10). There are two ways of doing this: by adding to the video signal before correction uncorrelated pseudorandom numbers that have a uniform distribution and are obtained from a special generator (Sect.6.3), or by not rounding off the numbers after calculation. The latter method is used if, as usually happens, the video signal values being corrected are derived from calculation in a computer with a longer word-length than the number of bits assigned for the quantized video signal, or if the values are presented as floating-point numbers. Both methods give practically the same results. The addition of fictional random quantization levels suppresses the false quantization contours, and the visual quality of the picture improves.

The enhancement of picture contrast can serve as an example of amplitude correction. Contrast stretching is used when for one reason or another the digital video signal occupies only part of its allowed range of values. This would arise for a system with an amplitude characteristic similar to the line w shown in Fig.7.14 for cases when the video signal only has values between the levels 12 and 26 out of the possible range of 0 - 31. The correction characteristic for this case is shown in Fig.7.14 by the line w_c. Because the gradient of the line w_c is necessarily greater than unity, the correction causes an increase in the quantization error (noise). The result correcting the picture shown in Fig.7.15a is presented in Fig.7.15b. The latter figure clearly shows the false contours (rings) resulting from the

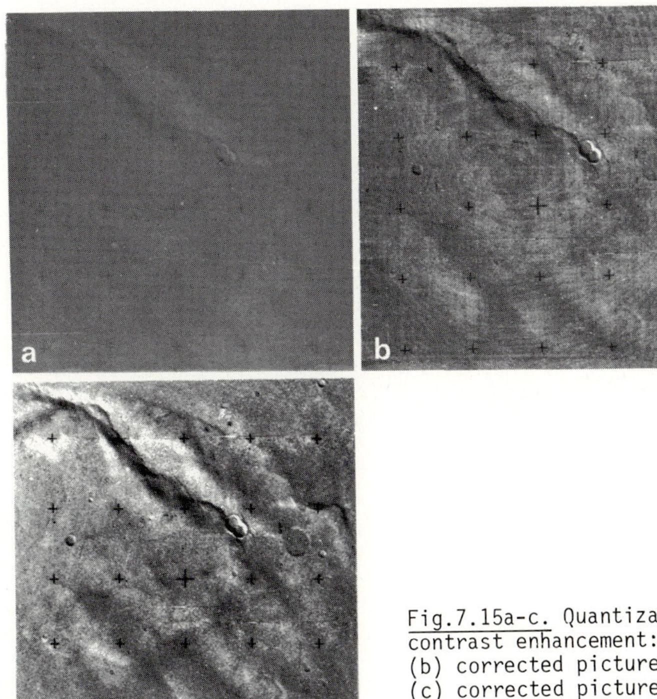

Fig.7.15a-c. Quantization effects during contrast enhancement: (a) original pictures (b) corrected picture with false contours; (c) corrected picture with false contours destroyed by the introduction of false quantization levels

described quantization effect. Figure 7.15c illustrates how the introduction of false quantization levels destroys the false contours.

The simplest, fastest and most commonly used method of digital amplitude correction is to provide a look-up table of the corrected quantized signal values in the memory device of the processor. The value of the transformed signal can be simply used as the address in this table from which the corresponding value of the corrected signal is found. This table requires as many memory cells as the values that the transformed signal can adopt, but on the other hand, the transformation itself is carried out by one call to the memory per signal sample. This method is also applied in special-purpose processors, e.g., display processors[4]. If, however, the required memory volume is too large, only certain values are tabulated and not the entire table. The intermediate values are found by interpolation.

---

[4] A display processor is a device permitting visual observation of the picture at the output of the digital processor. It enables the user to perform interactive processing.

8 Picture Enhancement and Preparation

Pictures carry information for visual perception. To help the user to appreciate this information, it is frequently necessary to prepare the picture in some way. This chapter describes the methods by which this is achieved.

In Sect.8.1 the idea of preparation is explained and a classification of picture preparation methods is given. The adaptive quantization of pictures based on mode analysis of histograms is presented in Sect.8.2. Section 8.3 describes methods of adaptive non-linear transformation of signals such as histogram equalization and its generalizations. Linear methods of picture preparation are presented in Sect.8.4 as methods of optimal adaptive linear filtering, and Sect.8.5 reviews methods of converting pictures into computer graphics. Geometrical transformations of pictures are briefly discussed in Sect.8.6.

8.1 Preparation Problems and Visual Analysis of Pictures

As has already been noted, it is often not sufficient in practice to represent an object to the observer using only ideal image-forming systems. In solving complex interpretation probelms requiring meticulous picture analysis (surveying, object identification, determination of various types of qualitative characteristics, generalized description, etc.) it is desirable to "reinforce" the eye by providing the observer with tools facilitating picture interpretation and information extraction. These are firstly technical tools, and secondly methods of video signal processing. This processing, which acts as an auxiliary tool for visual picture interpretation, is called *picture preparation*.

In contrast to the correction of imaging systems, preparation can be regarded as the correction of the interaction between the video signal sensor and the object. For example, in medical radiography, special X-ray opaque substances are injected into the patient to increase the contrast in the X-ray photograph. In principle, the desired information is contained in a photograph obtained without the contrast-forming substance, but this information is difficult to discern. This inadequacy can be compensated by special processing, or preparation, of the photograph. Preparation can also be treated as matching the video signal with the characteristics of the decision-making device, which in the given case is the visual system of the person involved in picture interpretation.

Preparation as processing designed to facilitate visual picture interpretation has two aspects: preparation for the collective user, as, for example, the viewers in television broadcasting, cinematography and the printing industry, and preparation for the individual user. In the first case the methods of preparation can be based on information about the properties of the viewing systems, which are studied in mass psychophysiological experiments. Preparation for the collective user is synonymous with the concept of "improving visual quality", or image enhancement. Preparation for the individual user must allow not only for the psychophysiological properties of viewing that are common to everyone, but also for specific professional experience. The only way of doing this is to incorporate the user into the processing system by organizing the picture processing in an interactive scheme which makes it possible for the user to manage processing algorithms and their parameters.

Preparation has long been used in scientific and even artistic photography as well as in techniques of visualizing information. Examples are graphic arts methods, solarization in preparing photographic prints, "dodging", the construction of isophotes (isodensities), the reproduction of pictures in pseudocolours [8.1,2]. Naturally, preparation methods are to a large extent determined by the technical means available. The digital technology of signal processing, with its great variety of possible preparation methods and the convenience with which it allows the required interactive regime — feedback from the user to the processing equipment — to be organized can be said without exaggeration to be unequalled in its universality and flexibility.

The development of computer technology and of means of visualizing information at the output of digital processors has opened up the possibility of creating interactive systems of picture processing. With them the observer, or user, employs advanced software (mathematical programming) with which the picture can be subjected to various transformations at will; the necessary decisions are then based on direct observation of the picture during processing. In essence, we can speak of the creation of a robot that increases not only the purely optical properties of sight, but also its analytical properties.

Picture preparation methods can be divided into two major classes: feature processing, and geometric transformation.

8.1.1 Feature Processing

Feature processing is the extraction, measurement and visualization of those video signal characteristics or features, that provide the eye with the greatest amount of information in the given problem of analysis. For example, the edges of objects are features that are necessary in order to recognize objects in pictures. Other features include brightness, colour, the degree of inhomogeneity of an object and texture.

The choice of features is determined by the nature of the picture analysis problem and the characteristics of the object being studied, as in problems of pattern recognition [8.3-5].

In selecting methods of measuring and transforming features in digital image-processing systems, it is best to proceed on the basis of the rational organization of the software of such systems. It is from this point of view that the basic transformation classes used to construct processing algorithms must be selected. In accordance with the principles formulated in Sect.3.1, the following classes of feature-processing methods can be defined:

a) nonlinear element-by-element transformations;
b) linear transformations;
c) combined transformations, i.e., any sequence of linear and element-by-element transformations.

We shall consider a few feature-processing methods of each type: adaptive quantization of modes; increasing local contrast by non-linear transformation of the video signal value; and methods of contouring and of constructing graphic representations.

8.1.2 Geometric Transformations

The geometric transformations include all processing associated with manipulating the plane projections of 3-D bodies. The most characteristic tasks of this type are encountered in processing aerial and space photographs in the composition of maps. Sometimes geometric picture transformation is necessary in order to compare two or more pictures of the same body obtained under different circumstances.

The use of digital processors for geometric transformation has no particular advantages over the analogue devices (optical or television) used for this purpose. The main advantage of digital over analogue processors in picture processing is that the transformation algorithm can be rapidly changed. However, this does not compensate for the inconvenience of performing numerous transfers of large amounts of data which are usually located in slow external bulk memory, or for the complexity of computation involved in ensuring high interpolation accuracy.

8.2 Adaptive Quantization of Modes

One trend in digital picture preparation is the use of decision-making algorithms similar to those encountered in problems of classifying and recognizing images [8.3,6,7]. The pictures obtained with such picture preparation algorithms form, as it were, classes resulting from decisions based on certain features, generally vectorial ones, which are presented to the user for further analysis.

Such decision-making algorithms can be used, for example, when areas having certain properties must be identified in a picture so that the configurations of these areas, their mutual distribution, size, and other properties can be visually analyzed. Such problems are encountered in geology, metallography, and medical and industrial diagnostics based on pictures. The features to be identified are usually known only qualitatively: they include the video signal value in various spectral ranges of the recorded radiation, the modulus or direction of the gradient of the video signal, and the intensity of the signal in various ranges of spatial frequencies. The quantitative values of these features corresponding to certain parts or details which are to be distinguished from one another and separated on the picture are usually unknown; they are found using the methods of cluster analysis.

These methods are based on the idea that various portions of the picture can be grouped according to the value of the features, which represent certain properties of the objects depicted. The grouping of picture elements on the basis of these features is apparent in the histogram of the distribution of the picture elements in terms of the features (or of the components of the features, if the latter are vectorial): around certain values there is a concentration (condensation), or mode, also known as a cluster, whereas other values occur seldom or not at all. By separating these modes, one can establish the borders between them; this gives the threshold values of the features, on the basis of which the picture can then be divided into parts.

Let us imagine, for example, that we are observing a colour picture of a sandy beach dotted with clumps of plants. Clearly, if we look at the distribution of the picture elements by colour we find three modes: one in the region of blue-green (the colour of the sea), another yellow-green (the plants) and a third silver-yellow (the sand). These modes will be concentrated to a greater or lesser extent depending on the homogeneity of the colouring of the sea, sand and plants. By determining the colour of each element of the picture, we can discover to which of the three modes it belongs and thus classify, or divide, the picture elements by colour.

The separation of modes in the histogram of feature values can be called *adaptive mode quantization*. Its quality depends on how clearly the peaks are separated, and the criteria for dividing them is therefore important. The separability of modes depends on their degree of concentration, i.e., on the homogeneity of the picture parts that must be considered one type within the context of the given problem. Suppose that in the example considered, we wished to separate the parts with plants from the sea and sand. Plants may be homogeneous in colour, but can also have variations (variegated leaves, shine on the leaves, stalk colour, various types of plants, etc.). These variations dilute the mode, and it may overlap with that belonging to the sand, whose colour can also have various shades. This means that the dividing boundaries must be chosen accurately, and if there are wide variations, a feature such as colour might possibly have to be rejected altogether.

To develop quantitative criteria for the choice of optimal borders between peaks it is necessary to have a description of the reason for the dilution of the modes and for loss resulting from incorrect classification. The loss function is determined by the demands of the subsequent visual analysis of the picture. In picture preparation problems, the most important requirement is that the number of inaccurately classified picture elements be minimal. Other possible requirements are that the edges of the separated parts be smooth, that there be no small marginal incursions within the part having the largest area, etc.

The extent to which the modes are diluted is determined by the properties of the object in relation to the chosen feature. These properties are usually difficult to describe formally, and therefore more or less likely models are constructed based on a priori knowledge of how the property of interest is reflected in the observed feature and how the observed picture is formed. For example, the picture being prepared can be treated as a result of the transformation, by random operators and/or noise, of the initial field, in which only "pure" modes were present (i.e., a field in which the histogram relating to the features being considered consists of a set of δ functions). Then, by applying the statistical decision theory [8.3], one can construct decision rules which are optimal, say, for the criterion of the minimum rate of errors in the classification of the picture elements. The decision field obtained in this case can be treated as the estimate of the original picture on the assumption that the picture being prepared was obtained by the distortion of a certain initial field with "pure" modes.

The simplest models with random operators acting on "ideal" pictures and with additive or multiplicative noise for which exact decision algorithms can be found, do not usually correspond satisfactorily to the actual interaction between the properties of the object that are distinguished and the features that are measured. Therefore other factors, often difficult to formalize, must be considered.

Thus, the distribution modes of features can be significantly diluted owing to averaging over the whole of the observed picture; the observed picture often cannot be treated as the result only of noise or only of a linear operator acting on the "ideal" signal with pure modes; and picture elements grouped by modes usually form extended areas, or at any rate such areas must be separated and the minor ones ignored.

Similar factors can be taken into account using the following supplementary approaches:

1) Limiting the size of the pictures to be analyzed. The picture is divided into regions in which the modes are only negligibly diluted, and processing takes place fragment by fragment. The latter can be arranged side by side or may overlap. In the latter case, only the non-overlapping parts are processed. This approach makes it possible to process pictures with a spatially inhomogeneous structure.

2) Division according to the type of mode dilution. In finding the modes and determining their boundaries – the quantization boundaries – the analysis of the distribution of values for the feature takes place only on those parts of the picture where the diluted mode can be considered to be the effect of noise alone or of the linear operator alone.

This approach is elucidated in Fig.8.1 for a 1-D feature, for example, the value of the video signal of the picture. The solid lines in Fig.8.1a show a video signal in the form of a step and those in Fig.8.1b, the corresponding histogram, which consists of two δ functions. Because of noise in the video signal, the δ function peaks are somewhat broadened, but if the noise is fairly weak the peaks do not overlap (see the dashed line in Fig.8.1b). The solid lines in Fig.8.1c show the same video signal at the output of a low-pass linear filter which smooths the step, and those in Fig.8.1d the histogram of this signal with two peaks mixed; the dashed lines show these same functions in the presence of noise. Clearly, if the histogram is measured from the parts of the video signal to the left and to the right of points y and y in Fig.8.1c, we again obtain excellent separated modes, as in Fig.8.1b.

3) Rejection of modes on the basis of "population". If the number of picture elements in a mode is below a certain threshold which is set on the basis of the minimum allowed area for the separated parts of the picture, the mode is ignored.

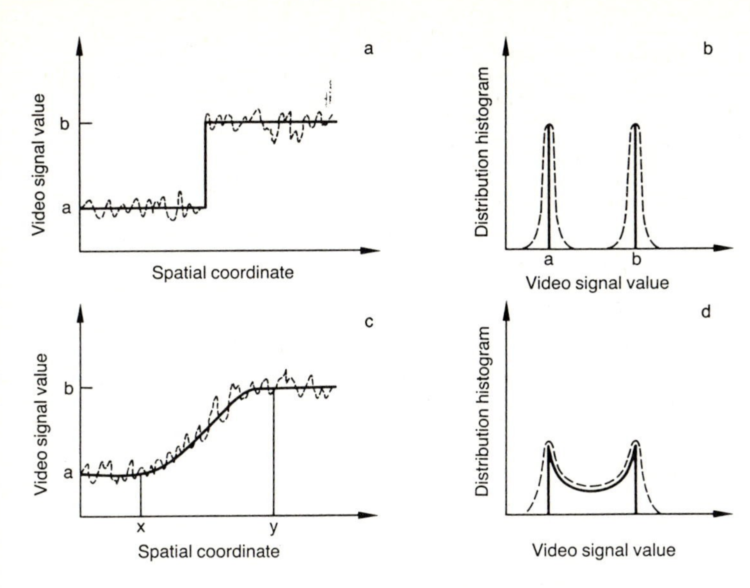

Fig.8.1. (a) Stepped video signal plus noise and its value distribution histogram, (b) the same for the signal at output of low-pass filter

4) Rejection of minor details. Filtering is conducted by joining to the dominant mode picture elements which belong to a mode other than the surrounding ones and which form a configuration having an area below the given threshold.

To illustrate the method of adaptive quantization, we show the results that are obtained when it is employed to separate picture parts according to the magnitude of the video signal [8.8,9]. The magnitude of the video signal is the simplest feature that can be measured in a picture, but precisely this magnitude is often associated with the basic physical properties of the objects represented.

A flow diagram of the package of program modules used for picture preparation using adaptive mode quantization of the video signal is shown in Fig.8.2 [8.9]. The rectangles denote the separate program blocks, the circles the parameters set by the user, and the ovals the processing stages at which picture preparation is obtained.

The "fragment separation" block sets the type of processing (either over the whole frame or over separate fragments). In the latter case the user states the size of the fragments and the degree of overlap. No precise a priori recommendations can be given for the choice of these sizes and the user selects the parameters on the basis of personal experience and visual evaluation of the picture being processed. This reservation also applies to the other parameters set by the user.

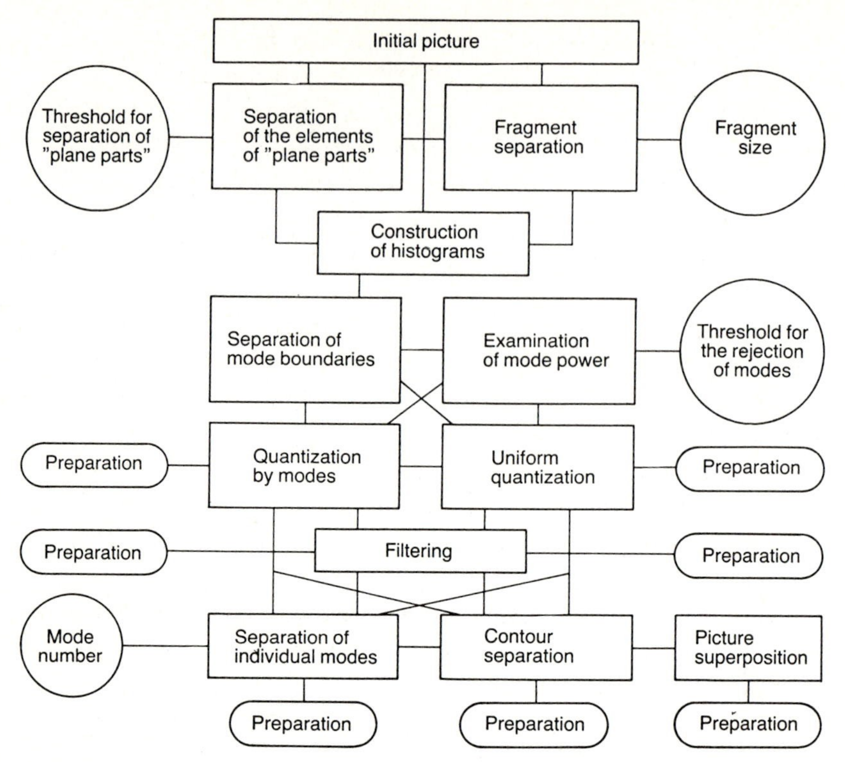

Fig.8.2. Diagram of the program package for picture preparation by adaptive quantization of modes

In the block "separation of flat zones", those picture elements are separated for which the maximum modulus of their difference from the neighbouring eight raster elements (left, right, above left and right, below left and right) does not exceed the threshold set by the user. Thus, in the block "construction of histograms" only those video signal values appear that correspond to picture parts with weak changes in the video signal.

The block "separation of mode boundaries" determines the position of the modes of the histograms and the quantization boundaries. If the values of the histogram differ from zero between neighbouring peaks, the boundary between them is set at minimum point of the histogram; if there is a region with zero values, the boundary is drawn half-way between the maxima being separated.

In the block "examination of mode power", peaks are rejected on the basis of "population". The area of the histogram is calculated under each of the peaks (mode power). Those for which it is smaller than a threshold set by the user are united with the closest stronger mode.

In the "quantization by modes" and "uniform quantization" blocks, the video signal values are quantized on the basis of the boundaries chosen in the preceding stages. With quantization by modes, picture elements within the borders of the given mode of the video signal value are assigned the magnitude corresponding to the maximum of the histogram in this mode. In uniform quantization these elements are ascribed a value equal to Δi, where i is the order number of the mode and Δ is the ratio of the given video signal range to the total number of modes separated. In quantization by modes a greater similarity is retained between the prepared picture and the initial one. Uniform quantization allows an increase in the relative contrast of details. With the appropriate devices for the reproduction of colour pictures, the result of uniform quantization can be shown in pseudocolours.

In the "filtering" block, picture parts that are small in area and do not belong to the same mode as their neighbours are rejected. To do this the picture is analyzed by fragments consisting of 6×6 elements, and the histogram of the element values belonging to the internal area of 4×4 elements of each of these fragments is examined. If there are modes in this histogram whose strength is below a certain given threshold, and the boundary elements of the 6×6 region do not belong to this mode, then the elements that do belong to this mode are ascribed the magnitudes corresponding to the strongest mode in the histogram.

The "separation of individual peaks" block allows only those picture parts belonging to one or several particular peaks to be shown on the picture preparation.

The described program package also includes the blocks "counter separation" and "picture-superposition", which are designed to display the boundaries of objects and superpose them on the initial picture (Sect.8.5).

In addition to picture preparation, this program package can give the user relevant numerical information which may be useful in picture analysis and decision making. Thus, in the method of adaptive mode quantization of the video signal, one can display the video signal histogram, the separation thresholds, the number and strength of the modes (i.e., the relative area occupied by details belonging to each mode), etc.

Some results achieved with this program package are presented in Fig.8.3. Part (a) shows an aerial photograph used in experiments on quantization, and Parts (b,c) show the results of uniformly quantizing the photograph with two different threshold for rejecting modes on the basis of their strength: 4 and 10 percent respectively. As a result there were 11 and 3 quantization levels (modes) respectively. A comparison of these figures reveals that with

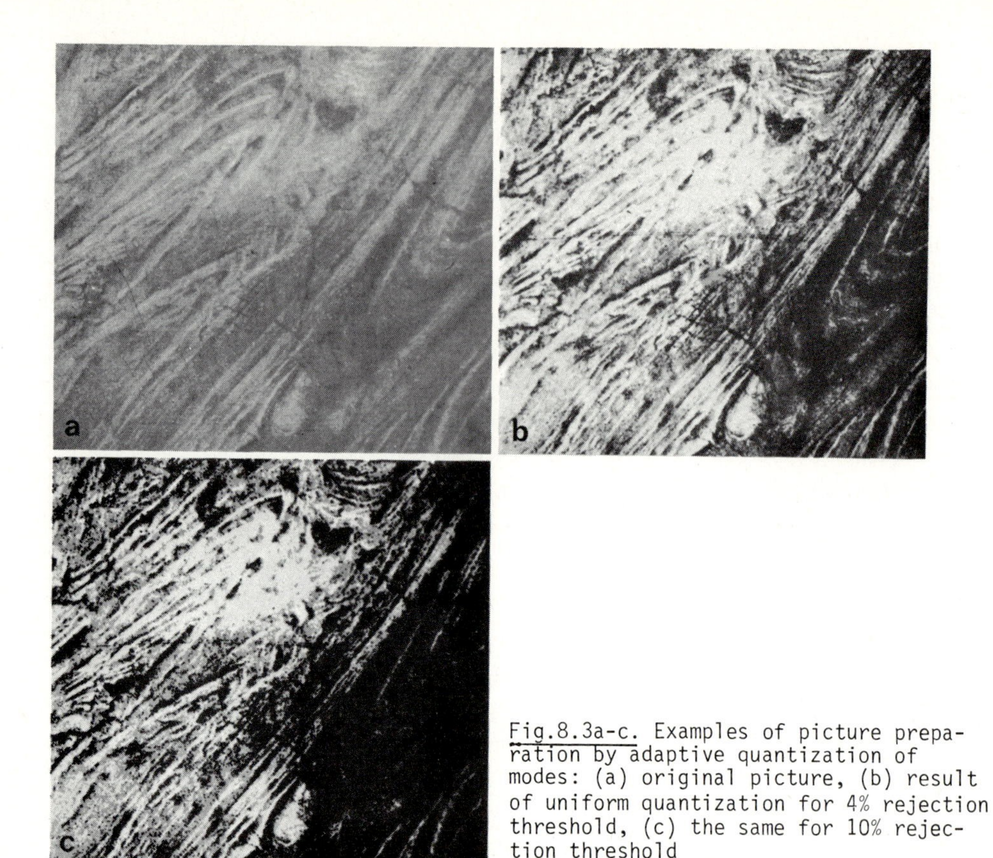

Fig.8.3a-c. Examples of picture preparation by adaptive quantization of modes: (a) original picture, (b) result of uniform quantization for 4% rejection threshold, (c) the same for 10% rejection threshold

an increase in the threshold for mode rejection, details disappear and a kind of generalization of the preparation takes place. Details belonging to each mode can be extracted from the other details. Examples of such preparation are shown in Figs.8.4a-c, which were obtained by the extraction of details belonging to various modes of the picture preparation with a threshold of 10% (Fig.8.3c).

Fragment-by-fragment quantization and global quantization (analysis of the entire frame) are compared in Fig.8.5. Part (a) is the initial picture of the micro-section, and Part (b) is the result of global quantization on three levels. Part (c) shows the result of fragment-by-fragment quantization without overlap. The fragment boundaries are shown by the grid. In Part (b) only the coarse structure of the section remained whereas in Part (c) many of the details of the original are retained, the photograph itself is sharper, and the boundaries between inclusions look better than in the initial photograph.

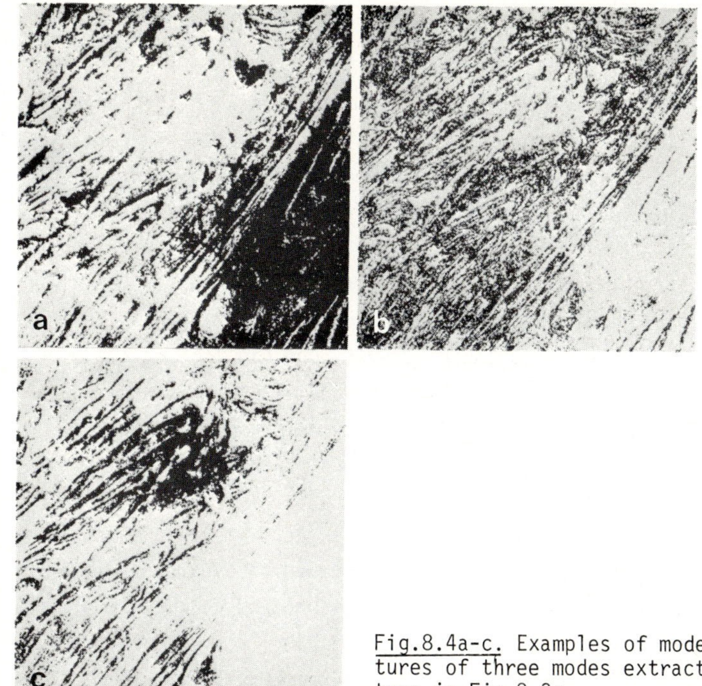

Fig.8.4a-c. Examples of mode extraction: pictures of three modes extracted from the picture in Fig.8.3c

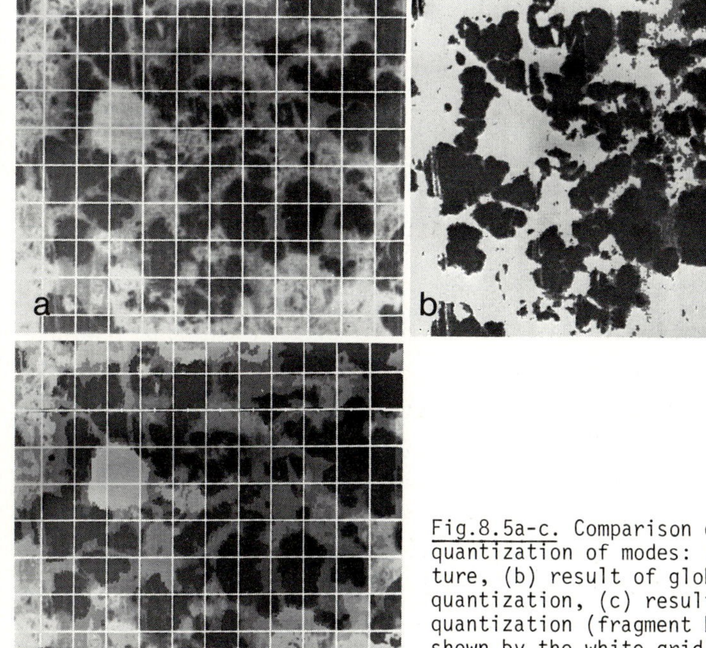

Fig.8.5a-c. Comparison of global and local quantization of modes: (a) original picture, (b) result of global adaptive mode quantization, (c) result of fragment quantization (fragment boundaries are shown by the white grid)

8.3 Preparation by Nonlinear Transformation of the Video Signal Scale

Nonlinear transformation of video signal values is applied when the contrast between weakly differentiated picture details must be increased. The simplest example is the amplitude window method. A certain portion of the range of video signal values is stretched over the entire scale of permissible values using the transformation shown in Fig.8.6a, which is applied to every picture element. In this way the contrast between the details within this range of values increases. Details outside the chosen range are suppressed. By moving this window along the range of video signal values, one can look at the picture successively as if through a contrast-magnifying glass. This stretching of the scale can also be carried out simultaneously on several portions of the range of video signal values by using the sawtooth transformation shown in Fig.8.6b. If the picture contains large details with slowly changing video signal values, such a transformation hardly impairs the picture as a whole, and yet it significantly increases the contrast between the low-contrast details. Data are available on the application of this method of picture preparation in biomedical research [8.10-12].

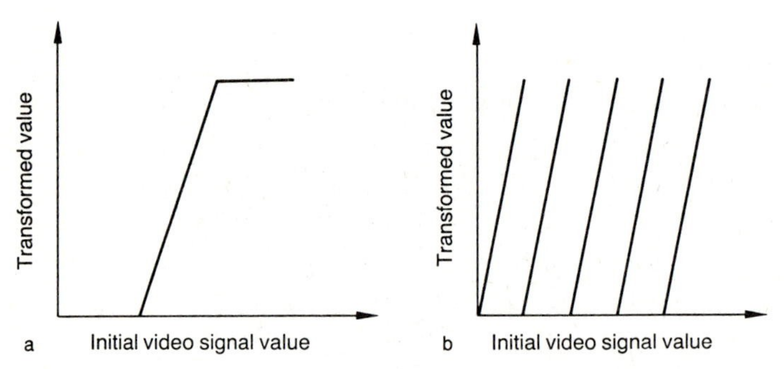

Fig.8.6a,b. Preparation by non-linear transformation of the video signal value scale: (a) amplitude window, (b) sawtooth transformation

A variant of the amplitude window method is "blocking" of binary digits in the video signal code [8.13]. In the simplest case, if a binary digit in a particular position is one, say, the entire signal is replaced by a predetermined value ("black") while if the same digit is zero, the signal is replaced by another value ("white"). The changes in one or several binary digits in the code of a video signal are blocked, and during the transformation of such a digital signal into analogue form, the analogue signal is stretched to the full range. It is easy to see that if the most significant digits of the code are blocked, the characteristic of the transformation takes the form shown in Fig.8.6b, with a number of portions equal to 2 to the power of the number of blocked digits. If the lower digits are blocked, the transformation characteristic becomes stepped. If one or more of the medium-order digits are blocked, the characteristic becomes more complex. The chief advantage of this method is the simplicity with which it can be realized in digital processors.

The amplitude window method has long been known and can be realized not only by digital but also by purely analogue means, e.g., photography or television. When using it in digital form, one should remember that since it is the quantized signal that is transformed, only the distance between the quantization levels in the stretched portion changes, but not their number. As a result of stretching, false contours can appear at the edges of the change between quantization levels. This can sometimes even be convenient, since these contours are lines of equal intensity and as such are of interest.

In using the amplitude window method the user must assign the parts of the video signal value range that are to be stretched. Usually no information can be obtained about them, and by just looking at the picture one sees the ranges of values only very approximately. To obtain a highly informative picture, the edges of the ranges must be chosen by trial and error. Therefore it is much more convenient to use adaptive amplitude transformations which are constructed on the basis of measurement and analysis of the histogram of the video signal values. Of these, the so-called equalization method [8.8,10,13-17] and a related method of hyperbolization [8.18] have achieved the greatest popularity. In equalization the transformation function of the video signal is chosen in such a way that the transformed signal has a uniform histogram (hence the name "equalization"). In hyperbolization the histogram of the values of the logarithm of the video signal is equalized. Note, however, that in the majority of video processing systems the logarithm of the video signal is taken before quantization and input into the computer, and this operation is compensated in the picture synthesizer at the output of the processing system. Therefore in practice, equalization of the digital signal frequently corresponds to hyperbolization of the picture brightness observed. The method of finding the required transformation function in equalization is best explained for a continuous signal. Let $h(a)$ be the histogram of the video signal being transformed ($a_{min} \leq a \leq a_{max}$), $\mathbf{w}(a)$ be the transformation function sought, and $h(\hat{a})$ be the histogram of the transformed signal $\hat{a} = \mathbf{w}(a)$, which is clearly equal to $1/[\mathbf{w}(a_{max}) - \mathbf{w}(a_{min})]$.

Of all the possible functions $\mathbf{w}(a)$ giving a uniform distribution of the transformed signal, we shall only look for monotonic functions preserving the one-to-one property of the transformation. Considering $w(a)$ to be monotonical growing (monotonically diminishing corresponds to transmission as a "negative"), from the obvious equality

$$h(\hat{a})d\hat{a} = d\hat{a} / [\mathbf{w}(a_{max}) - \mathbf{w}(a_{min})] = h(a)da] \quad , \tag{8.1}$$

we obtain

$$d\hat{a} / da = dw(a) / da = [\mathbf{w}(a_{max}) - \mathbf{w}(a_{min})]h(a) \quad , \tag{8.2}$$

upon solving which we find that

$$[\mathbf{w}(a) - \mathbf{w}(a_{min})] / [\mathbf{w}(a_{max}) - \mathbf{w}(a_{min})] = \int_{a_{min}}^{a} h(a_1)da_1 \quad . \tag{8.3}$$

With digital processing we can use the discrete analogue of this transformation:

$$\hat{m} = \left\{(M - 1) \left[\sum_{k=0}^{m} h(k) - h(0)\right] / h(0)\right\} \quad , \tag{8.4}$$

where m is the quantized value of the signal being transformed; h(k) is the corresponding histogram; $k = 0, 1, \ldots, M - 1$; $\hat{m}$ is the transformation value; and the braces denote rounding up to the nearest integer.

Sometimes the effectiveness of equalization is explained by the fact that it results in the maximum entropy of the quantized video signal. However, the real reason is that the histogram equalization operation leads to an increase in the contrast in the parts of the picture with the most frequently encountered video signal values. Indeed, the transformation gradient, the derivative of the function w(a), is greater, the larger h(a) is, as follows from (8.2). If the histogram h(a) is unimodal, i.e., if the majority of the picture elements have similar brightness values, forming a background for the other elements, then the mode widens as a result of equalization, and the contrast between details and the background becomes greater, the smaller it was before transformation. If the distribution histogram has several modes (for example, bimodal histograms are characteristic of landscape pictures with deep shadows), equalization results in the widening of each mode, and the range of values which each peak will occupy after transformation is proportional to the strength of the peak. This selectivity of equalization in relation to the frequency of the video signal values is the major advantage of this method over the simple linear contrast stretching described in Sect.7.2.

Equalization can be carried out over the whole picture or fragment-by-fragment, and the picture fragments may overlap [8.8]. In fragment-by-fragment equalization with overlap, the histogram is constructed for entire fragment, but only the central portion of it equal in size to that of the non-overlapping parts is transformed. If each successive fragment is shifted relative to its predecessor by one element, the transformation is called sliding. The table of sliding equalization transformation changes from element to element according to changes in the histogram of the picture regions surrounding these elements. Like adaptive mode quantization the fragment-by-fragment and sliding regimes of equalization are also a way of fitting to the spatial inhomogeneity of the picture.

Equalization is in a certain sense the antithesis of adaptive peak quantization, which was discussed in the previous section. Indeed, equalization widens the peaks in the distribution of video signal values, whereas adaptive peak quantization narrows them to δ functions. After quantization, the signal shown in Fig.8.7a becomes a pure step; equalization destroys the step in such a signal but increases the contrast in the deviations from the previous two values (Fig.8.7b).

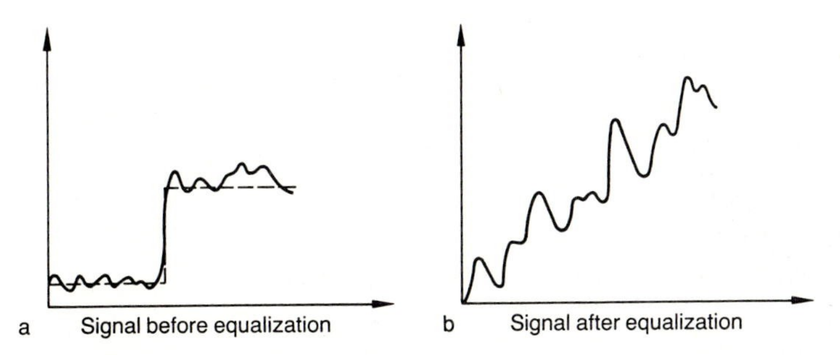

Fig.8.7a,b. Video signal transformation by histogram equalization: (a) initial video signal; (b) transformed video signal

The effect of equalization is graphically illustrated in Fig.8.8a. It shows the result of equalization of the aerial photograph shown in Fig.8.3a on fragments consisting of 30×30 elements with a 15×15 element step. It can be seen from Fig.8.8a that unlike the quantized photograph, the amount of detail in the picture is sharply increased. If this aerial photograph is equalized not by regions, but as a whole (Fig.8.8b), its general contrast also increases, but the fine details are much less clear than in Fig.8.7a.

Two further examples of fragment-by-fragment equalization are shown in Figs.8.9a-d. They show a panorama of the surface of Venus obtained from the "Venus 9" and "Venus 10" space probes. Parts (a,c) present the pictures before processing, and Parts (b,d) after equalization. It can be seen that equalization made many small, low-contrast details in the dark and light parts of the

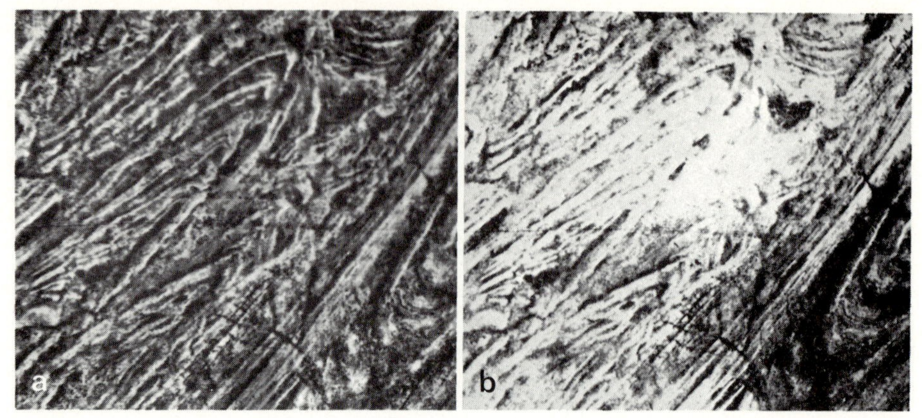

Fig.8.8a,b. Examples of picture preparation by histogram equalization: (a) result of sliding fragment-by-fragment histogram equalization of the picture shown in Fig.8.3a, (b) result of histogram equalization of the same picture this time as a whole

Fig.8.9a-d. Application of fragment-by-fragment equalization to panoramas of the surface of Venus: (a) and (c) show the initial pictures, (b) and (d) the equalized pictures

panorama more visible, and also emphasized the volume. This result was obtained with a fragment size of 15×15 elements and a step size of 3×3 elements (the total size of the picture is 128×512 elements). The cell structure in certain portions is connected with insufficient overlapping of the individual fragments. Sliding equalization does not yield such a structure.

Note that with fragment-by-fragment and sliding equalization the number of operations required to compile transformation tables can become exceedingly large if recursive algorithms similar to those described in Sect.6.1 for the local histogram are not used [8.19].

The equalization method can be applied in combination with other preparation methods. It is especially convenient to use following linear preparation methods, which are described in the next section. In this case equalization can be made spatially dependent (or context dependent [8.14]). Interesting possibilities are opened up by equalization which is controlled by preparation. In it the transformation law is constructed from the picture preparation histogram, and the initial picture is transformed.

In conclusion we will mention possible generalizations of the equalization procedure. One of them consists in considering equalization to be a particular instance of amplitude transformations which convert the observed signal into a signal with a given distribution law. In equalization this is a uniform law. Such a transformation towards a certain preset distribution law is used as a means of standardizing various pictures, for example in composing a photographic mosaic out of them [8.14,15], or in textural analysis [8.17].

This transformation can be carried out using the equalization algorithm in the following way:

1) From the observed video signal histogram the transformation look-up table giving histogram equalization is constructed.

2) A similar table is constructed for the given histogram.

3) The table for the required transformation of the histogram into the given one is constructed by joining these two tables in such a way that for every output value of the first table, the value corresponding to it at the input of the second table is found.

Another interesting possibility of generalization consists in changing the relationship between the steepness of the nonlinear transformation of the signal and its histogram. With equalization the transformation gradient is proportional to the histogram value (Sect.8.2). We can equally require it to be proportional to a certain power p of the histogram values [8.8]:

$$dw(a)/da = [w(a_{max}) - w(a_{min})]h^p(a) \quad . \tag{8.5}$$

We then obtain the following transformation formula:

$$\frac{\mathbf{w}(a) - \mathbf{w}(a_{min})}{\mathbf{w}(a_{max}) - \mathbf{w}(a_{min})} = \frac{\int_{a_{min}}^{a} h^p(a_1)da_1}{\int_{a_{min}}^{a_{max}} h^p(a_1)da_1} \quad . \tag{8.6}$$

This may be called the power intensification method. With $p > 1$, weak peaks in the histogram will be more suppressed the greater p is as a result of transformation, and stronger peaks will stretch out over the entire range. Linear stretching of the scale of video signal values corresponds to $p = 0$. With $p < 0$, the greater p is, the greater the narrowing of peaks which will take place as a result of transformation. With large negative p, transformation (8.6) will approach adaptive quantization of modes with quantization boundaries in the minima between modes.

Equation (8.6) is fully analogous to (3.77), which described the optimal law of predistortion of the signal when it is being quantized. This analogy elucidates adaptive mode quantization, discussed in the preceding section, and also equalization.

The above-discussed methods of preparation by means of non-linear transformations concerned scalar signals. They can be applied to the video signal and its spectrum (Sect.8.4) as well as to individual components of vector signals, for example the various components of multispectral photographs used in space and aerial photography. A variant of nonlinear transformation of vector signals is the method of pseudocolours. The signal being transformed is mapped onto a three-component video signal, which is reproduced in the form of a colour photograph. The eye is much more sensitive to colour contrast than to brightness contrast. Therefore the pseudocolour method is very convenient when the weak variations of a signal must be rendered visible. In the processing of a scalar signal the pseudocolour method can be considered a generalization of the amplitude window method.

8.4 Linear Preparation Methods

Linear picture preparation methods are essentially an extension of methods of correcting linear distortions in imaging systems. They can be treated as optimal linear signal filtering with noise, if noise is understood to mean details not essential to the given problems of picture analysis.

The most commonly used method is the suppression of low and the amplification of high spatial frequencies of the DFT of the signal. This operation gives good results in those cases when slow signal changes in the picture occupy the entire dynamic range of values allocated for the signal, and small details have low contrast. The suppression of low spatial frequancies allows a weakening of the slowly changing signals, and owing to this the contrast and thus the visibility of small details is significantly increased. The most convenient way of realizing such processing is by single or multiple signal filtering (parallel or series filtering; see Sect.4.3) by means of a 2-D separable recursive filter which is a modification of the filter (4.29):

$$\hat{a}_{k,\ell} = g_1 a_{k,\ell} + g_2\Bigg(a_{k,\ell} - \frac{1}{(N_{11}+N_{12}+1)(N_{21}+N_{22}+1)} \sum_{m=-N_{11}}^{N_{12}} \sum_{n=-N_{21}}^{N_{22}} \times a_{k+m,\ell+n}\Bigg) + (1-g_1)\bar{a} \quad , \tag{8.7}$$

where $a_{k,\ell}$ is the initial video signal; g_1 is the coefficient determining the degree of weakening of low-frequency (slowly changing) signal components; and g_2 is the constant defining the degree of enhancement of high-frequency components that are responsible for transmitting small details. The parameters N_{11}, N_{12}, N_{21}, N_{22}, which define the size of the rectangular region, characterize the domain over which signal averaging takes place, and are chosen so that the size of the rectangle is roughly equal to the average size of the details that must be extracted. Parameter $\bar{a}$ determines the average (background) signal value after processing.

The filter impulse response (8.7) is rectangular in form and thus the filter is suitable for the extraction of quasi-isotropic, vertically or horizontally oriented details. By means of multiple parallel filtering one can form arbitrarily oriented impulse responses, which are related to the orientation of the picture details. Series processing allows the formation of a smoother, and in particular more isotropic impulse response.

Figures 8.10,11 present examples of processing X-ray images of the ribcage and brain blood vessels in this way. Part (a) is the initial picture, and Part (b) the result of emphasizing the higher spatial frequencies. The processing parameters for Fig.8.10 were with a total picture size of 512×512 elements $g_1 = 0$, $g_2 = 2$, $\bar{a} = (a_{max} - a_{min})/2$, $N_{11} = N_{21} = N_{12} = 32$; and for Fig.8.11: $g_1 = 1$, $g_2 = 8$, $\bar{a} = 0$, $N_{11} = N_{21} = N_{12} = N_{22} = 10$.

An interesting pseudorelief effect is obtained if the summation in (8.7) is asymmetrical in (k,ℓ), i.e. $N_{11} \neq N_{12}$, $N_{21} \neq N_{22}$. This is illustrated in Fig.8.12 for $N_{11} = N_{12} = 0$, $N_{21} = N_{22} = 1$, $g_1 = 1$, $g_2 = 4$.

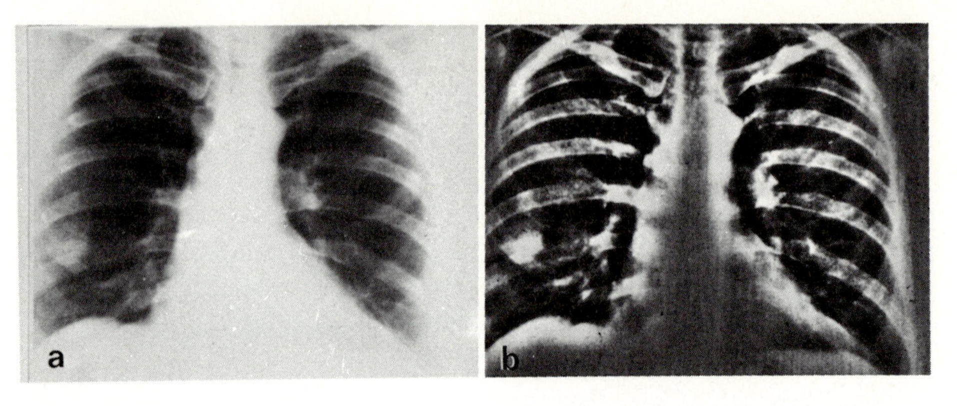

Fig.8.10a,b. Picture preparation by linear filtering: (a) initial picture, (b) result of emphasizing the higher spatial frequencies

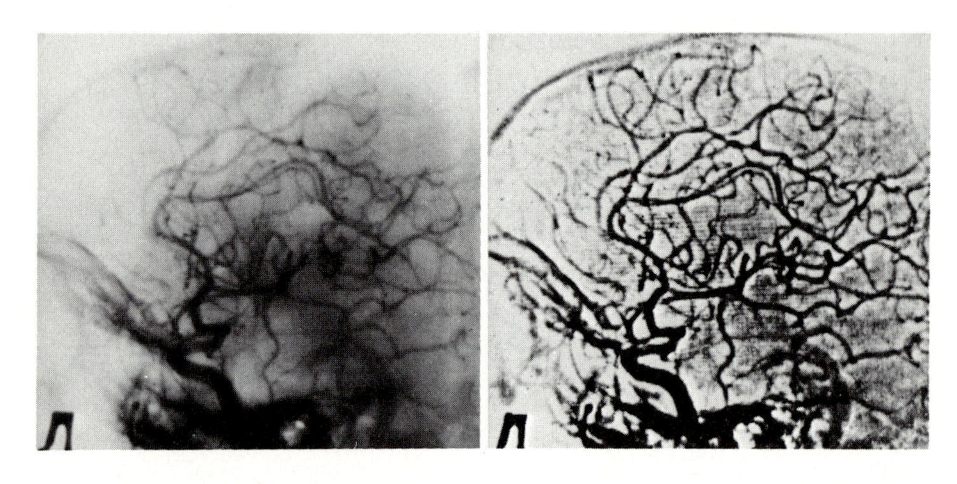

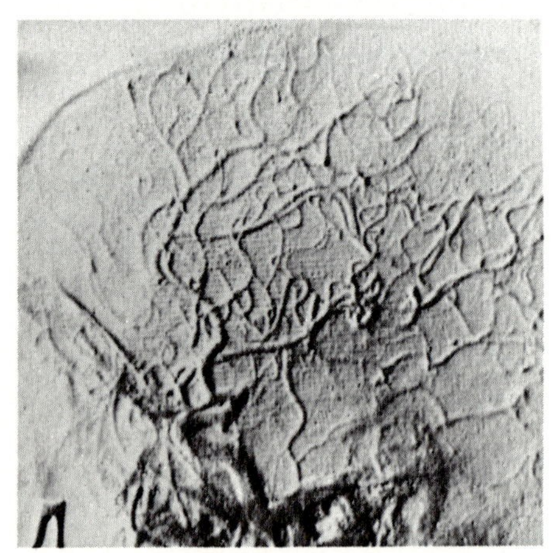

Fig.8.11a,b. Picture preparation by linear filtering: (a) initial picture of brain vessels, (b) result of emphasizing the higher spatial frequencies

Fig.8.12. Pseudorelief effect for the picture shown in Fig.8.11, obtained with an asymmetrical high-frequency filter

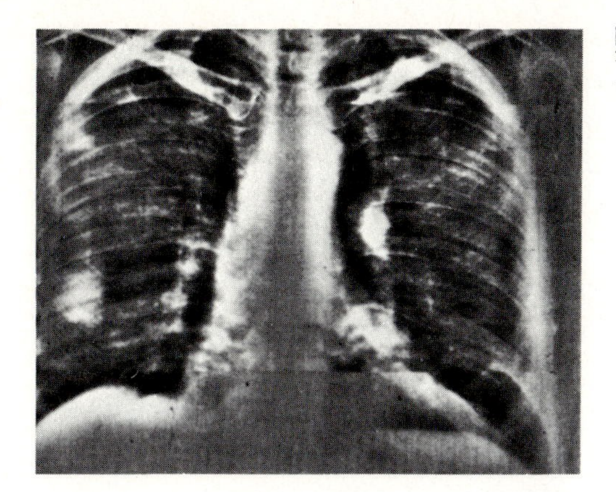

Fig.8.13. Removal of the ribs from the picture shown in Fig.8.10b by linear filtering

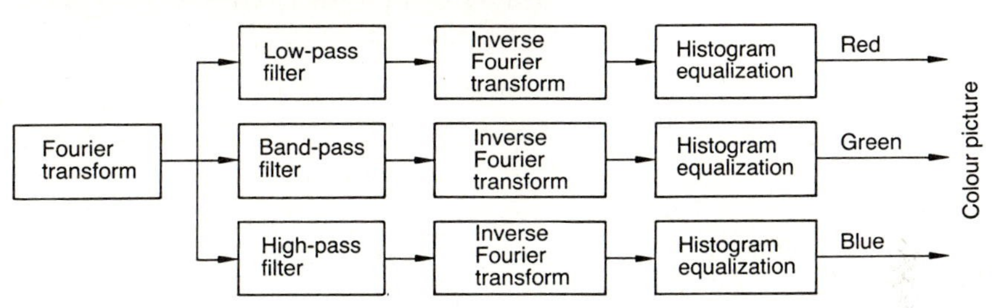

Fig.8.14. Flow chart of picture preparation procedure [5.3] with combination of linear and non-linear element-by-element transformations

In order completely to suppress individual components or narrow sections of the signal spectrum, it is convenient to filter in the frequency domain. This method was used to obtain the results shown in Fig.8.13; the periodic appearance of the ribs in the X-ray image of the rib cage shown in Fig.8.10 is weakened. The decrease in the contrast of the ribs here allows an increase in the contrast of details in the spaces between the ribs.

In the spectral domain, more complicated nonlinear forms of processing, designed to change the relationship between various spectral components of the signal, are also possible. Thus, good results are obtained by applying in the frequency domain a nonlinear, logarithmic transformation, which suppresses large values and strengthens small ones. This kind of transformation can be applied to Fourier or Walsh spectrum components [8.13,20]. Its visual effect is an increase in picture sharpness: the picture looks as if it had been retouched.

Still another example of spectrum transformation is the procedure described in [8.13]. The flow chart for it is shown in Fig.8.14. It is evident from the figure that as a result of the processing, the initial picture is converted into three pictures, which carry information about three separate parts of the spatial spectrum. In order to retain the unity of the images and at the same time extract this information, visualization in pseudocolours is used. These procedures are also interesting as an example of the combined application of several preparation methods: linear processing, nonlinear signal transformation and representation in pseudocolours. They can be useful in analyzing texture.

The intensity of the spectral components of the picture can in general be considered to be an attribute which better reflects the spatial structure of the picture than the initial video signal. Thus, isolated peaks in the spectrum correspond to periodic structures on the picture. Therefore manipulations of the spectrum such as the extraction, suppression or accentuation of individual portions are very effective methods of picture preparation.

Methods of adding and subtracting several pictures can also be regarded as linear preparation methods. They are commonly used to overlay on one photograph a signal which has been extracted either from another picture or is result of processing the given photograph. They may also be used to visualize changes in an object (e.g. to extract moving objects from successive frames of cinefilm, to reveal the progress of disease in organs with time in medical X-rays pictures [8.10], etc.).

8.5 Methods of Constructing Graphical Representation: Computer Graphics

One of the most common approaches to picture preparation is to convert the picture into a plot, a diagram composed of lines and special symbols and signs. The lines typically follow the edges of the various objects and their features. Signs and symbols are used to indicate the types of objects and their quantitative characteristics. Graphical representations can also be spatial drawings. In this case a form of projection employed in descriptive geometry is used to visualize picture features as a function of two coordinates.

The drawing of graphical representations is widely used in manual picture processing and, of course, in digital processing as well. The simplest computer graphics are lines of equal value of the video signal or one of its features (e.g., local averages, local variance, etc.). The concrete values to which these lines correspond are usually chosen as uniform steps on the value scale of the mapped magnitudes or as distribution quantiles of these values[1]. For convenience of interpretation, lines are sometimes superimposed

[1] A quantile is an area of part of the signal histogram, confined between two of values of the signal. The video signal values are distributed uniformly by quantiles if the same fraction of the picture elements falls between any two of these values.

Fig.8.15. Graphical preparation of a fragment of the picture shown in Fig. 8.9a. Lines of equal video signal values are overland on the picture

on the picture itself, as in Fig.8.15, which shows an enlarged region of the picture in Fig.8.9a overlaid with lines of equal video signal values uniformly following 10% quantiles.

The peculiarity of the digital construction of lines of equal value is that owing to the discrete nature of the picture, simply marking the picture elements where the video signal has a set value results in a network of broken or thickened lines that are wider than one element. If this is undesirable for one reason or another, a more complicated marking procedure must be adopted which takes decisions based not only on the values at the given point but also on the values in its immediate neighbourhood.

It is easy to obtain a graphical representation after the adaptive mode quantization of a picture (Sect.8.2). Picture elements differing from neighbouring ones are easily recognized. By selecting at the same time the video signal values in the selected elements, one can extract the edges of the portions belonging to one or several modes. Such a representation is shown in Fig.8.16a. It shows the edges of the portions of the aerial photograph of Fig.8.3a that were extracted in Fig.8.4c. Figure 8.16b, in which these edges are superimposed on the initial picture, clearly shows how they pick out the lightest details on the photograph. Note that if the purpose of preparation is to obtain a graphical representation, the procedure of adaptive mode quantization shown in Fig.8.2 need not be carried out to the end. It is sufficient to find the video signal values from the histogram separating one mode from another, and then construct lines which follow these values.

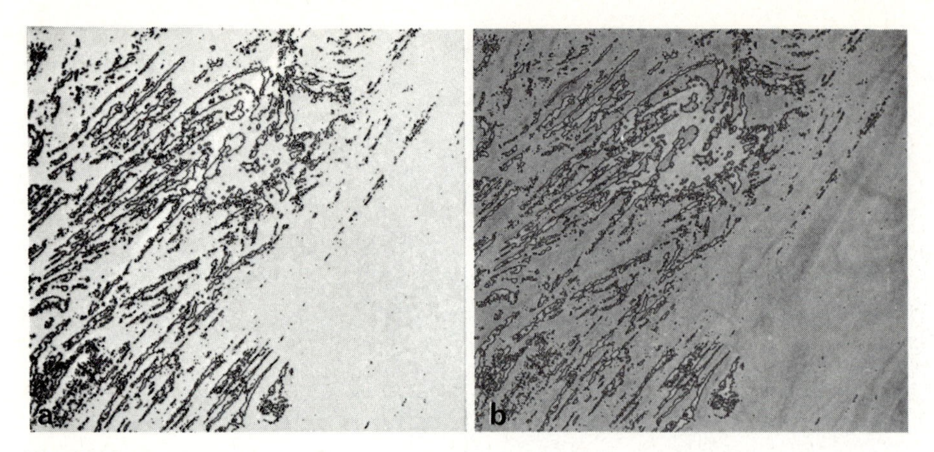

Fig.8.16a,b. Graphical preparation of the picture shown in Fig.8.3a. (a) edges of modes from the picture shown in Fig.8.4a, (b) the same, overlaid on the initial picture

The edges of objects can also be extracted directly from the original picture by noting which picture elements differ from their surroundings. The difference can be measured using one of the various digital differential operators. The following are the most commonly used:

1) The discrete analogue of the Laplace operator

$$\Delta a_{k,\ell} = a_{k,\ell} - 0.25(a_{k-1,\ell} + a_{k,\ell-1} + a_{k+1,\ell} + a_{k,\ell+1}) \tag{8.8a}$$

or its more isotropic variants

$$\Delta a_{k,\ell} = a_{k,\ell} - 0.125(a_{k-1,\ell-1} + a_{k-1,\ell} + a_{k-1,\ell+1} + a_{k,\ell-1} + a_{k,\ell+1} + a_{k+1,\ell-1} + a_{k+1,\ell} + a_{k+1,\ell+1}) \tag{8.8b}$$

$$\Delta a_{k,\ell} = a_{k,\ell} - 0.15(a_{k-1,\ell} + a_{k+1,\ell} + a_{k,\ell-1} + a_{k,\ell+1}) - 0.1(a_{k-1,\ell-1} + a_{k-1,\ell-1} + a_{k-1,\ell+1} + a_{k+1,\ell-1} + a_{k+1,\ell-1}) \quad . \tag{8.8c}$$

2) The modulus of the Laplace operator $|\Delta a_{k,\ell}|$.

3) The modulus of the Gradient

$$|\Delta a_{k,\ell}| = \sqrt{(a_{k-1,\ell} - a_{k,\ell})^2 + (a_{k,\ell-1} - a_{k,\ell})^2} \tag{8.9a}$$

or its more simply computable analogue

$$|\Delta a_{k,\ell}| = |a_{k-1,\ell} - a_{k,\ell}| + |a_{k,\ell-1} - a_{k,\ell}| \quad . \tag{8.9b}$$

In Fig.8.17a,18a, extraction of edges by means of the modulus of the gradient and the modulus of the Laplace operator is illustrated using the proces-

Fig.8.17a,b. Graphical preparation of the picture shown in Fig.8.9c. (a) module of gradient, (b) the same, overlaid on the initial picture

Fig.8.18a,b. Another example of graphical preparation on the picture shown in Fig.8.9c. (a) module of the Laplace operator, (b) the same, overlaid on the initial picture

sing in Fig.8.9c. Adding these edges to the initial picture can also give a desirable visual effect (Fig.8.17b,18b).

In order to recognize edges with preset directions, one can calculate the difference in the direction, perpendicular to that of interest. For example, the horizontal edges occur where the difference values on the vertical are large:

$$\Delta_8 a_{k,\ell} = |a_{k,\ell-1} - a_{k,\ell}| \quad . \tag{8.10}$$

If there is much fine detail in the picture (owing to noise in the video signal sensor, for example), these simple differential or difference operators give a signal that markedly differs from zero in nearly every picture element. As a result the line pattern formed by the edges is spoiled, having numerous short lines or separate spots. Additional methods are applied to clean it. These include retaining only those points where the signal difference exceeds a certain threshold; finding the difference between average values over neighbouring regions; and introducing line tracing algorithms [8.21]. There are quite a few of them, but the concept of edges is somewhat ambiguous, and the flexibility of digital methods opens up wide possibilities for invention.

The extended lines obtained from such an operation can be very jagged again, for example, as a result of noise from the video signal sensor. Various methods are used to smooth them, in particular the drawing of smooth curves by the least-squares method (see [8.22], in which the line smoothing is described for blood vessels).

Let us note in conclusion that the task of extracting contours can be approached in the same way as the task of measuring the coordinates of a given object in the picture, (this is analyzed in the next chapter). From this viewpoint it is easier to order all the various methods of extracting edges and explain rationally why they give results and how they relate to optimal methods. Moreover, analyzing the problems involved in localizing objects in pictures gives a deeper understanding of what picture contours are and why they are so important in viewing systems (Sect.9.4).

8.6 Geometric Picture Transformation

The following classes of problems requiring geometrical transformation of a picture can be defined:

1) Compilation of photographic maps and mosaics from aerial and space photographs of the surface of the earth and the other planets; superimposition of coordinate grids on photographs.

2) Correction of geometric distortion introduced by imaging systems (objective lenses, television scanning systems, etc.).

3) Comparison of photographs of 3-D objects taken with various settings of the photographic device.

4) Combining various pictures of one and the same object taken in various spectral ranges or at different times, to obtain colour pictures or reveal differences.

Geometric picture transformation can also be a stage in correction aberrations in imaging systems. Prior to linear filtering, the photograph is geometrically transformed and the imaging system thus converted into a spatially homogeneous one. After filtering, the reverse transformation is carried out, thereby restoring the previous coordinate system [8.23,24]. In principle, the task of geometrial transformation reduces to determining from the given picture $a(t_1, t_2)$ the value of the video signal at every point of the picture in the new coordinates $u_1(t_1, t_2)$, $u_2(t_1, t_2)$. Note that as we are discussing the pictures of continuous objects on a plane, the law of transforming the coordinates $u_1(t_1, t_2), u_2(t_1, t_2)$ is usually expressed in a continuous representation. Depending on the type of picture and the means of expressing the coordinate transformation law, there are two possible approaches to solving this problem.

1) For every element of the initial picture with the coordinates (k,ℓ), its coordinates in the new system $u_1(k\Delta t_1, \ell\Delta t_2)$, $u_2(k\Delta t_1, \ell\Delta t_2)$ are found, and the value of the video signal in that element is assigned to the element of the transformed picture with the coordinates $(\{[u_1(k\Delta t_1, \ell\Delta t_2)] / \Delta u_1\}, \{[u_2(k\Delta t_1, \ell\Delta t_2)] / \Delta u_2\})$, where Δt_1, Δt_2, Δu_1, Δu_2 are the discretization steps of the picture on the coordinates t_1, t_2, u_1, u_2, and the braces denote the integer part of quantities in them.

2) For every element of the transformed picture with the coordinates (k', ℓ'), its coordinates $[t_1(k'\Delta u_1, \ell'\Delta u_2), t_2(k'\Delta u_1, \ell'\Delta u_2)]$ are found in the initial picture $a(t_1, t_2)$, and the video signal value equal to $a[t_1(k'\Delta u_1, \ell'\Delta u_2), t_2(k'\Delta u_1, \ell'\Delta u_2)]$ is assigned to this element. Since, in digital processing, the initial picture is defined only at the sample points $(k\Delta t_1, \ell\Delta t_2)$, the quantity $a[t_1(k'\Delta u_1, \ell'\Delta u_2), t_2(k'\Delta u_1, \ell'\Delta u_2)]$ must be found using one of the various methods of interpolating the discrete signal $\{a_{k\ell}\}$.

The first approach is convenient in cases in which the picture being transformed is sparsely filled, i.e., contains few non-zero elements (e.g., when it is composed of lines, as in the superposition of coordinate grids). If it is used for the transformation of ordinary grey-scale pictures, which differ from zero in every element, then, firstly the error in mapping the coordinates will be large because they are rounded off during renumbering and secondly, empty (unfilled) places will remain in the transformed picture, and these will have to be filled by interpolation. As they will be distributed irregularly, this interpolation can be a laborious operation.

The second approach is applied for the transformation of "full" pictures. Interpolation on the initial picture can sometimes be avoided, and the transformed coordinates can be rounded up. This decreases the accuracy of measurement of the coordinates of the object in the transformed picture, owing to deformation of the shape of the object resulting from rounding errors. But if the curvature of coordinate grid is not large, this distortion is acceptable, since the avoidance of exact interpolation significantly simplifies the transformation procedure [8.25]. For a more faithful transformation, it may still prove necessary to interpolate the values of samples of the initial picture. Exact recommendation for correct interpolation is difficult to give, since at present it is hard to describe the link between accuracy of the coordinate transform and that of subsequent measurements using the transformed picture. Therefore, we shall limit ourselves to describing two practical interpolation methods.

8.6.1 Bilinear Interpolation

The 2-D analogue of 1-D piecewise interpolation, the bilinear interpolation, is one of the simplest means of interpolation for computing. Its underlying basis is shown diagrammatically in Fig.8.19, in which a_1, a_2, a_3, a_4, are the given values at four neighbouring picture samples, and $\hat{a}$ is the interpolated value in the intermediate sample at a point with the coordinates (α, β) relative to the first sample. The distance between the initial samples is taken to be unity. As can be seen from the figure, $\hat{a}$ is found by linear interpolation between two values, which in turn are found by linear interpolation between a pair of samples (a_1, a_4) and (a_2, a_3):

$$\hat{a} = [a_1(1 - \alpha) + a_4\alpha\ (1 - \beta) + a_2(1 - \alpha) + a_3\alpha]\beta \quad . \tag{8.11}$$

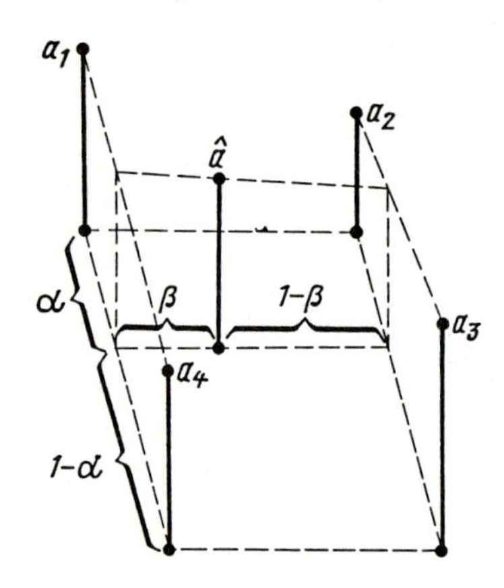

Fig.8.19. The method of bilinear interpolation

8.6.2 Interpolation Using DFT and SDFT

This method of interpolation is based on the properties of the DFT and SDFT (Sects.4.6,7) and is discussed in Sect.4.8. It is the discrete analogue of optimal continuous signal interpolation based on sampling theory (Sect.3.4), and thus in principle corresponds better to the picture structure. However,

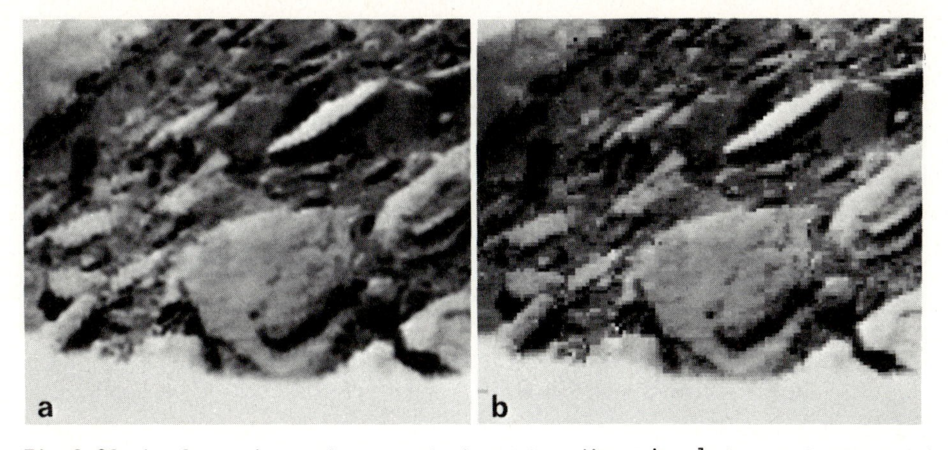

Fig.8.20a,b. Comparison of two methods of two-dimensional interpolation: (a) interpolation by the discrete analog of the sampling theorem, (b) interpolation by simple fourfold repetition in each direction of every sample of the fragment of the initial picture shown in Fig.8.9a

it is more complicated to calculate. Figure 8.20a shows a part of the view of Venus in Fig.8.9a as an example of interpolating in this way. The number of samples in this region is four times greater (in every direction) than the number of samples in the initial picture. For comparison, Fig.8.20b shows the same picture, interpolated by simple fourfold repetition of every sample of the initial picture in Fig.8.9a.

The operation of recalculating the coordinates and interpolating the video signal values is generally speaking fairly cumbersome, and if it is carried out for all the video signal samples much computer time is required. To decrease this, methods of reduced description are used which can turn a "full" picture into a "sparse" one. These include methods of picture coding based on the transformation of contours or of clearly marked discontinuities of the signal in the picture [8.26-28]. Indeed, insofar as the transformation of coordinates is topological, it is sufficient to extract only a few reference points in the picture from which it can be restored with sufficient accuracy. Only they need then be transformed, and the rest of the picture can be restored from the reference points.

Thus, in [8.29] contours are employed as the reference points. The contours used are those which arise when elements whose quantized value differs from the value of the previous elements in the scan are extracted from the initial picture. The coordinates of the elements chosen are transformed, and the rest, rather than being transformed, are found by means of linear interpolation between reference points.

MARCH and GARDNER [8.29] limited themselves to linear interpolation and were therefore forced to ensure that the separated elements were not excessively far from each other. They were guided by the following rule: a curve can be replaced by a straight line in the new coordinate system if its deviation from the straight line does not exceed half the distance between the elements of the transformed picture. It was discovered that for the type of projection investigated (Mercator), this criterion was met in the majority of cases by points spaced 6 to 16 elements from one another. It was thus possible to decrease by approximately one order of magnitude the number of operations directly necessary for the calculation of the transformed coordinates. As analysis of the task of measuring the position of objects on a plane shows (see Chap.9), there are grounds for believing that with accurate transmission of the coordinates of abrupt changes of signals, and more crude transmissions of the other elements, the required precision of measurement in the transformed pictures should not suffer too much.

9 Measuring the Coordinates of Objects in Pictures

This chapter is devoted to one of the characteristic problems of automatic picture interpretation, namely the localization of objects in a picture. A mathematical formulation of the problem is given in Sect.9.1, and in Sect.9.2 the optimal adaptive algorithm for localization of a precisely known object in a spatially homogeneous picture is derived and illustrated. Section 9.3 gives modifications of the algorithm for use when there is some uncertainty about the object and when the picture is inhomogeneous and blurred. In Sect. 9.4 the connection between optimal localization and picture contours is discussed, and an algorithm for automatic choice of the most easily detectable objects is suggested and illustrated. In Sect.9.5, the automatic detection and extraction of bench marks on aerial and space photographs is presented as a practical application of the methods developed.

9.1 Problem Formulation

One of the most important purposes of pictures is to carry information about the spatial distribution of objects in relation to one another. There are a number of practical tasks which require that objects be found and their coordinates measured (localization). These include detecting and measuring the coordinates of objects in photographic interpretation; measuring the mutual arrangement of corresponding points on stereo pairs of pictures, finding and measuring the position of benchmarks in aerial and space photographs; detecting faults and alien formations in medical and industrial diagnostics using pictures; finding set objects and symbols in information and retrieval systems; etc. These tasks can be fulfilled either "manually" by visual picture analysis, or with automatic digital or analogue (optical or electronic) computers by processing the corresponding 2-D signal.

To solve problems of automatic detection and/or localization (measurement of the coordinates) of objects in pictures, a description of the signal corresponding to the object sought must be available; otherwise the whole task becomes meaningless! Moreover, the means whereby this signal is represented in the observed picture must be also described.

The simplest additive model of the object sought and the picture observed is commonly used [9.1,2]. In it the picture observed is considered to be the sum of the signal sought, which is known with an accuracy up to a coordinate shift, and additive, signal-independent normal noise with a known correlation function. This model gives the well-known result that the optimal estimator of the coordinates of the signal sought, which yields the maximum a posteriori probability estimate, will be a linear estimator. The latter is composed of the linear optimal filter and the decision-making device defining the coordinates of the absolute maximum or a number of principal maxima (for the localization of several objects) of the signal at the filter output [9.3,4]. The linear optimal filter has the frequency response

$$H(f_1, f_2) = \frac{\alpha_0^*(f_1, f_2)}{P_n(f_1, f_2)} \quad , \tag{9.1a}$$

where f_1, f_2 are the spatial frequencies; $\alpha_0^*(f_1, f_2)$ is the complex-conjugate of the spectrum of the signal to be found taken at the origin of coordinates; and $P_n(f_1, f_2)$ is the power spectral density of the additive noise.

If the additive Gaussian noise is "white", i.e., if it has a uniform power spectrum [$P_n(f_1, f_2)$ = const], the optimal filter becomes the matched filter, or correlator [9.4,5]:

$$H(f_1, f_2) = \alpha_0^*(f_1, f_2) \quad . \tag{9.1b}$$

However, for many practical tasks of localizing objects in pictures the additive model is false, and in general the relationship between the signal sought and the picture against whose background it is to be found cannot be described deterministically. Thus, for example, in aerial photographs, signals from separate objects are not summed in the observed general signal, but "mortice" into it. Moreover, the observed signal of the object sought is determined not only by the object itself, but also by the neighbouring objects (e.g., their shadows), the lighting conditions during filming, weather conditions, camera noise, and other random factors that are difficult to formalize. Therefore the problem of localization objects in pictures should be solved with more realistic assumptions about the form in which the object is represented in the observed picture, and also with the least possible number of limitations.

There is yet another approach to detecting and localizing objects in pictures. It can be termed heuristic, and is characteristic of research on pattern recognition. It is based on the use of the Schwarz inequality, which tells us that the normalized correlation coefficient of two signals, say a(t) and b(t), cannot exceed unity:

$$\frac{\int\limits_{-\infty}^{\infty} a(t)b^*(t)\, dt}{\left[\int\limits_{-\infty}^{\infty} |a(t)|^2 dt \int\limits_{\infty}^{\infty} |b(t)|^2 dt\right]^{1/2}} \leqslant 1 \quad . \tag{9.2}$$

Equality is achieved only if the signals coincide exactly apart from a multiplicative constant $a(t) = kb(t)$. This approach also leads to a system composed of a correlator and a decision-making device that compares the signal at the output of the correlator with the threshold, which is proportional to the square root of the energy of the observed picture [9.6]. It is known, however, that this system has a large probability of false detection or identification, even in simple pictures consisting of letters, symbols and numbers. Therefore, several improvements have been proposed: predistortion of the sample objects before correlation, various types of "contour" extraction, quantization, etc.

The drawback of such improvements is that they have been invented without clear account being taken of the distinctive features and limits of the tasks. For this reason it is unclear, firstly, whether further improvements are possible, and secondly, when which suggestion is preferable.

The correlation detector-estimator is a variant of the idea of the linear-detector-estimator. In the latter a decision regarding the presence of the object sought and its coordinates is taken on the basis of the value of the signal at each point of the field at the output of a certain linear filter acting on the observed picture.

The purpose of the linear filter in such a device is to transform the signal space so that afterwards the decision can be based not on all of the signal, but independently on its individual coordinates in the transformed space. The structure of 1-D decision devices is not very complicated; it must perform only the operation of comparing two values. The task of synthesizing the estimator therefore reduces to synthesizing only its linear block, which is much simpler. Moreover, the division into independent linear and nonlinear threshold blocks significantly simplifies not only the analysis, but also the implementation of such devices in digital and analogue processors. This, in particular, explains the popularity of the correlation method of detecting and localizing objects in pictures.

Let us find the form for the linear estimator filter of the object coordinates which ensures optimal measurement quality.

The quality of measurement of the object coordinates is determined by two types of errors: those resulting from incorrect identification of the object in the observed picture, and those of measuring the coordinates close to their true value. Errors of the first type are determined by large deviations which exceed the size of the object sought in the results of coordinate measurement. In detection, these are errors of the false-alarm and object-passing (false-negative) type. We shall call them anomalous. Errors of the second type have a value of the order of the size of the objects and are associated with imprecise determination of the coordinates within the limits of the object itself. These we shall call normal.

Normal errors are primarily with signal distortion from the object sought. They are perfectly well described by the additive model, which takes only the noise of the video signal sensor into account. Therefore in terms of minimum normal error (normal errors can be considered to be characterized by their standard deviation), the optimal estimator is one with filter (9.1a), although it will cause many anomalous errors. The probability of anomalous error and the threshold property associated with it have been discussed for such an estimator in [9.3].

Let us find the characteristics of the linear filter of the estimator which is optimal in relation to anomalous errors. Let us also define the idea of optimality. In order to allow for spatial inhomogeneity of the picture, we will consider it to be divided into N fragments with areas S_n, $n = 0, 1, \ldots, N-1$.

Let $h^{(n)}(b, t_1^0, t_2^0)$ be the density distribution of the values of the video signal $b(t_1, t_2)$ at the output of the filter, measured for the nth fragment from points not occupied by the object, with the condition that the object is situated at the point with the coordinates (t_1^0, t_2^0). Let b_0 be the quantity of the signal at the point of the filter output at which the object is localized. (Without limiting the applicability, we can consider that $b_0 > 0$). Then, since the linear estimator determines the coordinates of the object sought from the coordinates of the absolute maximum of the signal at the output of the linear filter, the integral

$$Q_n(t_1^0, t_2^0) = \int_{b_0}^{\infty} h_n(b, t_1^0, t_2^0)db \tag{9.3}$$

is the proportion of the points of the nth fragment whose coordinates may be erroneously adopted by the decision device as the coordinates of the object.

Quantity b_0 should generally speaking be considered random, since it is influenced by noise from the video signal sensor, the conditions of filming and illumination, the orientation of the object during filming, the presence of neighbouring objects, and other random factors. To allow for these, let us introduce a function $q(b_0)$, the a priori probability density of b_0. The coordinates of the object must also be considered random. Moreover, in localization problems the weight of measuring errors can be unequal for various parts of the picture. To allow for these factors we introduce the weight functions $w^{(n)}(t_1^0, t_2^0)$ and W_n, which characterize the a priori significance of errors in determining the coordinates within the nth region and for every nth region respectively:

$$\iint_{S_n} w^{(n)}(t_1^0, t_2^0)dt_1^0dt_2^0 = 1 \quad , \quad \sum_{n=0}^{N-1} W_n = 1 \quad . \tag{9.4}$$

Then the quality of coordinate measurement by the estimator under consideration can be described by the integral (9.3) weighted by $q(b_0)$, $w^{(n)}(t_1^0, t_2^0)$ and W_n. Thus

$$\overline{Q} = \int_{-\infty}^{\infty} q(b_0)db_0 \sum_{n=0}^{N-1} W_n \iint_{S_n} w^{(n)}(t_1^0, t_2^0)dt_1^0dt_2^0 \times \int_{b_0} h_n(b, t_1^0, t_2^0)db \ . \tag{9.5}$$

If we are interested in the quality of the estimator averaged over a certain set of pictures, then the quantity $\overline{Q}$ must be averaged over this set.

We shall consider the estimator ensuring the minimum $\overline{Q}$ to be optimal [9.7].

9.2 Localizing a Precisely Known Object in a Spatially Homogeneous Picture

Let the object sought be precisely known. In the given case this means that the response of any filter to this object can be precisely calculated, i.e., $q(b_0)$ is a δ function:

$$q(b_0) = \delta(b_0 - \overline{b}_0) \quad . \tag{9.6}$$

Then (9.5), which determines the quality of the localization, becomes

$$\overline{Q} = \sum_{n=0}^{N-1} W_n \iint_{S_n} w^{(n)}(t_1^0, t_2^0)dt_1^0dt_2^0 \int_{b_0}^{\infty} h_n(b, t_1^0, t_2^0)db \tag{9.7}$$

or, if we write

$$\overline{h}_n(b) = \iint_{S_n} w^{(n)}(t_1^0, t_2^0)\, h_n(b, t_1^0, t_2^0)dt_1^0dt_2^0 \quad , \text{ we have} \tag{9.8}$$

$$\overline{Q} = \sum_{n=0}^{N-1} W_n \int_{b_0}^{\infty} \overline{h}_n(b)db \quad . \tag{9.9}$$

We consider the picture to be spatially homogeneous if $\overline{h}_n(b)$ and W_n are independent of n. We denote the histogram of a homogeneous picture by $\overline{h}(b)$:

$$\overline{h}(b) = \iint_S w(t_1^0, t_2^0)\, h(b, t_1^0, t_2^0) dt_1^0 dt_2^0 \quad , \tag{9.10}$$

where S is the area of the picture, and $w(t_1^0, t_2^0)$ is the weight function of errors in determining the coordinates in the picture.

With localization in this picture

$$\overline{Q} = \int_{b_0}^{\infty} \overline{h}(b) db \quad . \tag{9.11}$$

Let us first find the frequency response $H(f_1, f_2)$ of the optimal filter for the spatially homogeneous picture. The choice of $H(f_1, f_2)$ influences both the quantity $\overline{b}_0$ and the histogram $\overline{h}(b)$. As $\overline{b}_0$ is the magnitude of the response of the filter at the point where the object is located, it can be found if the spectrum of the object $\alpha_0(f_1, f_2)$ is known:

$$\overline{b}_0 = \int_{-\infty}^{\infty} \alpha_0(f_1, f_2)\, H(f_1, f_2) df_1 df_2 \quad . \tag{9.12}$$

The link between $\overline{h}(b)$ and $H(f_1, f_2)$ generally speaking has a complicated character. The dependence on $H(f_1, f_2)$ can only be described explicitly for the second moment of the histogram $\overline{h}(b)$. This is done by using Parseval's relation for the Fourier transform (see Tables 2.2,3, line 17):

$$m_2 = \left[\int_{-\infty}^{\infty} b^2 \overline{h}(b) db\right]^{1/2} = \left[\iint_S w(t_1^0, t_2^0) dt_1^0 dt_2^0 \int_{-\infty}^{\infty} b^2 h(b, t_1^0, t_2^0) db\right]^{1/2}$$

$$= \left[1/S_1 \iint_S w(t_1^0, t_2^0) dt_1^0 dt_2^0 \iint_{S_1} b^2(t_1, t_2) dt_1 dt_2\right]^{1/2}$$

$$= \left[1/S_1 \iint_S w(t_1^0, t_2^0) dt_1^0 dt_2^0 \int_{-\infty} \int_{-\infty} |\alpha_b(f_1, f_2)|^2\, |H(f_1, f_2)|^2 df_1 df_2\right]^{1/2}$$

$$= \left[1/S_1 \int_{-\infty}^{\infty} \int_{-\infty}^{\infty} \overline{|\alpha_b(f_1, f_2)|^2}\, H(f_1, f_2)\,^2\, df_1 df_2\right]^{1/2} \quad , \tag{9.13}$$

where S_1 is the area of the picture under analysis without allowance made for the area occupied by the signal of the sought-for object at the output of the filter; $\alpha_b(f_1, f_2)$ is the Fourier spectrum of the picture, in which the signal values in the part occupied by the object sought are replaced by zero values; and

$$\overline{|\alpha_b(f_1, f_2)|} = \left[\iint_S w(t_1^0, t_2^0)\, |\alpha_b(f_1, f_2)|^2\, dt_1^0 dt_2^0\right]^{1/2} \quad . \tag{9.14}$$

Therefore we will use the Chebyshev inequality, well known in probability theory, which for histograms takes the following form:

$$\overline{Q} = \int_{\overline{b}_0}^{\infty} \overline{h}(b)db \leqslant m_2^2 / b_0^2 \quad , \tag{9.15}$$

and we will require that the relation

$$g = m_2^2 / \overline{b}_0^2 \tag{9.16}$$

be minimal[1].

This condition is satisfied when the following quantity reaches a maximum:

$$g_1 = \frac{\overline{b}_0^2}{S_1 m_2^2} = \frac{\int_{-\infty}^{\infty}\int_{-\infty}^{\infty} \alpha_0(f_1, f_2)\, H(f_1, f_2) df_1 df_2}{\int_{-\infty}^{\infty}\int_{-\infty}^{\infty} |\alpha(f_1, f_2)|^2 \, |H(f_1, f_2)|^2 df_1 df_2} \quad . \tag{9.17}$$

To find the maximum of g_1 from $H(f_1, f_2)$ we use the Schwarz inequality (9.2):

$$\frac{\int_{-\infty}^{\infty}\int_{-\infty}^{\infty}\left(\frac{\alpha_0(f_1, f_2)}{\overline{|\alpha_b(f_1, f_2)|}}\right) [H(f_1, f_2)\overline{|\alpha_b(f_1, f_2)|}]\, df_1 df_2}{\int_{-\infty}^{\infty}\int_{-\infty}^{\infty} \frac{|\alpha_0(f_1, f_2)|^2}{\overline{|\alpha_b(f_1, f_2)|^2}} df_1 df_2 \int_{-\infty}^{\infty}\int_{-\infty}^{\infty} |H(f_1, f_2)|^2 \, \overline{|\alpha_b(f_1, f_2)|^2}\, df_1 df_2} \leqslant 1 \quad , \tag{9.18}$$

from which it follows that the maximum value of g_1

$$g_{1max} = \int_{-\infty}^{\infty}\int_{-\infty}^{\infty} \frac{|\alpha_0(f_1, f_2)|^2}{\overline{|\alpha_b(f_1, f_2)|^2}} df_1 df_2 \tag{9.19}$$

is reached for

$$H_{opt}(f_1, f_2) = \frac{\alpha_0(f_1, f_2)}{\overline{|\alpha_b(f_1, f_2)|^2}} \quad . \tag{9.20}$$

This formula is similar to (9.1), but the meaning of the denominator is different. Here it is the power spectrum of that part of the observed picture that does not contain the object and thus appears as "noise".

---

[1] Strictly speaking, in view of the fact that the Chebyshev inequality (9.15) is a very crude evaluation of $\overline{Q}$, the requirement that g be minimal is a necessary and sufficient condition for minimum $\overline{Q}$ only if $\overline{h}(b)$ is a normal (Gaussian) distribution density. However, experiments show that in real pictures the histogram $\overline{h}(b)$ of the signal at the output of the filter, which is found from the condition of minimum g, is close to Gaussian. This is a consequence of the normalizing effect of the linear filter.

Let us express $|\alpha_b(f_1, f_2)|^2$ in terms of the spectrum of the observed picture, $\alpha_p(f_1, f_2)$, and the spectrum of the object sought, $\alpha_0(f_1, f_2)$. Clearly,

$$\alpha_b(f_1, f_2) = \alpha_p(f_1, f_2) - \alpha_0(f_1, f_2) \exp[-i2\pi(f_1 t_1^0 + f_2 t_2^0)] \quad , \qquad (9.21)$$

i.e.,

$$|\alpha_b(f_1, f_2)|^2 = |\alpha_p(f_1, f_2)|^2 + |\alpha_0(f_1, f_2)|^2 - \alpha_p(f_1, f_2)\, \alpha_0^*(f_1, f_2) \times \exp[i2\pi(f_1 t_1^0 + f_2 t_2^0)] - \alpha_p^*(f_1, f_2)\, \alpha_0(f_1, f_2) \exp[-i2\pi(f_1 t_1^0 + f_2 t_2^0)] \ . \qquad (9.22)$$

Then, substituting (9.22) into (9.14), we obtain

$$\overline{|\alpha_b(f_1, f_2|^2} = |\alpha_p(f_1, f_2) - \alpha_0(f_1, f_2)\, W(f_1, f_2)|^2 \quad , \qquad (9.23)$$

where

$$W(f_1, f_2) = \iint_S w(t_1^0, t_2^0) \exp[-i2\pi(f_1 t_1^0 + f_2 t_2^0)]dt_1^0 dt_2^0 \qquad (9.24)$$

is the spectrum of the weight function $w(t_1^0, t_2^0)$.

Usually the area of the picture occupied by the object sought is much smaller than the area of the picture itself. Therefore in practice the approximation $|\alpha_b(f_1, f_2)|$ can often be used:

$$\overline{|\alpha_b(f_1, f_2)|^2} \approx |\alpha_p(f_1, f_2)|^2 \quad . \qquad (9.25)$$

Clearly, if the optimal filter for a set of pictures must be constructed, then the result of averaging over the picture spectra of the given set, rather than $|\alpha_p(f_1, f_2)|$, must be substituted into (9.23) and (9.25).

The resulting optimal filter can be realized fairly easily by optical means [9.8] and gives very good results [9.9].

In the realization of this filter in digital processors it is most natural to use signal processing in the frequency domain, since the frequency response (9.20) of the optimal filter is based on measurements of the Fourier spectrum of the observed picture.

It is interesting to examine the results of simulating the optimal linear estimator and comparing it with the traditional correlator. Figure 9.1 shows a picture consisting of 512×512 elements on which experiments were conducted [9.10]. The coordinates of 20 dark, square test marks consisting of 5×5 elements were measured, and the marks were superposed on the picture. The manner in which they were arranged is shown in Fig.9.2. The numbers in the boxes in Fig.9.2 show the serial numbers of the corresponding spots in the list of the 20 principal signal maxima at the output of the optimal filter. As can be seen

Fig.9.1. Picture with test marks used in an experiment on optimal localization

from the arrangement, the test objects were superposed on parts of the aerial photograph having different structures, thus permitting the performance of the correlator and of the optimal linear estimator to be evaluated under different conditions. The contrast in the marks amounted to approximately 25 percent of the range of the video signal on the aerial photograph. The ratio of the ampli-

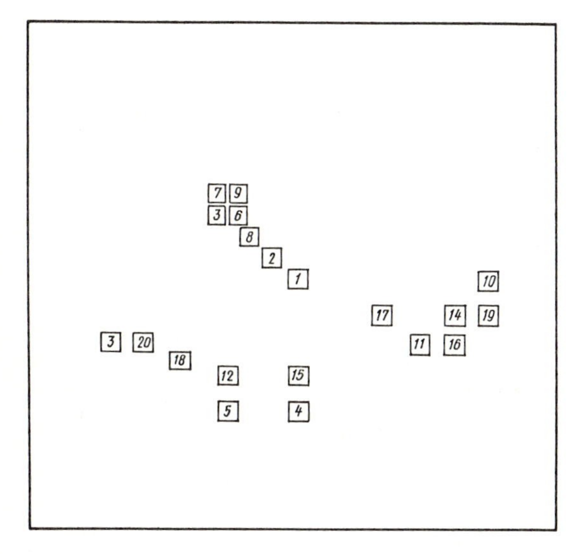

Fig.9.2. Arrangement of the test marks in the picture shown in Fig.9.1

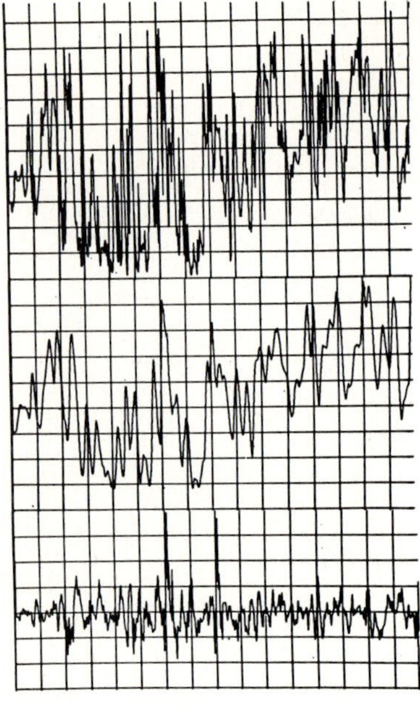
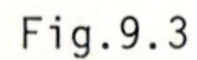
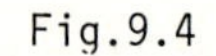

Fig.9.3. Comparison of optimal linear filter and conventional correlator. The top plot is the cross-section of the initial signal through marks 12 and 15 of the picture shown in Fig.9.1, the middle plot is the same cross-section of the output of the correlator, and the bottom plot is the same cross-section of the output of the optimal filter
Fig.9.4. Pattern of decisions at the output of the correlator

tudes of the marks to the mean-square value of the video signal in the background of the picture is about 1.5.

Figure 9.3,4 show the results of simulation. Figure 9.3 shows (from top to bottom) the plot of the cross-section of the initial signal, of the output of the correlator, and of the output of the optimal filter passing through the center of the marks 12 and 15 (Figs.9.1,2). The plot of the output signal of the correlator clearly shows the autocorrelation peaks of the test marks and the peaks of the mutual correlation function of the test marks with details of the background picture, including those that exceed the autocorrelation peaks and consequently give false decisions. The map of decisions taken by comparing the signal with a threshold is shown in Fig.9.4. The threshold was set so that all marks should be found without exception. As can be seen from Fig.9.4, the number of false detections is very great. If the lower plot in Fig.9.3 is compared with the middle plot evaluation can be made of the extent to which the optimal filter facilitates the localization problem for the decision device. In the given photograph the coordinates of all 20 test spots could be detected faultlessly by this means.

9.3 Uncertainty in the Object and Picture Inhomogeneity. Localization in "Blurred" Pictures

Let us now look at cases in which the object is not known with absolute accuracy, i.e., $q(b_0)$ cannot be considered a δ function. As before, we consider the picture to be spatially homogeneous. Now the optimal estimator should ensure that the integral

$$\overline{Q}_1 = \int_{-\infty}^{\infty} q(b_0)db_0 \int_{b_0}^{\infty} \overline{h}(b)db \quad , \tag{9.26}$$

passes through a minimum, where $\overline{h}(b)$ is defined by (9.10).

9.3.1 "Exhaustive" Estimator

Let us divide the interval of possible values of b_0 into subintervals within which $q(b_0)$ can be considered constant. Then

$$\overline{Q}_1 \approx \sum_i q_i \int_{b_0^{(i)}}^{\infty} h(b)db \tag{9.27}$$

where $b_0^{(i)}$ is representative of the ith interval, and q_i the area under $q(b_0)$ in the ith interval.

Since $q_i \geqslant 0$, $\overline{Q}_1$ is minimal if

$$\overline{Q}_1^{(i)} = \int_{b_0^{(i)}}^{\infty} h(b)db \tag{9.28}$$

is minimal.

Thus the problem leads back to the previous situation, in which a precisely known object was localized. It differs only in that now one must construct the estimator with the filter

$$H_{opt}^{(i)}(f_1, f_2) = \frac{\alpha_0^{*(i)}(f_1, f_2)}{|\alpha_b(f_1, f_2)|^2} \tag{9.29}$$

separately for every "representative" of the object from all of its possible variants, i.e., one must consider there to be not one given object, but several, which differ from one another by the values of the unknown parameters. This, of course, leads to an increase in processing time, needed for exhaustive searching.

9.3.2 Estimator Seeking an Averaged Object

If the spread of parameters is not large, the problem can be solved as if the object were precisely known. This is achieved, at the expense of a certain increase in the rate of anomalous errors, by modifying the optimal filter, to take the spread of the object parameters into account. To find the corrected filter transfer function, let us substitute the variable $b_1 = b - b_0$ into (9.26) and change the order of integration:

$$\overline{Q}_1 = \int_0^{\infty} db_1 \int_{-\infty}^{\infty} q(b_0)\, \overline{h}(b_1 + b_0) db_0 \quad . \tag{9.30}$$

The internal integral in (9.30) is the convolution of two distributions. Let us denote the resulting distribution by $\overline{h}_q(b_1)$. Its mean value is equal to the difference between the means b_0 and b_{ave} of the distributions $q(b_0)$ and $\overline{h}(b)$, and the variance is the sum of the variances of these distributions, i.e., $[m_2^2 - (b_{ave})^2] + \sigma_q^2$, where σ_q^2 is the variance of the distribution $q(b_0)$. Therefore

$$\overline{Q}_1 = \int_0^{\infty} h_q(b_1) db_1 = \int_{\overline{b_0}}^{\infty} h_q(b_1 - \overline{b}_0) db_1 \quad . \tag{9.31}$$

Thus, the problem reduces to the one examined in Sect.9.2, and by analogy with (9.20) the frequency response of the optimal filter can be written as

$$\overline{H}_{opt}(f_1, f_2) = \frac{\overline{\alpha}_0^*(f_1, f_2)}{\overline{|\alpha_b(f_1, f_2)|^2} + |\alpha_e(f_1, f_2)|^2} \quad , \tag{9.32}$$

where $\overline{\alpha}_0^*(f_1, f_2)$ is the complex conjugate of the spectrum $\alpha_0(f_1, f_2)$ of the object, averaged over the set of its unknown parameters [averaged over $q(b_0)$ in (9.30)], and

$$|\alpha_e(f_1, f_2)|^2 = \overline{|\alpha_0(f_1, f_2) - \overline{\alpha}_0(f_1, f_2)|^2} \tag{9.33}$$

is a similar mean-square difference.

Hence it is clear that the optimal filter changes its form somewhat in comparison with the deterministic case in which the object is precisely known. The filter is constructed on the basis of the "averaged" object and the power spectrum of the background picture, corrected by the mean square of the power spectrum of object variations.

On the basis of considerations relating to the evaluation of the power spectrum of the background picture that were introduced in Sect.9.2 in connection with (9.23), and because the variance $q(b_0)$ must be sufficiently small in order for the use of the averaged object as a reference to be meaningful, one can assume that the correction in the denominator (9.23) is not large when the area of the object is small compared with the area of the picture, and that the correction can be ignored.

The use of the above assumption regarding the spatial uniformity of the picture, i.e., the independence of the histogram $h(b)$ of the region on which it is measured, is rarely fulfilled in practice. It is usually more correct to consider the picture not to be spatially homogeneous in the sense stated in Sect.9.2. Therefore we turn to the general formula (9.5).

Depending on the implementation limitations, one can choose one of two possible ways of achieving minimum $\overline{Q}$.

9.3.3 Adjustable Estimator with Fragment-by-Fragment Optimal Filtering

With given W_n, minimum $\overline{Q}$ is achieved with the minimum of

$$\overline{Q}_2^{(n)} = \int_{-\infty}^{\infty} q(b_0)db_0 \iint_{S_n} w^{(n)}(t_1^0, t_2^0)dt_1^0dt_2^0 \int_{b_0}^{\infty} h_n(b,t_1^0, t_2^0)db \quad . \qquad (9.34)$$

for all n.

This means that the linear filter which transforms the picture must be adjustable, and that picture processing is carried out by fragments within which the picture can be considered spatially homogeneous. The optimal frequency response is found for each region using (9.18 or 32) on the basis of measurements of the locally observed power spectra of the fragments (allowing for the above reservation with regard to the influence which the spectrum of the object has on the observed picture spectrum). In accordance with (9.5) the transfer from fragment to fragment is effected in jumps. But based on (9.5) it is not difficult to understand that in principle a sliding algorithm arises from it which is based on the evaluation of the current local power spectrum of the picture, since the weight of the errors can be expressed by continuous functions. Let us also note that with fragment-by-fragment and sliding processing by means of the adjustable filter, the frequency response of the filter does not depend on the weights W_n or the corresponding continuous function.

The local spectrum can be evaluated by the recursive algorithm described in Sect.5.7.

9.3.4 Non-Adjustable Estimator

If it is impossible to implement the adjustable filter with fragment-by-fragment or sliding processing, the estimator must be based on the power spectrum of the picture fragments averaged over W_n. Indeed, from (9.5) it follows that

$$\overline{Q} = \int_{-\infty}^{\infty} q(b_0)db_0 \int_{b_0}^{\infty} \left[\sum_{n=0}^{N-1} W_n \iint_{S_n} w^{(n)}(t_1^0, t_2^0)dt_1^0dt_2^0 h_n \quad (b, t_1^0, t_2^0)dt_1^0dt_2^0\right]db$$

$$= \int_{-\infty}^{\infty} q(b_0)db_0 \int_{b_0}^{\infty} h(b)db \quad , \tag{9.35}$$

where $h(b)$ is the histogram $h_n(b)$, averaged over $\{W_n\}$ and $w^{(n)}(t_1^0,t_2^0)$. Hence by analogy with (9.20,32), we can conclude that

$$H_{opt}(f_1, f_2) = \frac{\alpha_0^*(f_1, f_2)}{\overline{\overline{|\alpha_b(f_1, f_2)|^2}} + |\alpha_e(f_1, f_2)|^2} \quad , \qquad \text{where} \tag{9.36}$$

$$\overline{\overline{|\alpha_b(f_1, f_2)|^2}} = \sum_{n=0}^{N-1} W_n \overline{|\alpha_b(f_1, f_2)|^2} \quad . \tag{9.37}$$

Thus, in this case the frequency response of the optimal filter depends on the weights $\{W_n\}$.

9.3.5 Localization in Blurred and Noisy Pictures

Sometimes the observed picture containing the object sought is defocused or blurred owing to the conditions of photography or the imperfection of the objective lens. As already noted in Chap.7, this effect can usually be described as the result of a certain linear system acting on the focused field. For simplicity let us look at spatially-invariant (isoplanatic) system. Such a system is fully characterized by its frequency response. Let us denote it by $H_s(f_1, f_2)$. The optimal estimator must be adjusted to the object, subjected to the same transformation as the observed picture; i.e. the frequency response of its filter must be defined by the relation[2]

$$H_{opt}(f_1, f_2) = \frac{H_s^*(f_1, f_2)\, \alpha_0^*(f_1, f_2)}{\overline{|\alpha_b(f_1, f_2)|^2}} \quad . \tag{9.38}$$

[2] For simplicity we shall proceed from the formula (9.20) for a precisely known object and a spatially homogeneous picture.

Depending on how much easier it is to realize this filter and in what form the reference object is given, various modifications of the formula are possible. For example, representation of $H_{opt}(f_1, f_2)$ in the form

$$H_{opt}(f_1, f_2) = \frac{H_s^*}{|H_s|(\overline{|\alpha_b|^2})^{1/2}} \frac{\alpha_0^*}{(\overline{|\alpha_b|^2})^{1/2} / |H_s|} \tag{9.39}$$

corresponds to an estimator in which the observed blurred picture spectrum undergoes "whitening" by means of the filter $H_s^* / |H_s|(\overline{|\alpha_b|^2})^{1/2}$ (rendering its power spectrum almost uniform, whence by analogy with the term "white noise" the term "whitening" has arisen) and then is correlated with the reference $\alpha_0^* / (\overline{|\alpha_b|^2})^{1/2} / |H_s|$. Relation $(\overline{|\alpha_b|^2})^{1/2} / |H_s|$ can be considered to be the spectrum of the picture at the output of a filter that is inverse to the defocusing filter, i.e., to be the spectrum of the picture corrected by the inverse filter. With this approach a link appears between the problem of localization in blurred pictures and correction of such images (Sect.7.3). The link can be even better established if one considers the case in which the observed picture is the sum of the defocused picture and additive independent noise.

Let $P_n(f_1, f_2)$ be the power spectrum of such noise, and $\overline{|\alpha_{b_0}(f_1, f_2)|^2}$ the spectrum of the focused picture without noise, so that the power spectrum of the observed background picture, averaged over additive noise, can be written as

$$\overline{|\alpha_b(f_1, f_2)|^2} = |H_s(f_1, f_2)|^2 \overline{|\alpha_{b_0}(f_1, f_2)|^2} + P_n(f_1, f_2) \quad . \tag{9.40}$$

Then (9.38) can be rewritten as

$$H_{opt}(f_1,f_2) = \left(\frac{1}{H_s(f_1, f_2)} \quad \frac{\overline{|\alpha_b(f_1, f_2)|^2} - P_n(f_1, f_2)}{\overline{|\alpha_b(f_1, f_2)|^2}}\right) \frac{\alpha_0^*(f_1, f_2)}{\overline{|\alpha_{b_0}(f_1, f_2)|^2}} \; . \tag{9.41}$$

Thus the optimal filter consists of two successive filters: the optimal adaptive filter analogous to filter (7.23) [the first factor in (9.41); see Sect.7.3] and the optimal filter of type (9.20) for the focused picture. It is interesting to note that when the noise intensity $P_n(f_1, f_2)$ is low in comparison with the background picture, the reliability of localization is almost independent of the degree to which the picture is defocused (blurred). Indeed, with a blurred and noisy picture, (9.19), which determines the reliability of localization, changes to

$$g_{1\max} = \int_{-\infty}^{\infty}\int_{-\infty}^{\infty} \frac{|H_s(f_1, f_2)|^2 |\alpha_0(f_1, f_2)|^2}{|H_s(f_1, f_2)|^2 |\alpha_b(f_1, f_2)|^2 + P_n(f_1, f_2)} df_1 df_2 \quad , \qquad (9.42)$$

from which the above conclusion follows.

9.4 Optimal Localization and Picture Contours. Choice of Reference Objects

To understand the point of the operations performed on observed pictures by the optimal filter of the linear estimator, it is convenient to present the frequency response (9.20) of this filter in the form

$$H_{opt}(f_1, f_2) = H_1 H_2 = \frac{1}{[\overline{|\alpha_b(f_1, f_2)|^2}]^{1/2}} \; \frac{\alpha_0^*(f_1, f_2)}{[\overline{|\alpha_b(f_1, f_2)|^2}]^{1/2}} \quad . \qquad (9.43)$$

In this representation the action of the optimal filter reduces to whitening of the background picture (filter H_1), which was mentioned in Sect.9.3, and to subsequent correlation of the whitened picture with the objects sought (filter H_2), which are transformed in the same way.

An interesting peculiarity of the optimal filter (9.42) is that the action of the whitening component $H_1(f_1, f_2) = 1/[\overline{|\alpha_b(f_1, f_2)|^2}]^{1/2}$ usually (but not necessarily) leads to contouring of the observed picture since higher spatial frequencies are enhanced. The reason for this is that the power spectrum of the picture is usually a quite rapidly decreasing function of spatial frequencies and consequently, $H_1(f_1, f_2)$ is an increasing function. This conclusion is illustrated by Fig.9.5, which shows the results of whitening the picture shown in Fig.9.1 and by Figs.9.6,7[3].

This accounts adequately for the empirically supported recommendations which certain researchers have made about the correlation method of localization. They find that in order to achieve more reliable localization, it is advisable to subject the picture to contouring using methods of spatial differentiation or to quantize the picture roughly to obtain sharp edges, before correlation. Moreover, this result once again clarifies what picture contours are and why they are so important in visual systems. The concept of contours is frequently encountered, and it is defined in various ways in works on picture processing and pattern recognition [9.6,11-15]. From the standpoint of object localization in pictures by means of the optimal linear estimator, "contours" are what is obtained by picture "whitening". The stronger this

[3] Approximation (9.25) was used in obtaining the results.

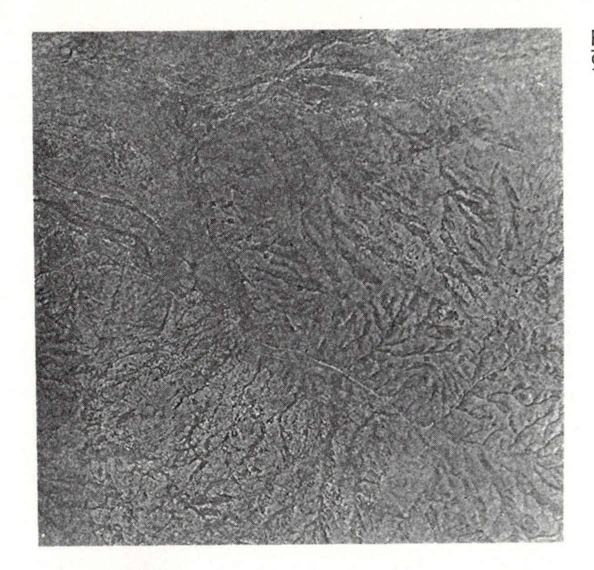

Fig.9.5. Picture shown in Fig. 9.1 after whitening

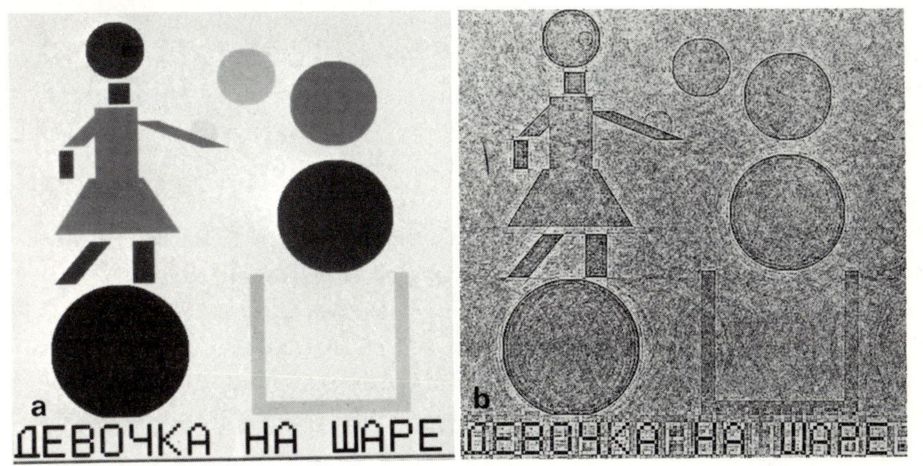

Fig.9.6a,b. Whitening: (a) initial picture, (b) whitened picture

"contour" part of the signal described the object is (specifically, the sharper the picture of the object subjectively is), the more reliable is localization.

This explains the well-known psychophysical fact that the visibility of interference and distortions is lower near abrupt changes in brightness (edges, objects) than in places where the brightness changes smoothly, i.e., where the intensity of the "contour" signal is low.

Fig.9.7a,b. Whitening: (a) initial picture, (b) result of whitening. Note that inclined elements of letters are not suppressed

Note that when reference is made to the extraction of contours, what is usually meant are isotropic procedures of extraction. The whitening which is optimal for localization is not necessarily isotropic, as it is determined by the background pictures (and in the spatially non-uniform estimator, by picture fragments), on which the search for the given object must be carried out. Furthermore, for this very reason it is adaptive, i.e., the characteristic of the whitening filter is adjusted to suit the observed picture, and whitening affects different pictures in different ways. Thus, with rectangles and parallelograms on a circular background, the corner points are emphasized (Fig.9.6b), while in pictures of text the vertical and horizontal parts of the letters are contoured (in effect only their corners remain), but the sloping parts hardly change, since they are rarely encountered (Fig.9.7b).

The entire preceding analysis is based on the assumption that the object to be located is in some way given. There are, however, many problems in which it is not given and must be chosen. Thus in stereogrammetry, regions must be selected from one picture of a stereo pair and located in the other. The question is how this choice can best be made. This is the problem of so-called characteristic points of the picture which is encountered in stereogrammetry and in certain tasks relating to artificial intelligence [9.15-16].

In works on stereogrammetry it is usually recommended that features having sharply defined local characteristics such as cross-roads, meandering rivers, or individual buildings, be chosen as these reference objects. ZAVALISHIN and MUCHNIK [9.16] recommended choosing as these reference objects parts in which certain local differential operators give extreme values. Similar qualitative recommendations are also discussed in other works on pattern recognition.

The above analysis shows the precise meaning of such recommendations. Indeed, from the formula (9.19) for the maximum value of the signal-to-noise ratio that can be achieved at the input of the decision unit of the linear estimator, it follows that the best references are those picture regions where the power of the whitened spectra, $\alpha_0(f_1, f_2) / \overline{[|\alpha_b(f_1, f_2)|^2]}^{1/2}$, is maximal. Such references give a greater response at the output of the optimal filter, and consequently keep the number of false identifications to a minimum.

The following recommendation on the choice of reference objects, say, in stereogrammetry can hence be made. One of the pictures of a stereo pair should be divided into regions which are sufficiently small in area, and then the ratio of each of their spectra $\alpha(f_1, f_2)$ to the modulus of the spectrum $\overline{|\alpha_b(f_1, f_2)|}$ of the second picture should be found. Following this, integral (9.19) is calculated for each region (in digital processing, the corresponding sum is calculated)[4], and the appropriate number of regions giving the largest results is selected.

Since the picture spectrum is usually a function that decreases rapidly towards the boundaries of the frequency domain, the best references are those

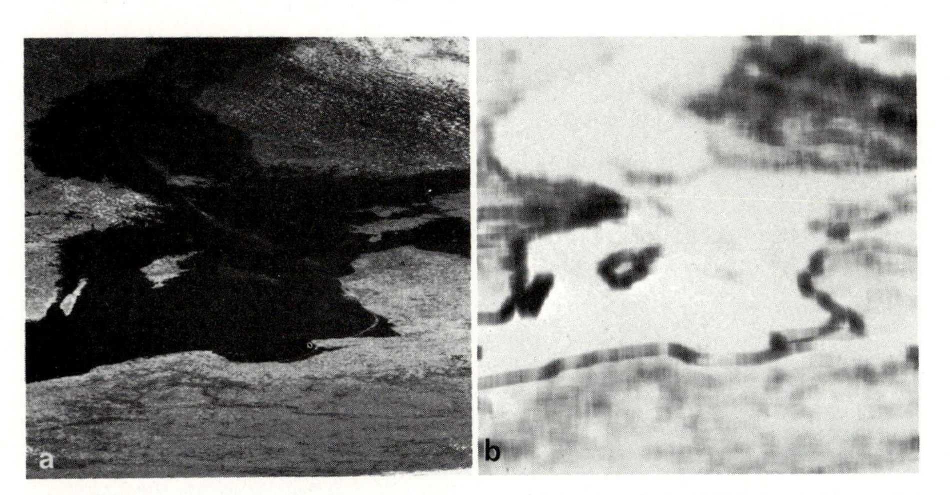

Fig.9.8a,b. Automatic choice of reference objects: (a) initial picture, (b) pattern of optimal reference objects. Object quality is represented by the degree of shadow

[4] Owing to Parseval's relation this quantity can be obtained by integrating the square of the modulus of the "whitened" signal.

that have a slowly falling spectrum, i.e., those parts of the picture that are seen to contain intense contours. This is clearly visible in Fig.9.8 [9.17], of which Part (a) is the initial picture, and Part (b) the result of fragment-by-fragment sliding processing by the algorithm described. The degree of shading in Fig.9.8b corresponds to the value of the integral (9.19). The picture clearly shows that the darkest parts correspond to sharply accentuated local features of Fig.9.8a: abrupt changes in brightness and alterations in texture, for example.

9.5 Algorithm for the Automatic Detection and Extraction of Bench-Marks in Aerial and Space Photographs

In photographic and television recording systems used in aerial photography and in surveying the earth and the other planets with artificial satellites and space probes, the pictures are exposed through a mask containing opaque bench marks in the form of points or crosses. These marks are used in the geometric calibration of the picture during interpretation. One of the tasks involved in automating this interpretation process is the detection and extraction of these marks on the photograph. These tasks are also necessary when processing photographs to improve their visual quality, since bench marks are foreign objects and their presence can lead to artefacts as a result of picture processing (Sect.7.5).

An algorithm for the automatic detection and extraction of marks using computers is described below, and the results of using it to process the pictures transmitted from "Mars 4" and "Mars 5" to extract bench marks, as well as small defects and spots present in some photographs [9.18] are presented. This algorithm is based on the ideas discussed above in the analysis of the optimal detector-estimator. In the present case the estimator was constructed in such a way as to find the bench marks at the output of the linear filter by comparing the signal with the threshold, and also to extract them as far as possible, without distortion, and at the same time to locate small spots and other defects in certain photographs. Thus the filter was adjusted not only for bench marks in the form of crosses, but also for any small details (several dozen picture elements).

The spectrum of small details as compared to the spectrum of the whole picture can on the average be considered to be almost uniform. The filter is therefore constructed on the assumption that the average object spectrum, $\alpha_0(f_1, f_2)$, in (9.32) is constant for an imprecisely set object. In order to prevent the filter from distorting the bench marks to any serious extent, filter (9.32) was approximated only in the low-frequency domain. It is convenient to use a 2-D recursive filter of type (4.29) for this.

In principle, the indicated approximations decrease the noise immunity of the estimator, but they also make it possible to extract the marks virtually without distortion by means of a simple comparison of the filtered signal with the threshold. The threshold can be chosen so that not a single mark is overlooked. It is true that certain contrast elements of the relief in the picture are thereby also extracted; for this reason possible false detection was taken into account in constructing the entire procedure of picture processing. In this way the contrast elements of the relief of the surface of Mars, which were extracted with the bench marks and spots could be restored in the processed picture.

To do this, picture elements in which marks or defects were found to be present (because the signal exceeded the threshold in these elements) were noted, and the video signal value in these elements was replaced by value interpolated from neighbouring innocuous elements. An auxiliary picture was formed whose elements had values equal to the difference between the initial video signal at the extracted elements and the interpolated values where appropriate. The other elements of the auxiliary picture were set to zero. By adding this auxiliary picture to the processed picture, from which the bench marks and spots had been removed, it was possible to restore both the bench marks and the small contrast elements of the relief. With the aid of a special program, unwanted parts of the auxiliary picture (e.g., spots) could be eliminated before this picture was added to the processed photographs [9.18,19].

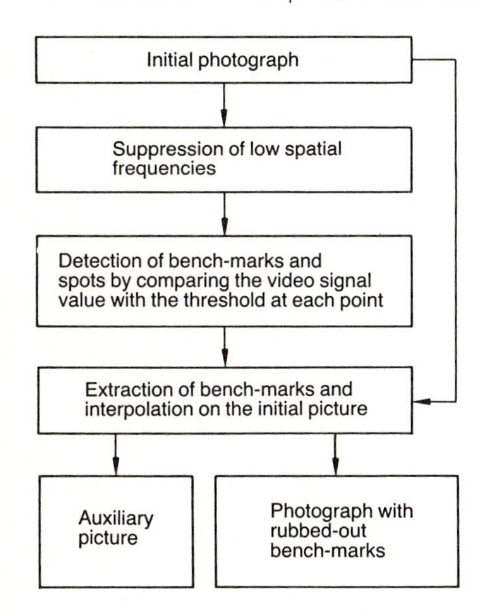

Fig.9.9. Flow chart of the algorithm of automatic bench-mark detection

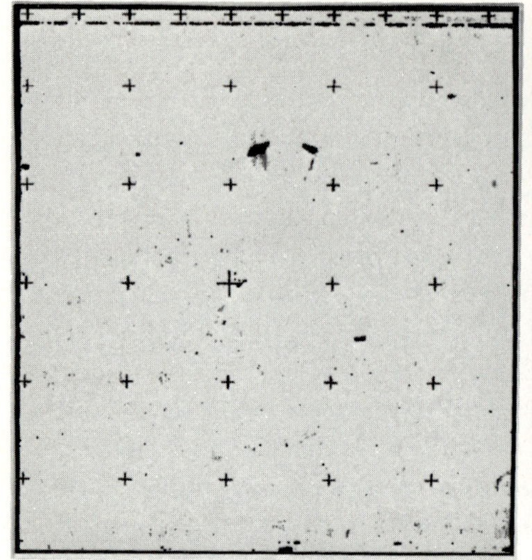

Fig.9.10. Pattern of bench-marks and foreign spots extracted from the picture shown in Fig.7.4

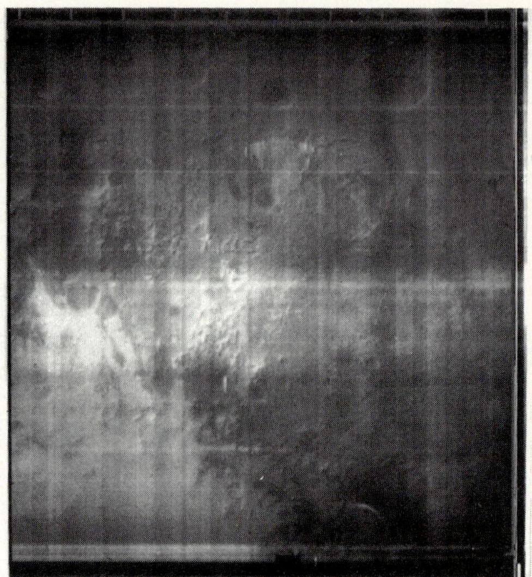

Fig.9.11. Picture shown in Fig.7.4 after extraction of bench-marks and foreign spots and interpolation

The procedure by which bench marks are detected and extracted is shown in Fig.9.9.

Figures 9.10,11 show the results of extracting marks from the picture shown in Fig.7.4. Figure 9.10 is the auxiliary picture obtained by extracting marks and spots, and Fig.9.11 is the picture after extraction and interpolation.

10 Conclusion

The way in which the possibilities afforded by digital picture processing are implemented is determined by the level of hardware and software. At present the main tendency is to create special-purpose computers with dedicated equipment and programs for picture processing.

The main features of picture processing are the large volume of processed information (up to several dozen megabits for one picture); the two-dimensionality of the data array and the vector nature of the data; and an intrinsic lack of formal criteria for the visual quality of the end result, as a consequence of which an interactive processing regime is required, i.e., the processing must be directly supervised and guided by the user. This also determines the architecture of the special-purpose circuitry and software.

At present the specialized computer complexes for the digital processing of pictures are usually based on minicomputers which act as the central processing unit and contain

- Devices interfaced to the following storage and peripherals, which transform the video signal into a digital signal, and devices which record the results photographically (input-output picture devices). The nature of this equipment is determined by the specific purpose of the complex and the form in which the original video information is represented.
- Large-capacity memory with rapid access (usually magnetic discs).
- Archival memory (usually magnetic tape).
- Means for outputting the numerical results and records of the work (alphanumeric printers, plotters, etc).
- Devices for the dynamic visualization of pictures (displays and display processors).
- Special-purpose digital processors (Fourier processors, array processors, etc.).

The interactive processing regime is made possible by display and display processors. The main purpose of display is to reproduce high-quality black-and-white or colour pictures, as well as all accompanying comments, alphabetical, digital and plot information, in a form convenient for direct visual perception and in real time; i.e., in such a way that displaying the picture does not retard the dialogue.

Today's microminiaturized computing technology has opened up the possibility of creating new types of display devices, namely display processors. These permit certain types of picture processing to be performed by hardware and create a flexible association between the user and the central processing unit via the display processor and the "language" of joystick and keyboard, which is more natural for man than the language of programs and numbers. A display processor, supplied with such devices as a light pen and/or cursor, makes it possible for the user to participate actively in the processing and to transmit his decisions in the form of pictures to the central processor for further processing.

Display processors have three main features which distinguish them from passive display: the analogue means of communication with the user (joystick, light pen, tracker ball, control knobs, switches, etc.); devices for processing the picture without the participation of the central processing unit; and two-way communication with the central processor.

The structure of the software for special-purpose digital picture processing systems is determined by the fact that the programs must:

1) Free the user from concern about storing great masses of picture data and keeping them accessible. The user must be able to refer by means of natural coordinate addressing to the massive array representing the picture, with the addressing corresponding to the coordinates in which the picture is expressed in the input-output device and on the display screen.

2) Ensure flexible access to the special-purpose equipment of picture input-output and provide for user dialogue and, in systems with display processors, data exchange between the central and display processors.

3) Offer the means of automating the programming procedure for picture processing. In principle, the user should merely have to provide sequences of high-level macroinstructions acting on a library of programs, together with numerical parameters, links and other service words (as in high-level languages such as FORTRAN or BASIC). The number of minimal special symbols not having their own usual meaning must be.

4) Have specialized means of debugging: programs that generate test signals and pictures, as well as means by which numerical information and other results from processing test signals can be rapidly visualized in a form convenient for observation.

5) Include as an integral feature a library of standard picture-processing programs.

Experience shows that this library should consist of programs from the following basic categories:

- element-by-element video signal transformation
- linear transformation in the spatial domain
- linear transformation in the spectral domain
- preparation by nonlinear algorithms
- statistical measurement
- geometric transformations
- programs for the coding and decoding of pictures for archival storage.

The software of picture processing should also include general-purpose program packages: program assemblers, loaders, debugging programs, translators of programming languages, and, if necessary, a supervisor to organize simultaneous access by several users and a library of the standard programs of computing mathematics.

Digital picture processing is one of the most rapidly developing fields in technical cybernetics, and has a wide range of practical applications. In this book we have discussed only a few of the basic problems of digital processing and presented some examples illustrating the possibilities and results of applying it. These examples show that many practical problems can be solved with the existing methods and technological means.

References

Chapter 1

1.1 E.E. David: Proc. IRE **49**, 319 (1961)
1.2 R.E. Graham, J.K. Kelly: IRE WESCON Conv. Rec. **4**, 41 (1958)
1.3 L.P. Yaroslavsky: Vopr. Elektron. Ser.9, Tekh. Telev. **6**, 35 (1965)
1.4 A.M. Kukinov, D.S. Lebedev: Izv. Akad. Nauk SSSR, Tekh. Kibern. **2**, 17 (1963)
1.5 D.G. Lebedev, D.S. Lebedev: Izv. Akad. Nauk SSSR, Tekh. Kibern. **1**, 88 (1965)
1.6 M.D. McFarlane: Proc. IEEE **60**, 768 (1972)
1.7 Special Issue: *Computer Graphics*, Proc. IEEE **62** (April 1974)
1.8 T. Pavlidis: *Algorithms for Graphics and Image Processing* (Springer, Berlin, Heidelberg, New York 1982)
1.9 J.D. Foley, A. Van Dem: *Fundamentals of Interactive Computer Graphics* (Addison Wesley, London 1982)
1.10 M.P. Extrom, B.H. Mayall: Computer **7**, 72 (May 1974)
1.11 L.P. Yaroslavsky: Radiotekhnika **32**, 11. 72 (1977)
1.12 L.P. Yaroslavsky: "Some Topics of Digital Picture Processing" in Proc. ICO-11 Conf., Madrid (1978)
1.13 E. Person, K.S. Fu: Computer **5**, 70 (1976)
1.14 C.L. Patterson, G. Buechler: Computer **7**, 46 (May 1974)
1.15 Prospecting for Minerals with Mini-Computers: Aviat. Week Space Technol. **5**, 48 (1975)
1.16 G.W. Wecksung, K. Campbell: Computer **7**, 63 (May 1974)
1.17 Special Issue on Digital Image Processing, Proc. IEEE **69** (January 1981)
1.18 L. Bolc, Z. Kulpa (eds.): *Digital Image Processing System*, Lecture Notes Comp. Sci., Vol.109 (Springer, Berlin, Heidelberg, New York 1981)
1.19 Special Issue on Computer Architectures for Image Processing, Computer **12** (January 1983)

Chapter 2

2.1 L.E. Franks: *Signal Theory* (Prentice Hall, Englewood Cliffs, NJ 1969)
2.2 A.M. Trakhtman: *Spektral'naya teoriya signalov* [Spectral Theory of Signals] (Sovetskoe Radio, Moscow 1972)
2.3 H.F. Harmuth: *Transmission of Information by Orthogonal Functions* (Springer, Berlin, Heidelberg, New York 1970)
2.4 A.M. Trakhtman: Avtom. Telemekh. **4**, 81 (1974)
2.5 B.R. Frieden: *Probability, Statistical Optics, and Data Testing*, Springer Ser. Inform. Sci., Vol.10 (Springer, Berlin, Heidelberg, New York 1983)
2.6 E.C. Titchmarsh: *Introduction to the Theory of Fourier Integrals* (Oxford, New York 1948)
2.7 A. Papoulis: *Systems and Transforms with Applications in Optics* (McGraw Hill, New York 1965)

Chapter 3

3.1 D.S. Lebedev, I.I. Zuckerman: *Televidenie i teoriya informatsii* [Television and Information Theory] (Gosenergoizdat, Moscow 1965)
3.2 C. Shannon: Bell Syst. Tech. J. **27**, 379 and 623 (1948)
3.3 B.R. Frieden: *Probability, Statistical Optics, and Data Testing*, Springer Ser. Inform. Sci., Vol.10 (Springer, Berlin, Heidelberg, New York 1983)
3.4 T.S. Huang, W.F. Schreiber, O.J. Tretiak: Proc. IEEE **60**, 1586 (1972)
3.5 L.E. Franks: *Signal Theory* (Prentice Hall, Englewood Cliffs, NJ 1969)
3.6 N.K. Ignat'ev: Izv. Vyssh. Uchebn. Zaved. Radiotekh. **6**, 684 (1961)
3.7 L.P. Yaroslavsky: "Diskretisatsiya s propuskami" [Sampling with 'Gaps'] in *Tezisy dokladov pervoy Vsesoyusnoy Konferentsiy "Metody i sredstva preobrasovaniya signalov"* [Abstracts of the First All-Union Conference "Methods and Technical Means of Signal Transformations"] (Zinatne, Riga 1978) p. 162
3.8 F.C. Billingsley: "Noise Consideration in Digital Image Hardware", in *Picture Processing and Digital Filtering*, ed. by T.S. Huang, Topics Appl. Phys., Vol.6, 2nd ed. (Springer, Berlin, Heidelberg, New York 1979)
3.9 N.K. Ignat'ev: Tekh. Kino Telev. **8**, 21 (1957)
3.10 L.P. Yaroslavsky: *Ustroistva vvoda-vyvoda izobrazhenii dlya ETSVM* [Picture Input-Output Devices for Computers] (Energiya, Moscow 1968)
3.11 A. Girarde: "Methode d'analyse d'image infrarouge par codage multiplex", Franco-Soviet Symp. on Optical-Spectral Devices for Picture Processing, Moscow, Sept. 1976
3.12 H.H. Barrett: J. Nucl. Med. **13**, 382 (1972)
3.13 N.J.A. Sloane, M. Harwit: Appl. Opt. **15**, 107 (1976)
3.14 R.D. Swift, R.B. Wattson, J.A. Decker, R. Paganetti, M. Harwit: Appl. Opt. **15**, 1595 (1976)
3.15 H.J. Landau, H.O. Pollak: Bell Syst. Tech. J. **41**, 1295 (1962)
3.16 W.B. Davenport, W.L. Root: *An Introduction to the Theory of Random Signals and Noise* (McGraw Hill, New York 1958)
3.17 V.A. Garmash: Elektrosvyaz **10**, 11 (1957)
3.18 J. Max: IRE Trans. IT-**4**, 7 (1960)
3.19 W.M. Goodal: Bell Syst. Tech. J. **30**, 33 (1951)
3.20 R. Nuna, K.R. Rao: "Optimal Quantization of Standard Distribution Functions," in 12th Ann. Southeast Symp. on System Theory, Virginia Beach, VA 1980
3.21 V.R. Algazi: IEEE Trans. COM-**14**, 297 (1966)
3.22 L.G. Roberts: IRE Trans. IT-**8**, 145 (1962)
3.23 M.A. Kronrod: "Biblioteka programm B-71 dlya raboty s izobrazheniyami" [The B-71 Program Library for Working with Pictures], in *Ikonika: Tsifrovaya golografiya. Obrabotka izobrazhenii* [Ikonika. Digital Holography. Picture Processing] (Nauka, Moscow 1975)
3.24 L.P. Yaroslavsky, A.M. Fajans: "Issledovanie vozmozhnosti obrabotki i analiza interferogramm na EVM" [Investigation of the Possibilities of the Processing and Analysis of Interferrograms by Computer], in *Ikonika. Tsifrovaya golografiya. Obrabotka izobrazhenii* [Digital Holography. Picture Processing] (Nauka, Moscow 1975)
3.25 I.M. Bockstein, L.P. Yaroslavsky: Avtometriya **3**, 66 (1980)
3.26 L.P. Yaroslavsky, I.M. Bockstein: Soviet Patent No. 633043, Byulleten' izobretenii (Bulletin of Inventions) **42** (1978)
3.27 A. Habibi, G.S. Robinson: IEEE Trans. C-**7**, 22 (1974)
3.28 T.A. Wintz: Proc. IEEE **60**, 7 (1972)
3.29 N. Ahmed, N. Natarajan, K.R. Rao: IEEE Trans. C-**23**, 1 (1974)
3.30 W.K. Pratt, L.R. Welch, W. Chen: IEEE Trans. COM-**22**, 1075 (1975)
3.31 A. Habibi: IEEE Trans. COM-**22**, 614 (1972)
3.32 J.A. Roese, G.S. Robinson: SPIE Proc. **66**, 172 (1975)
3.33 D.A. Huffman: Proc. IRE **40**, 1098 (1952)

3.34 D.N. Graham: Proc. IEEE **55**, 336 (1967)
3.35 S. Werner: Comput. Graph. Image Proc. **6**, 286 (1977)
3.36 A. Netravali, J.U. Limb: Proc. IEEE **68**, 366 (1980)
3.37 A.K. Jain: Proc. IEEE **69**, 349 (1981)
3.38 T.S. Huang: *Image Sequence Analysis*, Springer Ser. Inform. Sci., Vol.5 (Springer, Berlin, Heidelberg, New York 1981)

Chapter 4

4.1 B. Gold, C.M. Rader: *Digital Processing of Signals* (McGraw-Hill, New York 1969)
4.2 L. Rabiner, B. Gold: *Theory and Applications of Digital Signal Processing* (Prentice-Hall, Englewood Cliffs, 1975)
4.3 R.E. Bogner, A.G. Constantinides: *Introduction to Digital Filtering* (Wiley, London 1975)
4.4 I.Sh. Pinsker: "Representation of the Functions of Many Variables in the Form of the Function Product Sums of one Variable", in *Mathematical Processing of Medical and Biological Information* (Nauka, Moscow 1976)
4.5 T.S. Huang, W.F. Schreiber, O.J. Tretiak: Proc.IEEE **60**, 1586 (1972)
4.6 G.M. Robbins, T.S. Huang: Proc. IEEE **60**, 862 (1972)
4.7 A.A. Sawchuk: Proc. IEEE **60**, 854 (1972)
4.8 T.S. Huang (ed.): *Two-Dimensional Digital Signal Processing I*, Topics Appl. Phys., Vol. 42 (Springer, Berlin, Heidelberg, New York 1981)
4.9 L.P. Yaroslavsky: In *Abstracts of the First All Union Conference on Methods and Means of Signal Transformation* (Zinatne, Riga 1978) p. 162
4.10 L.P. Yaroslavsky: Probl. Peredachi Inf. **15**, 102 (1979)
4.11 L.P. Yaroslavsky: "Shifted Discrete Fourier Transforms", in *Digital Signal Processing*, ed. by V. Cappellini, A.G. Constantinides (Academic, London 1980) p. 60
4.12 G. Bonnerot, M. Bellanger: Proc. IEEE **64**, 392 (1976)
4.13 J.L. Vernet: Proc. IEEE **59**, 1384 (1971)
4.14 N. Ahmed, N. Natarajan, K.R. Rao: IEEE Trans. C-**23**, 1 (1974)
4.15 N. Ahmed, K.R. Rao: *Orthogonal Transforms for Digital Signal Processing* (Springer, Berlin, Heidelberg, New York 1975)
4.16 A.K. Jain, E. Angel: IEEE Trans. C-**23**, 470 (1974)
4.17 L.P. Yaroslavsky: *Picture Input-Output Devices for Computers* (Energiya, Moscow 1968)
4.18 R.M. Mersereau: "A Two-Dimensional Fourier Transform for Hexagonally Sampled Data", in *Digital Signal Processing*, ed. by V. Cappellini, A.G. Constantinides (Academic, London 1980) p. 97
4.19 A.R. Butz: IEEE Trans. C-**28**, 577 (1979)
4.20 A.M. Trakhtman, V.A. Trakhtman: *Fundamentals of the Theory of Discrete Signals at Finite Intervals* (Sov. Radio, Moscow 1975)
4.21 H.F. Harmuth: *Sequency Theory*, Advances in Electronics and Electron Physics, Suppl. 9 (Academic, London 1977)
4.22 H.C. Andrews, J. Kane: J. Assoc. Comput. Mach. 17, 260 (1970)
4.23 I.J. Good: J. Roy. Stat. Soc. B**20**, 362 (1958)
4.24 R. Bellman: *Introduction to Matrix Analysis* (McGraw-Hill, New York 1960)
4.25 K.R. Rao, M.A. Narasimhan, K. Revuluri: IEEE Trans. C-**24**, 888 (1975)
4.26 H. Reitboeck, T.P. Brody: Inf. Control **15**, 130 (1965)
4.27 W.K. Pratt, L.R. Welch, W. Chen: IEEE Trans. COM-**22**, 1075 (1975)
4.28 B.J. Fino, V.R. Algazi: Proc. IEEE **62**, 653 (1974)
4.29 K.R. Rao, N.A. Narasimhan, K. Revuluri: Comput. Electr. Eng. **2**, 367 (1975)
4.30 K.R. Rao, K. Revuluri, M.A. Narasimhan, N. Ahmed: IEEE Trans. ASSP-**24**, 18 (1976)
4.31 P.A. Lux: Arch. Elektr. Übertrag. **31**, 267 (1977)
4.32 L.L. Boyko: "Generalized Fourier-Haar Transform on a Finite Abelean

Groop", in *Digital Signal Processing and its Applications*, ed. by L.P. Yaroslavsky (Nauka, Moscow 1980) p. 12
4.33 R. Haralick, K. Shanmugam: IEEE Trans. SMC-**4**, 16 (1974)
4.34 L.P. Yaroslavsky: "Elements of Matrix Formalism: Factorizing the Matrices of Orthogonal Transforms for Synthesis of Fast Algorithms", in Fifth International *Symposium on Information Theory*, Abstracts, Part 2 (Moscow 1979)
4.35 L.P. Yaroslavsky: Radiotekh. Elektron. **1**, 66 (1979)
4.36 L.P. Yaroslavsky: "A Common Representation of Orthogonal Matrices for Digital Signal Processing and Fast Algorithms", in *EUSIPCO-80. Signal Processing: Theory and Applications*, 1st European Signal Processing Conference, Lausanne, 1980 (Elsevier, Amsterdam 1980) p. 161
4.37 H.C. Andrews: "Two-Dimensional Transforms", in *Picture Processing and Digital Filtering*, ed. by T.S. Huang, Topics Appl. Phys., Vol. 6, 2nd ed. (Springer, Berlin, Heidelberg, New York 1975) Chap. 2
4.38 W.K. Pratt: IEEE Trans. C-**21**, 636 (1972)
4.39 W.K. Pratt: "Walsh Functions in Image Processing and Two-Dimensional Filtering", in Proc. Symp. Applications of Walsh Functions, 1972, p. 14
4.40 W.K. Pratt: *Digital Image Processing* (Wiley, New York 1978)

Chapter 5

5.1 N. Ahmed, K.R. Rao: *Orthogonal Transforms for Digital Signal Processing* (Springer, Berlin, Heidelberg, New York 1975)
5.2 H.C. Andrews, J. Kane: J. Assoc. Comput. Mach. **17**, 260 (1970)
5.3 H.C. Andrews: *Computer Techniques in Image Processing* (Academic, New York 1970)
5.4 B. Gold, C.M. Rader: *Digital Processing of Signals* (McGraw-Hill, New York 1969)
5.5 I.J. Good: J. Roy. Stat. Soc. B**20**, 362 (1958)
5.6 R.E. Bogner, A. Constantinides (eds.): *Introduction to Digital Filtering* (Wiley, London 1975)
5.7 A.M. Trakhtman: Avtom. Telemekh. **4**, 81 (1974)
5.8 D.K. Kahaner: IEEE Trans., AU-18,4, p.422 (1970)
5.9 T. Theilheimer: IEEE Trans., AU-17,2, p.158 (1969)
5.10 A.M. Trakhtman, V.A. Trakhtman: *Osnovy teorii diskretnykh signalov na konechnykh intervalakh* [Fundamentals of the Theory of Discrete Signals at Finite Intervals] (Sovetskoye Radio, Moscow, 1975)
5.11 L. Rabiner, B. Gold: *Theory and Applications of Digital Signal Processing* (Prentice Hall, Englewood Cliffs, NJ 1975)
5.12 L.P. Yaroslavsky, N.S. Merzlyakov: *Methods of Digital Holography* (Consultants Bureau, New York 1980)
5.13 L.P. Yaroslavsky: Radiotekhnika **32**, 15 (1977)
5.14 L.P. Yaroslavsky: IEEE Trans. ASSP-29,3, p.448 (1981)
5.15 I. Delotto, D. Delotto: Comput. Graph. Image Proc. **4**, 271 (1975)
5.16 W.B. Kend: IEEE Trans. C-**23**, 88 (1974)
5.17 K.J. Nussbaumer, P. Quandalle: IBM J. Res. Dev. **22**, 134 (1978)
5.18 H.J. Nussbaumer: *Fast Fourier Transform and Convolution Algorithms*, Springer Ser. Inform. Sci., Vol.2 (Springer, Berlin, Heidelberg, New York 1981)
5.19 I.S. Reed, N.M. Shao, T.R. Truong: Proc. IEEE **128**, Part E, 1 (1981)
5.20 J.H. McClellan, C.M. Rader: *Number Theory in Digital Signal Processing* (Prentice Hall, Englewood Cliffs, NJ 1979)
5.21 S. Winograd: Math. Comput. **32**, 178 (1978)
5.22 S. Zohar: IEEE Trans. ASSP-**27**, 4 (1978)
5.23 S. Zohar: In *Two Dimensional Digital Signal Processing II*, ed. by T.S. Huang, Topics Appl. Phys., Vol.43 (Springer, Berlin, Heidelberg, New York 1981) Chap.4

Chapter 6

6.1 C. Shannon: Bell Syst. Tech. J. **27**, 379 and 623 (1948)
6.2 J.M. Wozencraft, I.M. Jacobs: *Principles of Communication Engineering* (Wiley, New York 1965)
6.3 B. Saleh: *Photoelectron Statistics*, Springer Ser. Opt. Sci., Vol. 6 (Springer, Berlin, Heidelberg, New York 1978)
6.4 B.R. Frieden: *Probability, Statistical Optics, and Data Testing*, Springer Ser. Inform. Sci., Vol.10 (Springer, Berlin, Heidelberg, New York 1983)
6.5 A. Papoulis: *Probability, Random Variables and Stochastic Processes* (McGraw-Hill, New York 1965)
6.6 B.R. Frieden (ed.): *The Computer in Optical Research* (Springer, Berlin, Heidelberg, New York 1980)
6.7 W.K. Pratt: *Digital Image Processing* (Wiley, New York 1978)
6.8 D.S. Lebedev: *EUSIPCO-80. Signal Processing: Theory and Applications*, 1st European Signal Processing Conference, Lausanne, 1980 (Elsevier, Amsterdam 1980) p.113
6.9 D.S. Lebedev, L.I. Mirkin: "Two-Dimensional Picture Smoothing Using 'Component' Fragment Models", in *Ikonika. Digital Holography. Picture Processing* (Nauka, Moscow 1975)
6.10 N.I. Chentsov: Dokl. Akad. Nauk SSR **147**, 45 (1962)
6.11 J.S. Bendat, A.G. Piersol: *Random Data: Analysis and Measurement Procedures* (Wiley, Interscience, New York 1971)
6.12 W.A. Blankinship: IEEE Trans. ASSP-**22**, 76 (1975)
6.13 P.V. Lopresti, H.L. Suri: IEEE Trans. ASSP-**22**, 449 (1974)
6.14 G.S. Robinson: IEEE Trans. AU-**20**, 271 (1972)
6.15 D.I. Golenko: *Stimulation and Statistical Analysis of Pseudorandom Numbers on Computers* (Nauka, Moscow 1965)
6.16 F.F. Martin: *Computer Modelling and Simulation*, (John Wiley, New York 1969)
6.17 D.E. Knuth: *The Art of Computer Programming*, Vol.2, Seminumerical Algorithms (Addison-Wesley, London 1969)
6.18 L.I. Mirkin, M.A. Rabinovich, L.P. Yaroslavsky: Zh. Vychisl. Mat. Mat. Fiz. **12**, 5 (1972)
6.19 L.P. Yaroslavsky: "Automatic Image Distortions Diagnostics and Imaging Systems Adaptive Correction Methods", in 14th Int. Congress on High-Speed Photography and Photonics, Moscow, October 1980
6.20 M.G. Kendall, A. Stuart: *The Advanced Theory of Statistics*, Vol.2, Inference and Relationship (Griffin, London 1971)
6.21 J.W. Tukey: *Exploratory Data Analysis* (Addison-Wesley, London 1971)
6.22 E.R. Kretzmer: Bell Syst. Tech. J. **31**, 751 (1959)
6.23 L.I. Mirkin: Vopr. Kibern. Ikonika Tsifrovaya obrabotka i filtratsiya izobrazhenii **38**, 73 (1978)
6.24 L.I. Mirkin, L.P. Yaroslavsky: Vopr. Kibern. Ikonika Tsifrovaya obrabotka i filtratsiya izobrazhenii **38**, 97 (1978)

Chapter 7

7.1 T.S. Huang (ed.): "Introduction", in *Picture Processing and Digital Filtering*, Topics Appl. Phys., Vol.6, 2nd ed. (Springer, Berlin, Heidelberg, New York 1979)
7.2 T.S. Huang, W.F. Schreiber, O.J. Tretiak: Proc. IEEE **60**, 1586 (1972)
7.3 W.K. Pratt: *Digital Image Processing* (Wiley, New York 1978)
7.4 J.L. Vernet: Proc. IEEE **59**, 10 (1971)
7.5 B.R. Frieden: "Image Enhancement and Restoration" in *Picture Processing and Digital Filtering*, ed. by T.S. Huang (Springer, New York 1975)
7.6 J.M. Wozencraft, I.M. Jacobs: *Principles of Communication Engineering* (Wiley, New York 1965)

7.7 L.P. Yaroslavsky: "Investigation of the Effectiveness and Noise Immunity of Still Picture Transmission Systems", Ph.D. Thesis, Inst. for Energetics, Moscow (1969)
7.8 B. Gold, C.M. Rader: *Digital Processing of Signals* (McGraw-Hill, New York 1969)
7.9 G.M. Robbins, T.S. Huang: Proc. IEEE **60**, 862 (1972)
7.10 A.A. Sawchuk: Proc. IEEE **60**, 854 (1972)
7.11 G.W. Stroke, M. Halioua, F. Thon, D.H. Willasch: Proc. IEEE **65**, 39 (1977)
7.12 M.M. Sondhi: Proc. IEEE, **60**, 7 (1972)
7.13 I.M. Bockstein, L.P. Yaroslavsky: Avtometriya **3**, 66 (1980)
7.14 L.S. Gutkin: Theory of Optimal Methods of Radio Reception with Fluctuating Interference (Sovetskoe Radio, Moscow 1972)
7.15 A. Rosenfeld, A. Kak: *Digital Image Processing* (Academic, New York 1976)
7.16 W.K. Pratt: IEEE Trans. C-**21**, 636 (1972)
7.17 D.S. Lebedev, L.I. Mirkin: "Two-Dimensional Picture Smoothing Using 'Component' Fragment Models", in *Ikonika. Digital Holography. Picture Processing* (Nauka, Moscow 1975)
7.18 T.P. Belikova, M.A. Kronrod, P.A. Chochia, L.P. Yaroslavsky: Kosm. Issled. **6**, 898 (1975)
7.19 T.P. Belikova, P.A. Chocia, M.A. Kronrod, L.P. Yaroslavsky: "Digital Picture Processing", in *Surface of Mars* (Nauka, Moscow 1980)
7.20 L.P. Yaroslavsky: Geod. Kartogr. **10**, 37 (1976)
7.21 N.A. Avatkova, O.M. Sveshnikova, I.S. Fainberg, L.P. Yaroslavsky: "Synthesis of Colour Pictures of the Martian Surface", in *Surface of Mars* (Nauka, Moscow 1980)
7.22 D.S. Lebedev, L.P. Yaroslavsky: "Non-linear Filtering of Pulse Interference in Pictures", in *Ikonika* (Nauka, Moscow 1970)
7.23 B.V. Nepoklonov, G.A. Leikin, A.S. Selivanov, L.P. Yaroslavsky, E.P. Aleksashin, I.P. Bokstein, M.A. Kronrod, P.A. Chochia: "Processing and Topographical Interpretation of the Television Panoramas Obtained from the Landed Capsules of the Space Probes 'Venera-9' and 'Venera-10'", in *First Panoramas of the Surface of Venus* (Nauka, Moscow 1979)
7.24 D.S. Lebedev, A.V. Trushkin: "Using FFT to Restore Pictures with Distortions Such as Convolution", in 7th All-Union Conf. on the Theory of Coding and Information Transmission; Reports, Part 7: Methods of Reducing Redundancy, Moscow (1978)
7.25 A.S. Selivanov, M.K. Narayeva, I.F. Sinelnikova, B.A. Suvorov, V.Ya. Elenskii, G.M. Aleshin, A.G. Shabanov: Teknika Kino i Televideniya /Cinema and Television Engeneering/, 9, p.55 (1974)
7.26 H.C. Andrews, B.R. Hunt: *Digital Image Restoration* (Prentice Hall, Englewood Cliffs, NJ 1977)
7.27 D.S. Lebedev, O.P. Milyukova: Vopr. Kibern. Ikonika, Tsifrovaya obrabotka i filtratsiya izobrazhenii **38**, 18 (1978)

Chapter 8

8.1 W. Krug, H.G. Weide: *Wissenschaftliche Photographie in der Anwendung* (Akademischer Verlag, Leipzig 1972)
8.2 P.W. Hawkes (ed.): *Computer Processing of Electron Microscope Images* Topics Curr. Phys., Vol.13 (Springer, Berlin, Heidelberg, New York 1980)
8.3 R.O. Duda, P.E. Hart: *Pattern Classification and Scene Analysis* (Wiley Interscience, New York 1973)
8.4 H. Niemann: *Pattern Analysis*, Springer Ser. Inform. Sci., Vol.4 (Springer, Berlin, Heidelberg, New York 1981)
8.5 K.S. Fu (ed.): *Digital Pattern Recognition*, Communication and Cybernetics, Vol.10, 2nd ed. (Springer, Berlin, Heidelberg, New York 1980)
8.6 A. Rosenfeld: *Picture Processing by Computer* (Academic, New York 1969)

8.7 A. Rosenfeld (ed.): *Digital Picture Analysis*, Topics Appl. Phys., Vol.11 (Springer, Berlin, Heidelberg, New York 1976)
8.8 T.P. Belikova, L.P. Yaroslavsky: Vopr. Radioelektron. Ser. Obshchetekh. **14**, 88 (1974)
8.9 T.P. Belikova, L.P. Yaroslavsky: Tekh. Kibern. **4**, 139 (1975)
8.10 E.L. Hall, R.P. Kruger, S.J. Owyer, D.L. Hall, R.W. McLaren, C.S. Lodwik: IEEE Trans. C-**20**, 1032 (1971)
8.11 T.S. Huang (ed.): *Image Sequence Analysis*, Springer Ser. Inform. Sci., Vol.5 (Springer, Berlin, Heidelberg, New York 1981)
8.12 G. v. Bally, P. Greguss: *Optics in Biomedical Sciences*, Springer Ser. Opt. Sci., Vol.31 (Springer, Berlin, Heidelberg, New York 1982)
8.13 H.C. Andrews: Endeavour **31**, 88 (1972)
8.14 R.A. Hummel: Computer Graph. Image Proc. **4**, 209 (1975)
8.15 D.L. Milgram: Computer Methods for Creating Photomosaics. Tech. Rpt., University of Maryland, July 1974, p.313
8.16 R.H. Selzer: JPL Tech. Rpt., October 1968, p.1336
8.17 A. Rosenfeld, E.B. Troy: "Visual Texture Analysis," in Conf. Record of Symp. on Feature Extraction and Selection in Pattern Recognition, October 1970, IEEE Publ. 70C51-c, Argonne, IL, p.115
8.18 S. Werner: Comput. Graph. Image Proc. **6**, 286 (1977)
8.19 M.A. Kronrod: Vopr. Kibern. Ikonika, Tsifrovaya obrabotka i filtratsiya izobrazhenii **38**, 49 (1978)
8.20 W.K. Pratt: IEEE Trans., 1971, EMC-**13**, No 3, p.38
8.21 L.S. Davis: Comput. Graph. Image Proc. **4**, 248 (1975)
8.22 R.H. Selzer: "Computer Processing of Angiograms", in Symp. on Small Vessel Angiography, April 1972
8.23 G.M. Robbins, T.S. Huang: Proc. IEEE, **60**, 7, p.862 (1972)
8.24 A.A. Sawchuk: Proc. IEEE **60**, 854 (1972)
8.25 M.A. Kronrod: Geod. Kartogr. **6**, 77 (1975)
8.26 D.N. Graham: Proc. IEEE **55**, 336 (1967)
8.27 D.G. Lebedev, D.S. Lebedev: Izv. ANSSSR. Tekniches-Kaya Kiberuetika, **1**, 88 (1965)
8.28 D.S. Lebedev, I.I. Zuckerman: *Television and Information Theory* (Gosenergoizdat, Moscow 1965)
8.29 R.E. Mach, T.L. Gardner: IBM J. Res. Dev. **6**, 290 (1962)

Chapter 9

9.1 A. Van der Lugt: Proc. IEEE, **62**, 10, p.1300 (1974)
9.2 A. Van der Lugt: IEEE Trans. GE-**9**, 10 (1971)
9.3 L.P. Yaroslavsky: Radiotekh. Elektron. **4**, 714 (1972)
9.4 J.M. Wozencraft, I.M. Jacobs: *Principles of Communication Engineering* (Wiley, New York 1965)
9.5 L.S. Gutkin: *Teoriya optimal'nykh metodov radiopriema pri fluktuatsionnykh pomekhakh* [Theory of Optimal Methods of Radio Reception with Fluctuating Interference] (Sovetskoye Radio, Moscow 1972)
9.6 A. Rosenfeld: *Picture Processing by Computer* (Academic, New York 1969)
9.7 L.P. Yaroslavsky: Vopr. Kibern. **38**, 32 (1978)
9.8 L.P. Yaroslavsky: Soviet Patent No.536497, Byulleten' izobretenii (Bulletin of Inventions) **43** (1976)
9.9 V.N. Dudinov, V.A. Krishtal', L.P. Yaroslavsky: Geod. Kartogr. **1**, 42 (1977)
9.10 L.P. Yaroslavsky: Vopr. Radioelektron. **8**, 70 (1975)
9.11 L.S. Davis: Comput. Graph. Image Proc. **4**, 248 (1975)
9.12 D.S. Lebedev, I.I. Zuckermann: *Televidenie i teoriya informatsii* [Television and Information Theory] (Gosenergoizdat, Moscow 1965)
9.13 A. Rosenfeld, A. Kak: *Digital Image Processing* (Academic, New York 1976)
9.14 A. Rosenfeld: "Image Processing and Recognition", in *Advances in Computers*, Vol.18 (Academic, New York 1979)

9.15 R.O. Duda, P.E. Hart: *Pattern Classification and Scene Analysis* (Wiley Interscience, New York 1973)
9.16 N.V. Zavalishin, I.B. Muchnik: *Modeli zritel'nogo vospriyatiya i algoritmy analiza izobrazhenii* [Models of Visual Reception and Picture Reception Algorithms] (Nauka, Moscow 1974)
9.17 A.N. Belinsky, L.P. Yaroslavsky: Issled. Zemli iz Kosmosa **4**, 85 (1980)
9.18 L.P. Yaroslavsky: Geod. Kartogr. **10**, 37 (1976)
9.19 T.P. Belikova, P.A. Chochia, M.A. Kronrod, L.P. Yaroslavsky: "Tsifrovaya obrabotka izobrazhenii" ["Digital Picture Processing"], in *Poverkhnost' Marsa* [Surface of Mars] (Nauka, Moscow 1980)

Subject Index